LOCAL AND REGIONAL SYSTEMS OF INNOVATION

Economics of Science, Technology and Innovation

VOLUME 14

The titles published in this series are listed at the end of this volume.

LOCAL AND REGIONAL SYSTEMS OF INNOVATION

Edited by

JOHN DE LA MOTHE
University of Ottawa

and

GILLES PAQUET
University of Ottawa

KLUWER ACADEMIC PUBLISHERS
BOSTON / DORDRECHT / LONDON

Distributors for North, Central and South America:
Kluwer Academic Publishers
101 Philip Drive
Assinippi Park
Norwell, Massachusetts 02061 USA
Telephone (781) 871-6600
Fax (781) 871-6528
E-Mail <kluwer@wkap.com>

Distributors for all other countries:
Kluwer Academic Publishers Group
Distribution Centre
Post Office Box 322
3300 AH Dordrecht, THE NETHERLANDS
Telephone 31 78 6392 392
Fax 31 78 6546 474
E-Mail <orderdept@wkap.nl>

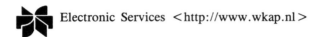 Electronic Services <http://www.wkap.nl>

Library of Congress Cataloging-in-Publication Data

A C.I.P. Catalogue record for this book is available
from the Library of Congress.

Printed on acid-free paper.

Printed in the United States of America

CONTENTS

CONTRIBUTORS

Zoltan Acs is a Professor of International Business and Management at the University of Baltimore

Nabil Amara is in the Department of Political Science at Laval University

John de la Mothe is an Associate Professor of Science and Government and Director of the Program of Research in International Management and Economy (PRIME) in the Faculty of Administration, the University of Ottawa

Jerome Doutriaux is an Associate Professor in the Faculty of Administration at the University of Ottawa

Felix R. FitzRoy is with the University of St. Andrews

Richard Florida is the Heinz Professor of Economic Development and Co-Director of the Center for Economic Development at Carnegie Mellon University

David Garkut is with the Program of Research on Globalization and Regional Innovation (PROGRIS) at the University of Toronto

Meric Gertler is a Professor in the Department of Geography at the University of Toronto

Hervey Gibson is with the Department of Economics at Glasgow Caledonian University

Adam Holbrook is with the Center of Policy Research on Science and Technology (CPROST) at Simon Fraser University

L.P. Hughes is with the Center for Policy Research on Science and Technology at Simon Fraser University

David Keeble is with the ESRC Centre for Business Research at the University of Cambridge

Réjean Landry is a Professor in the Department of Political Science at Laval University

Clive Lawson is with the ESRC Centre for Business Research at the University of Cambridge

Helen Lawton Smith is with the Centre for Local Economic Development at Coventry University

Barry Moore is with the ESRC Centre for Business Research at the University of Cambridge

Richard Nimijean is a Research Associate with the Program of Research in International Management and Economy (PRIME) at the University of Ottawa

Tim Padmore is with the Center for Policy Studies on Higher Education at the University of British Columbia

Gilles Paquet is a Professor of Economics and Management and Director of the Program of Research in International Management and Economy (PRIME) in the Faculty of Administration, the University of Ottawa

Jeffrey Roy is a Fellow of the Center of Governance at the University of Ottawa

Annalee Saxenian is an Associate Professor of City and Regional Planning at the University of California at Berkeley

Hans Schuetze is Director of the Center for Policy Studies in Higher Education at the University of British Columbia

Ian Smith is with the University of St. Andrews

Roger Voyer is with Coopers and Lybrand Consulting, Ottawa

Frank Wilkinson is with the ESRC Centre for Business Research at the University of Cambridge

David Wolfe is an Associate Professor of Political Science and Director of the Program of Research on Globalization and Regional Innovation Systems (PROGRIS) at the University of Toronto

Cliff Wymbs is a doctoral candidate in regional economic development and planning at Rutgers University

PREFACE

The initial purpose of the 1997 Workshop from which this book emerged was to provide an opportunity for a number of people who were professionally engaged in the practice of regional economic development and researchers who were interested in local and regional systems of innovation to take stock of what is known and not known about the spatial dynamics of the innovation process.

Over the past decade, there has been much ethnographic work done in various regions of Canada, the United States and Europe concerning high technology clusters, and there has been much speculation about the shape and contours of these sub-national innovation systems. However, we know little about their inner dynamics, their boundaries, or their development. Moreover, little has been collected in the form of comparable data that might be used to gauge the relative effectiveness of these different systems and various contexts.

One of the main reasons behind this research-reality lag has been the fixation that scholars have developed on 'national' systems of innovation. Every nation has found within its borders someone who is quite willing to celebrate the importance of the national innovation system by reference to some sort of inventory of national infrastructure features (human, physical, communications, etc.) and by asserting that through mysterious ways, an 'innovation system' not only existed but was co-terminus with the administrative contours of these infrastructures. While these vignettes provided a very limited insight into the *workings* of the presumed national 'system' and revealed nothing about the sources of their effectiveness, they more or less suggested explicitly that there was no real need to probe the sub-national levels.

It is only more recently, circa 1994, with the work of Thomas Courchene, Charles David, B.Ä. Lundvall, John Naisbitt, Kenichi Ohmae, Annalee Saxenian, and others - and building on a rich tradition in economic geography and regional science and on the pioneering efforts of Giacomo Becattini, Georges Benko, Gioacchino Garofoli, Allen Scott, Michael Storper, and many others - that increasing attention has been paid to the dynamics and workings of sub-national learning-oriented production and innovation systems.

This literature, generated over the past few years, has been quite rich and substantial. But it has been balkanized into segments rooted in quite different strategies. Some of the work has been based strictly on case studies and has not generated much in terms of generalizeable results. Other work has built on models of spatial or territorial interdependence, of patterns of proximity, of learning, or of cognitive-evolutionary processes. Finally, many monographs have focused on simple comparisons between two areas or districts, or have exploited specific data sets to elicit some general transversal co-relations between spatial, organizational and economic variables. But the results of these differing literatures remain unintegrated.

Our objective here is not to provide the definitive integrative framework that is capable of generating a new synthesis for the regional debate. Our ambitions are more modest. We want to suggest a few guidelines for the analysis of local and regional systems of innovation through a three-pronged approach that emphasizes the organizational, cognitive-evolutionary, and proximity dimensions. This is developed in Part A of the book and is designed to provide a broad workable framework encompassing the various approaches used in the book. Indeed, at the end of Part A, a taxonomy is provided to nest the chapters that follow.

To understand the central importance of sub-national systems of innovation as well as their core dynamics, we present two broad sets of perspective in Parts B and C. The first sample provides an overview of a variety of experiences from Europe and America. As such, Part B blends conceptual and empirical perspectives in order to characterize, in a broad way, some of the key features of learning regions on the two continents. Four broad features emerge from these papers:

1.   The new cognitive dynamics of the learning regions making up the local and regional systems of innovation (LRIS);
2.   the great importance of proximity (in a variety of senses) of the specificity of the cultural milieu in the explanation of the success or failure of the different types of LRIS;
3.   the key role played by the process of investment in the emergence, growth and maintenance of these LRIS; and
4.   the catalytic role of government-business relations in explaining both success and failure.

The second sample, in Part C of the book, provides a set of Canadian experiences to emphasize particularly the great diversity of settings within the broad administrative arrangements available in Canada. If ever there was a remnant of faith in the determining importance of the 'national' shell, these comparisons should put it to rest. Four broad lessons have emerged from these Canadian case studies, and there are reasons to believe that other national surveys of local and regional innovation systems would reveal a similar pattern:

e)   the extraordinary variety of the different regional experiences;
f)   the particular importance of the cultural/contextual factors in explaining those differences;
g)   the very idiosyncratic dynamics of the different local and regional systems of innovation; and
h)   the unlikely success of homogenous 'national' policies based on a 'one-size-fits-all' approach.

In Part D, we have tried to distill some clearer lessons and to identify some of the most important challenges facing local and regional systems of innovation. In closing, we formulate in closing some directions that would appear to be broadly identifiable for model builders, data gatherers and for economic/ethnographic researchers.

In sum, this volume is meant to be what one calls in French *un propos d'étape*: that is, a progress report on what we know and what we strive to know at this time. Our hope is that this volume will help trigger a broad new wave of interest in the socio-economic importance of location and regions.

John de la Mothe and Gilles Paquet

University of Ottawa

ACKNOWLEDGMENTS

This book is the result both of a symposium that was held in Ottawa, Canada on March 12-13 1997 and the year long conversation that ensued.

Our gratitude must be extended and expressed to Dr. Fred Gault of Statistics Canada who first articulated the need for such a meeting and who provided the necessary pecuniary spark to make the symposium a reality. We must also thank our colleagues from across the public, private and civic sectors of Canada as well as those researchers from the United States and Britain who accepted our invitation to come to Ottawa, to overlook Parliament Hill (albeit on a gray and bleak day), and to share their experiences and their results. Clearly, without the enlightened perspective of both our sponsor and the selfless cooperation of our colleagues, the symposium would not have brought forth this rich volume.

The organization and execution of the many tasks involved in hosting an international meeting and of bringing that event into a manuscript form has benefited greatly from the contribution of a number people. Some were either co-workers in the Faculty of Administration or colleagues from the public service. Some were students involved with the Program of Research in International Management and Economy (PRIME) at the University of Ottawa. And some, particularly in the latter stages of manuscript preparation, were involved with the newly created Center on Governance at the University of Ottawa of which PRIME is a member. In particular we would like to thank Frances Anderson (Statistics Canada), Diane Fontaine, Christine Holke, Jennifer Khurana, Anastasia Vakos, and Chris Wilson. Their collective efforts were indispensable and carried out with great skill, timeliness and humor. Additionally, all the contributors to this volume are indebted to Bo Carlsson, a fine scholar and an editor of this Kluwer series on the economics of science, technology and innovation who helped provide this book with an outlet. And finally, the historical record of producing this book would be revisionist in the extreme if we were not to mention, with thanks, the titanic patience and most pleasant demeanor of Ranak Jasani at Kluwer Academic Publishers. The many delays in submitting the final manuscript to the publishers - some due to the modern vagaries of incompatible computer programs or corrupted electronic copies; some due to the more traditional problems of the last minute updating and correction of figures and data, or the problem of simply finding more than 20 contributors who were scattered geographically and who travel regularly - must have made the commissioning editor of this volume wonder at times if she was dealing with ghosts.

Nevertheless, we feel that in the end, the final product is a book of high quality. We are most grateful to the authors for the time and care that they have given to the task of preparing their final manuscripts. All of our frustrations faded to memories once we had the pleasure of finally reading the complete and clean manuscript. What became clear is that we have been privileged to be able to cooperate with the contributors and to partake in this joint venture. We hope that readers will share our enthusiasm and our gratitude to the contributors.

PART A

IN SEARCH OF CONCEPTUAL FRAMEWORK

1 LOCAL AND REGIONAL SYSTEMS OF INNOVATION AS LEARNING SOCIO-ECONOMIES

John de la Mothe and Gilles Paquet
University of Ottawa

1.1 INTRODUCTION

It is becoming widely recognized by policy makers, statistical agencies and economic analysts alike that, in an era of techno-globalism, the innovative capacity of regions is of heightened importance. However, one cannot proceed with the development of a new array of policy-relevant tools or statistical indicators pertaining to sub-national systems of innovation until a more comprehensive and realistic picture of the real economic dynamics of local and regional systems of innovation has been better developed. This book provides such an attempt. Up 'til now, one of the major weaknesses of the very best work that has been done in this area is the emphasis on the various infrastructures (physical, human, communications, etc.) associated with these systems. These features are not unimportant, but they only provide a limited insight into the workings of innovation systems. Moreover, they tend to generate the false impression that "systems" are co-terminous with the administrative contours of these infrastructures.

This often means that the existence of a local or regional system of innovation, when it is recognized, is inferred from the existence of these infrastructures, and is in some way seen as being welded to them. And since national infrastructures are both more formalized and more celebrated than sub-national ones, this has led to a vast operation of co-relations between these national infrastructures and the presumed national systems of innovation.

From the point of view of a national statistical agency for example, this sort of classification (based on administrative infrastructures) is understandably a useful first phase in data gathering. But it is no more sophisticated than the classification of animals according to the number of legs – a classification that was in good currency in biology a very long time ago. Clearly, such simplistic classifications are unlikely to throw much light on the complex process of innovation and wealth creation *per se.*

Yet, how can one hope to do better without a meaningful dynamic evolutionary image of the innovation process?

1.2   PROCESS RATHER THAN STRUCTURE

Some attempts have been made at stylizing the *workings* of the innovation systems. For instance, Annalee Saxenian (1994) has insisted on the interactions between the structural features of corporate and industrial organization and local institutions and culture. A cognate effort has been undertaken by Richard Florida (1995) at the regional level. The Conseil de la science et de la technologie du Québec (1997) has proposed a three-tiered image of the innovation system at the regional/national level: at its core is the innovating firm embedded in a proximate environment of alliances and networks, and in a broader environment providing the legal, regulatory, economic and public support infrastructures defining the rules of the game. The coalescence of the interactions between those three levels is seen as being at the core of the innovation system.

While Saxenian's and Florida's ethnographic work and the provisional synthesis of the Quebec Council have proved helpful, this work remains on the whole largely descriptive. The major contribution of Saxenian, Florida and the Quebec Council has been to provide loose interpretative frameworks that emphasize structural features even though it puts at the center of the stage, and this is an important novelty, the nexus of interactions (at the local and regional levels) as the determining factor.

Our framework, which we might label *process rather than structure (PRTS),* focuses firmly on the process of innovation as its central concern. It recognizes *above* that the innovation process is not linear but is network-shaped and dynamized in a fundamental way by a complex multilogue which weaves the various partners together. Moreover it regards as utopian the hopes that by manipulating structures one may transform the process. What is centrally important and what has more heuristic power is an understanding of the dynamics of the process. Finally, it emphasizes the interaction between three major dimensions (the cognitive space, the capabilities/ absorptive capacity of firms, and the governance of the communities of practice) as playing key roles in the innovation process (Acs, de la Mothe, Paquet 1996).

We present the basic features of the PRTS approach in this chapter, and we suggest that it may well hold the key to the construction of a dynamic and evolutionary framework for the analysis of local and regional systems of innovation. The local milieu, rather than the regional/national structures, is the central focus of the PRTS approach. A subsidiary objective of this chapter is the identification of the crucial *process dimensions* of the innovation system and of the main *strategic externalities* generated by innovation races, for they will be of crucial importance in the construction of the next generation of statistical indicators.

The PRTS approach suggests a three-dimensional decoder of the learning process involved in generating new practical knowledge.

1. First, we build on the fact that an innovation is first and foremost new practical knowledge. Consequently, what is required is a stylized view of the innovation process in a *cognitive space*. The only template we have of such a world is the one suggested a few years ago by John Ziman (1991) but it has never generated the attention it deserved. Ziman has suggested nothing less than "some unpacking" of the traditional model of discovery and an exploration of the different knowledge domains in the cognitive space. Ziman focused on the cognitive roots of innovation and suggested that a neural net model of innovation was probably the most useful approach.

Ziman uses an old metaphor derived from the language of geography to define the cognitive space: it is used to identify transversal areas constructed from patterns of connections between different domains of the different layers of fields in the innovation process (sciences, technologies, marketing, etc.). Certain scientific ideas, practical techniques, and marketable commodities (each at different layers of the innovation process) are cognitively interconnected in the sense that they are related in a certain pattern of interdependence. This connectionist vision of the innovation process is very much like a neural network: what looks from the outside like a random set of connections is in fact a neural pattern capable of learning like a real neural net in the brain. The redundancy of connections allows information flows to circumvent a hole or a lesion. Indeed, the pattern of nodes and connections constitutes a system that is capable of experiencing learning as nodes or patterns of connections are modified.

The pattern of nodes and connections in this cognitive space need not be similar or identical as one looks from one system of innovation to another, for the ideas, techniques and commodities involved are not identical. The learning engine lies in the particular configuration or pattern of ideas, techniques and commodities *and* the specialized items of knowledge linking them. And the learning capability of this transversal network is its capacity to transform.

While the traditional view is that individual persons learn, what is suggested by more recent analyses is that the learning organization does not store its knowledge in separate heads but instead it stores it in the relationships that exist between stakeholders. It co-creates value by a dialogue of equals (Wikstrom and Normann 1994). What is required is a better understanding of that process. Already some progress has been made toward the development of cognitive economics but this remains to a great extent uncharted territory (McCain 1992; Paquet 1998a, 1998b).

2. Second, this new practical knowledge is generated by organizations throughout the cognitive space, and these organizations are defined by their *capabilities/absorptive capacities* (Langlois and Robertson 1995). One must therefore examine the factors defining the relative emphasis on exploration in search of the new and exploitation of the existing knowledge base that characterize the different communities of practice that make up these organizations.

In a rapidly evolving, surprise-generating context that is plagued with uncertainty and learning failures, the challenge is not simply to make the best use of existing resources and knowledge, to better *exploit* better the available possibilities, but it is also to *explore* new possibilities and opportunities. The strategic choice strikes a balance between exploitation and exploration: undue emphasis on the exploitation of available knowledge may trap an organization in sub-optimal situations while undue emphasis on exploration may prevent an organization from gaining much from successful experimentation (March 1991, Marengo 1993, Dosi and Marengo 1994).

Organizations will be the locus of on-going conversations that will produce new knowledge and value-added through networking and partnering. The main challenge then is to determine what is the best organization of knowledge production if the objective is to generate learning and innovation.

According to Marengo (1993), this will depend greatly on a balance between commonality and diversity of knowledge, between coherence and mutual learning, between exploitation and exploration. Ethnographic work on the daily life of organizations suggests that these features are not captured in the formal characteristics of organizations. Practice differs significantly from job descriptions or rules and procedures. The corporate culture embodies a communal interpretation that may have little to do with the documentation available. Corporate culture embodies generally unwritten principles meant to generate a relatively high level of coordination at low costs by bestowing identity and membership.

This corporate culture is nested at the organization level according to Orr (1990) who is extensively quoted in Brown and Duguid (1991), through the central features of work practice: stories of flexible generality about events of practice that act as repositories of accumulated wisdom, the evolution of these stories constituting collective learning of an evolving way to interpret conflicting and confusing data but also an on-going social construction of a community of interpretation.

3. Third, one must identify the features of the broader social process within which the organization is nested to ascertain the extent to which the workings of the innovation system depend on features of the *governance of the communities of practice* (proximity, trust, solidaristic values, etc.) (Hollingsworth 1993).

Actual practice in the workplace has a communal base. Learning has also a communal base. It is not about transmission of abstract knowledge from someone's head to another's. It is about learning to "function in a community", about the "embodied ability to behave as community members", about "becoming a practitioner". Learning is *legitimate peripheral participation* and it is fostered by facilitating access to, and membership of, the community-of-practice of interest (Brown and Duguid 1991).

Trust is at the core of both the fabric of the communities of practice and of its communal base. Trust is a way to transform "laborers into members", to convert an employment contract into a membership contract. The "concept of membership, when made real, would replace the sense of belonging to a place with a sense of belonging to a community" (Handy 1995). Belonging is one of the most powerful agents of mobilization. So what is required is an important "moral" component to the new employment contract, *a new refurbished moral contract* that is not mainly contractual but mainly moral: "a networks of civic engagement...which can serve as a cultural template for future collaboration... broaden the participants' sense of self... enhancing the participants'"taste" for collective benefits" (Putnam 1995; Paquet and Pigeon 1998).

But a situation in which such a membership contract would become hegemonic would risk degenerating into a situation in which a dominant macro-culture would prevail. If such were the case, in the long run, such an homogeneous and coherent culture would cease to be innovative. Innovation requires a certain diversity of knowledge and stems from the interplay of separate communities, from the interplay between the core community and the emergent communities at the periphery (the suppliers and the customers of a firm, for instance), from the organization and the environment it actively interacts with. Exploration calls for diversity, for separate stories to be in good currency. It happens at the interface between the organization and its environment and depends on the capacity for the non-canonical to prevail on the canonical.

It is therefore in the structure of the communities of practice that one must seek the levers that are likely to foster both learning and innovation by intervening with the work place. When the gap between the canonical and the actual practices widens too much, the only way to promote the growth of knowledge is to legitimize and support a number of enacting activities that may be disruptive, to foster a reconception of the organization as a community of communities and promote the view that communities of practice must be allowed to take some distance from the received wisdom.

The very coherence that makes learning easier is likely to make innovation more difficult. The central challenge, for the promotion of the growth of knowledge, is to find the requisite degree of dissonance that is necessary for a system to become innovative, and to identify the most effective schemes of decomposition of large organizations into quasi-isolated sub-systems likely to provoke the emergence of a workable degree of inconsistency and therefore of innovation.

The differences in the way the sub-organizations search for knowledge increase the scope of the search. So, as these differences are legitimized and the different ways of searching for new knowledge have maximum opportunity to rub against one another (as in industrial districts or more closely interconnected communities of practice), innovation will ensue.

1.3   MESO-SYSTEMS OF INNOVATION

In an economy dynamized by information flows, knowledge, competence and capabilities, and the communities of practices, the new relevant units of analysis have to be those that serve as the basis to understand and nurture innovation. Focusing either on the firm or on the national economy would appear to be equally misguided. Under the microscope, too much in a firm is idiosyncratic and white noise is bound to run high. Under the national macroscope, much of the innovation and restructuring going on is bound to be missed. One may therefore argue, we think persuasively, that the most useful perspective point is the Schumpeterian/Dahmenian meso-perspective focusing on development blocks, sub-national fora, etc. where the learning is really occurring (de la Mothe and Paquet 1994, 1996).

In an evolutionary model, the process of learning and discovery is only one blade of the pair of scissors. The other blade is the interactive mechanism with the context or environment through which selection occurs. This interactive mechanism "both provide[s] the source of differential fitness - firms whose R&D turn up more profitable processes of production or products will grow relative to their competitors - and also tend to bind them together as a community" (Dosi and Nelson 1994:162).

It is very important to realize that social proximity is bound to play a fundamental role on both sides of the equation. Both on the organization side and on the forum/environment side, proximity breeds interaction and socio-economic learning (Boswell 1990). Moreover these interactive mechanisms are fueled by dynamic increasing returns to agglomeration. In most cases, these agglomeration economies are bounded, and therefore do not give rise to monopoly by a single region or location, but they generate increasing returns snowballing (Arthur 1990).

Nelson and Winter (1977) have suggested that at the core of the processes of innovation, learning and discovery, and of the processes of diffusion of technical and organizational innovations, is the notion of "selection environment" which is defined as the context that "determines how relative use of different technologies changes over time" (p.61). This context is shaped by market and non-market components, conventions, socio-cultural factors, and by the broader institutional structure. This selection environment constitutes the relevant *milieu* (internal and external, broader or narrower) in explaining the innovative capacity of a sector/region.

The notion of milieu has been defined as "un ensemble territorial formé de réseaux intégrés de ressources matérielles et immatérielles, dominé par une culture historiquement constituée, vecteur de savoirs et savoir-faire, et reposant sur un système relationnel de type coopération/concurrence des acteurs localisés" (Lecoq 1989). Consequently, the notion of *milieu* connotes three sets of forces: (1) the contours of a particular spatial set vested with a certain unity and tonus; (2) the organizational logic of a network of interdependent actors engaged in cooperative innovative activity; and (3) organizational learning based on the dialectics between the *internal milieu* and the external milieu (Maillet 1992).

At the core of the dynamic milieu and of the innovation network are a number of intermingled dimensions (economic, historical, cognitive and normative) but they all depend to a certain degree on *trust and confidence*, and therefore on a host of cultural and sociological factors that have a tendency to be found mainly in localized networks and to be more likely to emerge on a background of shared experiences, regional loyalties, etc. This is social capital in Coleman's sense and such social and cultural capital plays a central role in both the dynamics and the capacity to learn and transform of meso-systems.

The innovation process depends as much on the features of a selection environment as on the internal milieu.

First, innovation is all about continuous learning and learning does not occur in a socio-cultural vacuum, either within the organization or in the environment. An innovation network is more likely to blossom in a restricted localized milieu where all the socio-cultural characteristics of a dynamic milieu are likely to be found. Moreover, it is most unlikely that this sort of *milieu* will correspond to the national territory. Therefore, if one is to identify *dynamic milieus* as likely systems on which one might work to stimulate innovation, they are likely to be local or regional systems of innovation.

Second, some geo-technical forces would appear to generate meso-level units where learning proceeds faster and better. As Storper argues, "in technologically dynamic production complexes ... there is a strong reason for the existence of regional clusters or agglomerations. Agglomeration appears to be a principal geographical form in which the trade-off between lock-in, technological flexibility (and the search for quasi-rents), and cost minimization can be most effectively managed because it facilitates efficient operations of a cooperative production network. Agglomeration in these cases is not simply the result of standard localization economies (which are based on the notion of allocative efficiency in minimizing costs), but of Schumpeterian efficiencies" (Storper 1992:84; Dosi, Andrea-Tyson, Zysman 1990).

Third, the deconstruction of national economies, the widespread devolution of central government powers, the rise of region-states and the growth of the new tribalism would tend to provide a greater potential for dynamism at the meso level. But Storper has argued that "codes, channels of interaction, and ways of organizing and coordinating behaviors" are what makes learning possible (p.85). He feels that the confluence of issues (learning, networks, lock-in, conventions and types of knowledge) must be rooted in political-economic cultures, rules and institutions and that in many countries these are highly differentiated at the regional level. Therefore one region may trigger technological learning and innovation networks in one sub-national area much faster than in others.

Canada, the USA and Mexico are such countries where one may reasonably detect a mosaic of political-economic cultures, rules and conventions with differential innovative potential (Maddox and McGee 1994). Consequently, one may say that there is a genuine "territorialization of learning" in such a Schumpeterian world.

1.4 ORGANIZATIONAL LEARNING, PROXIMITIES AND COORDINATION

At the core of the meso-innovation system are three profoundly intertwined processes: an adaptive and cumulative learning process, a system of interactions among proximate agents and groups, and a dynamic coordination process built on the other two. Together, they constitute the dynamics of the meso-innovation system.

1. Learning is not sheer adaptation. Adaptation is a process of adjustment to new circumstances on an *ad hoc* basis. Whatever new routine has evolved in this manner may be easily forgotten. Learning is quite different. It is a cumulative process through which new knowledge, however trivially different from what was already in store, gets embedded in new rules, conventions, routines and filters (Lazaric et Monnier 1995: ix). Indeed, learning is not restricted to a simple modification of routines and rules, it may also trigger a transformation of the representations, objectives, norms and strategies. But learning requires a reasonably well-defined internal milieu, acting as a sensitive surface, for without it the new experiences would not be registered or represented as new, and would not call for adjustment in the organizational pattern.

Routines, rooted in context and history, are embedded in a pattern of rationales, rules, conventions, and institutions that provide the process of decision-making and learning with unity and stability. In this world, the meaningful units of analysis are the parallel patterns of belief systems and mental representations *and* techno-organizational conventions and rules in which the process is nested.

For any agent or group, the environment is apprehended as a representation. And within a well-coordinated organization, there is much common representation. But there may be differences among the representations of different members. This is a source of learning since the organization may exploit such differences and diversity as a source of knowledge. A common representation may well be an amalgam or a synthesis of these partial representations. For organizational learning amounts to a reconfiguration of collective representation.

At any time, these representations and rules may be more or less fitting, i.e., they may be more or less effective socio-technical armistices between the evolving physical environment and the evolving values and plans of agents and groups. The degree of fitness is not invariant as circumstances change: it is rooted in the probability of survival and in the capacity to develop the requisite competences and capabilities to survive (Paquet 1998a).

Given the ever imperfect nature of mental representations, of conventions and of rules, and the mistake-ridden learning processes, the modification of both representations and rules constitutes the way in which the socio-economy evolves, transforms and learns. This learning may be more or less effective, depending on the nature of the challenges generated by the environment, and the nature of the competences, readinesses and reactive capacities of agents and organizations. For instance, the pattern of representations and rules may easily accommodate minor variations in the environment and adapt quickly to these new circumstances. However, it may be capable of only limited learning within a narrow band of circumstances. In the face of radical changes in the environment that call for a dramatic reframing of representations, rules and conventions, it may mainly generate a great deal of cognitive dissonance and dynamic conservatism, and be incapable of learning (Ciborra 1990:210).

Indeed, as Nelson and Winter have shown (1982), organizations tend to stick to their usual routine as long as performance remains above certain target levels. It is only when performance indicators fall below such levels that the organization searches for better alternatives. This balance between the exploitation of the available knowledge and the exploration and search for new knowledge and new possibilities, underlined by March (1991), is closely related to the mechanisms of selection and mutation: a procedure or routine that performs well being adopted by the system (i.e., a higher probability of survival being bestowed on it) in the case of selection, while some misfit between routines and the changing milieu may lead to a mutation in the routine or procedure (Dosi and Marengo 1994).

Evolution emerges from this process of mutual learning between agents and organizations, in the form of the parallel and interactive processes of selection and mutation. This learning may be faster or slower depending on the nature of the organization. For instance, hierarchical organizations may be able to filter out new local events and prevent learning of certain sorts: the design of organizations most efficient for knowledge exploitation purposes may lead to learning disabilities in exploration, to organizations that have lesser exploratory competences.

Toulmin has sorted change mechanisms into four categories (Toulmin 1981). Calculative change is triggered by rational choice as employed by mainstream economic theory; homeostatic change occurs in response to stimuli in accordance with fixed rules, as in single-loop learning in an organization (using new means); development change is typified by life-cycle theories and might correspond to instances of double-loop learning (learning new goals), i.e., to the restructuring of the selecting unit, its goals or mission; finally, populational change is triggered by changes in the environment and in the adoption of selective units (i.e., a change in its probability of being adopted and nurtured by the environment) - natural selection.

2. The notion of a learning economy (Lundvall and Johnson 1994) and of innovation as a form of learning have helped to dramatically reframe our perspectives on economic progress. The emphasis has been shifted from the economy as an allocative mechanism to the economy as a learning process. This does not mean however that learning proceeds inevitably or that it proceeds with as much momentum as would be desired at all times. In the same manner that in the old economy there were allocative failures, in the new economy, there are learning failures: the learning economy is often disconcerted (Paquet 1998c).

This process of collective learning, based on conversations with the situation and on interactions among stakeholders, is of necessity *situated* (Salais 1996:9), whether one is focusing on technological or organizational or institutional innovation. In each case, proximity of one sort or another is necessary for the conversation or interaction to proceed.

Proximity is not a unidimensional or solely spatial concept. Agents and groups may be proximate territorially, organizationally, institutionally, ideologically, etc. Moreover since everything happens in time and space, there may be a propensity to ascribe to geographical or historical linkages more causal importance than they really deserve for the simply reason that co-relations are more easily detectable in those dimensions. In fact, complementarities of knowledge, competences or capabilities, or organizational and technological complementary connections may well be the causal factors through which to explain faster learning through dynamic external economies (Richardson 1972; Haas 1995).

But learning remains a social cognitive process. Consequently, it requires some interaction and a major source of external benefits comes from the geographical closeness that generates not only maximal probability of contacts but also maximal probability of *learning by learning*, i.e. of developing new capabilities not only through agglomeration economies but through a greater density of situated cognition-driving interactions (Kirat and Lung 1995). In that sense, it is much less the spatial interactions *per se* than the mix of situated culture and institutions that characterizes the context and facilitates communication, cumulative informative exchange and community learning.

This is what has led some analysts (Abdeklmalki *et al* 1996) to insist on labeling those situated innovation systems as *territorialized* innovation systems: where the institutional density of the territory, its embedded loops of learning, and the sedimentation in the same space of multiplex institutionalized relationships (incorporated in cooperative projects, relational exchanges, and other initiatives and alliances) are welding the localized networks into a trustful community of practice.

Spatial contiguity may be important and even necessary for this sort of proximities and the sort of multiplex relationships that ensue to develop, but it is not sufficient. The fundamental fabric of this territorialized innovation system is the substantive logic of collective action that becomes instituted in the local culture and in the supportive collaborative framework, and underpins the different *readinesses* of the system to make the highest and best use of the existing capabilities but also to make adjustment in those capabilities (Langlois 1992).

3. The goodness of fit between the learning process, on the one hand, and the territorial and organizational clustering process that accompanies and underpins it, on the other, is a central feature of the governance of the learning economy. The territorialized system of innovation constitutes a way to bring about a certain framework that helps to coordinate the interdependent decisions of the various actors or groups. In that sense, territorialization provides some of the relevant coordinates and shapes a *world* of constraints that leads to a resolution of the indeterminacy generated by interdependence (Pecqueur 1996 Ch. 10).

Indeed, one of the important limitations of conventional neo-classical economics is the presumption that the interaction among rational individuals is fully intermediated by market mechanisms. Nobel economist Kenneth Arrow (1974) has shown that it is unlikely, but the exact nature of the visible and invisible institutions necessary to complement the market in the coordination arrangements have not been specified in a general way. They are quite different depending on the contexts. For instance, they may differ widely from one region to the next, from one culture to the next, from one period to the next. Individual rationality and mercantile contracts are not sufficient for coordination to occur. There are various non-market mechanisms at work: organizations, institutions, standards, norms and rules that are necessary for coordination to materialize (Orléan 1994).

Schelling has examined the obstacles to coordination in his analysis of pure coordination games (Schelling 1960). The archetypal situation is portrayed in the case of car operators who have to coordinate their activity by choosing on which side of the road they will drive. It matters little whether they choose to drive on the right or the left hand side of the road, but one must choose. It is a case where two acceptable Nash equilibria exist. And the indeterminacy can only be resolved by some convention or rule. In practice, the indeterminacy is lefted as a result of some contextual information that constitutes common knowledge and helps generate a focal point on which the protagonists will agree.

In the case of firms, the solution to such problems is found in the corporate culture: the firm as a community of practice is also a community of interpretation and finds in the set of its common values the source of a choice among the many possible solutions (Kreps 1990). This means that the coordination solution emerges as a *collective cognitive benchmark* which depends on the identity of the individuals partaking in the interaction. It depends on common knowledge; it presupposes a certain common cognitive capital, a certain experiential community.

The cognitive processes that lead to the emergence of a focal point materialize in the form of conventions on the basis of some interactive rationality. These conventions emerge from the social setting in which the economic game is nested. Indeed, the sort of social learning involved depends on the embedding network and the social proximity it provides, for this is the basis for the community of interpretation. These features of the coordination mechanism emphasize the necessary incompleteness of the market mechanisms and the need for *complementary collective cognitive devices* to ensure effective coordination

This is central in the process of innovation which is based on complex *intersubjective* and *intertemporal* complementarities among the stakeholders during the sequential process leading to an innovation (Arthur 1990; Bruno 1997). And Pecqueur (1997) has shown that, in this complex coordination game, the *territory* may well be the locus of coordination of knowledge production, the root of the convergence to the focal points. For Pecqueur, situated territorialization is not a matter of cost of transportation or spatial friction, but the basis of collective learning, a source of common knowledge, and a crucial component in the guidance toward focal points.

4. These three processes (learning, proximity, and coordination) underpin the basic process of innovation. The local and regional systems of innovation are spatial configurations that have emerged as covenants of identity and as frameworks for belonging and learning. But territory does not become automatically a *collective cognitive device* (CCD) in the sense of Favereau (1989). It becomes a CCD only if it becomes the source of common ideas, a clan, a culture of shared values. Only in such a case can it become "un guide de connaissance et de reconnaissance entre individus" (Pecqueur 1997:169).

Such a localized system of innovation is very fragile since it is based not on administrative structures but on a *frame of mind*, a common representation, a collective cognitive map that may well not be transferable from one generation to the next.

In some cases, the localized system of innovation can more easily develop permanent roots when it becomes based on a convention of quality pertaining to products or on some development of the territory as an effective learning world in the new cognitive division of labour (Moati and Mouhoud 1994). In such a world (in the sense of Storper and Salais (1997)) the logic of innovation and learning is fueled by proximity and generates a new form of dynamic coordination through relational exchange (Goldberg 1989).

It should be clear that the transversal importance of trust in this dynamic is crucial. But it is also a constructed dimension of the system of coordination: proximity, learning and sustainable evolutionary coordination mechanisms generate a self-re-inforcing dynamics of interdependence and trust (Brousseau et al 1997: 418) that others like Herbert Simon have compared to some form of altruism (Simon 1993) and which lead to limited opportunism. This in turn lends support to a much more robust learning system as learning goes much beyond the accumulation and development of technological protocols and competences into the accumulation and development of knowledge about the organization of relations among partners, the transformation and reconfiguration of their representations, and even the redefinition of interests and objectives. In such cases, the local system of innovation becomes a *trust system* that is likely to be maintained as an evolutionary sustainable convention (Laurent and Paquet 1998).

Evolutionary sustainable conventions are by definition robust. They are immune to minor deviance because of the fact that unless a substantial proportion of the population modifies its behaviour those who adhere to the prevalent convention will remain dominant. What will bring about a drift from the prevalent convention to another is nothing less than a major discontinuity. Robust localized social networks may therefore both ensure important inertia at times and allow the new convention to spread well and fast at other times (Boyer and Orléan 1994).

This sort of convention shift is however in the nature of a reframing of perspectives. It requires literally a move to a new "world" and not simply a marginal adjustment. This is the case because of the fact that conventions are part of an order that evolves as a collective: individual conventions are resilient until the order shifts. One may easily understand how different "worlds" may coexist in our modern economies where they pertain to different segments of the production arrangements (Storper and Salais 1997) but also how entrepreneurship, democratic action and the cultivation of solidarity may contribute significantly to the transformation of existing worlds or to the disclosing of new worlds (Spinosa *et al* 1997).

1.5   OUR RESEARCH STRATEGY IN THIS BOOK

There is not a single place where one may find a fully developed conceptual framework that encompasses all those dimensions and that is heuristically powerful enough to generate the *modest general propositions* that are necessary for the investigation to proceed. But there are many in which this sort of work is going on. The symposium on which this book is based has brought together a number of the important protagonists in this debate in order to develop a provisional statement of what is now well established and what areas require attention.

Some of our guests have taken advantage of the Ottawa meeting to address some of the conceptual issues that are still nagging researchers. Their papers are presented in Part B. Richard Florida explores in a general way the impact of the knowledge revolution on the capitalist system and the forces that have made regions the focal points for knowledge creation and learning. Annalee Saxenian examines the ways in which regional technical communities resolve their adjustment and learning problems via the strategic use of social and professional networks, and the impact of these new arrangements in bestowing a comparative advantage to firms through a blurring of their boundaries. Padmore and Gibson review and adapt two aggregate models – of innovation (Kline and Rosenberg 1986) in one case, and of competitiveness (Porter 1990) in the other – and examine their usefulness to guide the analysis and evaluation of regional systems of innovation.

Part C focuses on the analysis and comparison of various localized systems of innovation on different continents. Roger Voyer surveys broadly a sample of knowledge-based industrial clusters in different regions and develops a template to map out the major forces behind their emergence and performance. Acs, Fitzroy and Smith analyze, in a similar way but in a more quantitative manner, high-tech industrial clusters in different metropolitan areas in the United States. Lawton-Smith *et al* perform a detailed careful analysis of two regional innovation systems in the United Kingdom and correlate their performance to some features of the underlying social and professional networks. Finally, Cliff Wymbs studies the telecom industry in New Jersey and shows the importance of locational capabilities for investment decisions.

Part D presents a series of vignettes of different local and regional systems of innovation in Canada. These vignettes cover various localized systems in Western, Central and Eastern Canada, but deal also with selected aspects of the innovation process. These vignettes illustrate a multitude of aspects of the localized learning economies at different stages of their development (from the emergent system in New Brunswick to the more seasoned local system of innovation in Kitchener-Waterloo in Ontario), from diverse perspectives (from the multifaceted role of research parks to the behaviour of small firms), with a wide-ranging focus (from a small region of Quebec to the whole of Ontario), and using very different data sets (from official statistics and special survey data to very general qualitative overviews).

Part E is a very short synthesis of the lessons derived from these various perspectives, and a bold attempt at indicating new directions for the researchers and data gatherers interested in the research agenda for the next decade.

1.6   PROVISIONAL CONCLUSIONS

The focus on the analysis of the innovation process at the meso-level is promising but it is not a formulaic panacea. Even the fundamental problem of what the likely boundaries of localized systems of innovation are remains unresolved. From our own ethnographic work, we are led to believe that the "learning region" may be much more restricted than is usually presumed on the basis of a strict structural-institutional approach (Acs, de la Mothe and Paquet 1996) We see "regional" systems of innovation built on the notion of community of practice as corresponding more closely to "metropolitan" areas than to provinces or states. But this remains a presumption. It may simply be that there is no "optimal size" or "one-best-spatial-size" for learning regions.

The problems for statistical agencies when they take their distance from official administrative units in their data collection are very daunting. These problems become overwhelming when statistical agencies realize that the unit of interest when one refers to "territorialized systems of innovation" may vary considerably from place to place and from time to time.

There are various ways to approach the problem: some have been emphasized by the different authors of the papers in this book, either directly or indirectly, other approaches emphasizing the meso-level have been available for decades (Gillard 1972, 1975).

What is required at this time is a period of stock-taking, the realization that the data sets we have do not and cannot suffice, and some boldness in dealing with the challenges that are looking at us in the face. Understanding the complex dynamics of local innovative and economic growth require our recognition of this.

REFERENCES

Abdelmalki, L. et al (1996). "Technologie, institutions et territoires: le territoire comme création collective et ressource institutionnelle" in B. Pecqueur (ed.).

Acs, Z., de la Mothe, J., and Paquet, G. (1996). "Local Systems of Innovation: In Search of an Enabling Strategy". In P. Howett, *The Implications of Knowledge-Based Growth for Micro-Economic Policies*, (pp. 339-360). Calgary: University of Calgary Press.

Arrow, K. (1974). *The Limits of Organization*. New York, Norton.

Arthur, W.B. (1990). "Silicon Valley' Locational Clusters: When Do Increasing Returns Imply Monopoly?" *Mathematical Social Sciences*, 19, pp. 235-251.

Boswell, J. (1990). *Community and the Economy,* London: Routledge.

Boyer, R. et Orléan, A. (1994). "Persistance et changement des conventions" in A. Orléan (ed), pp. 219-247

Brousseau, E. *et al* (1997). "Confiance, connaissances et relations inter-firmes" in B. Guilhon *et al* (eds).

Brown, J.S. and Duguid, P. (1991). "Organizational Learning and Communities in Practice: Toward a Unified View of Working, Learning and Innovating" *Organization Science*, 2, 2.1, 40-57.

Bruno, S. (1997). "L'évaluation économique des choix comme convention institutionnelle: implications pour les décisions d'innovation" in B. Guilhon *et al* (eds) *Economie de la connaisance te organisations*. Paris: L'Harmattan, pp. 39-69.

Ciborra, C.U. (1990). "X-Efficiency, Transaction Costs, and Organizational Change" in K. Weiermair and M. Perlman (eds) *Studies in Economic Rationality - X-Efficiency Examined and Extolled: Essays Written in the Tradition of and to Honor Harvey Leibenstein,* Ann Arbor: The University of Michigan Press, pp. 205-222.

Conseil de la science et de la technologie du Québec (1997). *Pour une politique québécoise de l'innovation*. Quebec: Gouvernement du Québec.

de la Mothe, J. and Paquet, G. (eds) (1996). *Evolutionary Economics and the New International Political Economy,* London: Pinter.

de la Mothe, J. and Paquet, G. (1994). "The Dispersive Revolution" *Optimum* 25,1,pp.42-48.

Dosi, G. and Marengo, L. (1994). "Some Elements of an Evolutionary Theory of Organizational Competences" in R.W. England (ed). *Evolutionary Concepts in Contemporary Economics* (Ann Arbor: The University of Michigan Press), pp. 157-178.

Dosi, G. and Nelson, R.R. (1994). "An Introduction to Evolutionary Theories in Economics", *Journal of Evolutionary Economics*, *4*, 3, pp. 153-172.

Dosi, G., Zysman, J. and Tyson, L.D. (1990) "Technology, Trade Policy and Schumpeterian Efficiencies" in J. de la Mothe and L.M. Ducharme (eds) *Science, Technology and Free Trade*, London: Pinter, pp. 19-38.

Favereau, O. (1989). "Marchés internes, marchés externes" *Revue économique*, 40, 2, pp. 273-328.

Florida, R. (1995). "The Learning Region" *Futures*, 27, 5, pp. 527-536.

Gillard, L. (1975). "Premier bilan d'une recherche sur la méso-analyse" *Revue économique*, 26, pp. 448-516.

Gillard, L. (1972). "Six propositions pour transformer l'analyse sectorielle" *Revue d'économie politique*, 82, pp. 126-138.

Goldberg, V.P. (ed.) (1989). *Readings in the Economics of Contract Law*. Cambridge: Cambridge University Press.

Haas, S. (1995). "Economies externes technologiques, apprentissage et rendements d'agglomération" in N. Lazaric et J.M. Monnier (eds) pp. 180-205.

Handy, C. (1995). "Trust and the Virtual Organization" *Harvard Business Review*, 73, 3.

Hollingsworth, R. (1993). "Variation among Nations in the Logic of Manufacturing Sectors and International Competitiveness" in D. Foray and C. Freeman (eds) *Technology and the Wealth of Nations* (London: Pinter) pp. 301-321.

Kirat,T. et Lung, Y. (1995). "Innovations et proximités: le territoire, lieu de déploiement des processus d'apprentissage" in N. Lazaric et J.M. Monnier (eds) pp. 206-227.

Kline, S.J., and Rosenberg, N. (1986). "An Overview of Innovation", in R. Landau and N. Rosenberg (eds), *The Positive Sum Strategy: Harnessing Technology for Economic Growth*. Washington, National Academy Press.

Kreps, D. (1990). "Corporate Culture and Economic Theory" in J.E. Alt and K.A Shepsle (eds) *Perspectives on Positive Political Economy*. Cambridge: Cambridge University Press, pp. 90-143.

Langlois, R.N. and Robertson, P.L. (1995). *Firms, Markets and Economic Change*. London: Routledge.

Langlois, R.N. (1992). "Orders and Organizations: Toward an Austrian Theory of Social Institutions" in B.J. Caldwell and S. Boehm (eds) *Austrian Economics: Tensions and New Directions*. Boston: Kluwer Academic Publishers, pp. 165-183.

Laurent, P. et Paquet, G. (1998). *Economie et épistémologie de la relation ¥ coordination et gouvernance distribuée*. Lyon: Vrin.

Lazaric, N. et Monnier, J.M. (eds) (1995). *Coordination économique et apprentissage des firmes*. Paris: Economica.

Lecoq, B. (1989). *Réseau et système productif régional*, Dossiers de l'IRER, 23.

Lundvall, B. and Johnson, B. (1994). "The Learning Economy" *Journal of Industry Studies*, 1, pp. 23-42.

Maddox, J. and McGee, H. (1994). "Mexico's Bid to Join the World" *Nature*, 28 April, pp. 789-804.

Maillet, D. (1992). "Milieux et dynamique territoriale de l'innovation", *Canadian Journal of Regional Science*, 15,2:199-218.

March, J.G. (1991). "Exploration and Exploitation in Organizational Learning", *Organization Science*, 2, pp. 71-87.

Marengo, L. (1993). "Knowledge Distribution and Coordination in Organizations: On Some Social Aspects of the Exploitation vs Exploration Trade-off", *Revue internationale de systémique*, 7, 5.

McCain, R.A. (1992). *A Framework for Cognitive Economics*. Westport: Praeger.

Moati, P. et Mouhoud, E.L. (1994). "Information et organisation de la production: vers une division cognitive du travail" *Economie appliquée*, XLVI, 1, pp. 47-73.

Nelson, R.R. and Winter, S.G. (1982). *An Evolutionary Theory of Economic Change,* Harvard University Press, Cambridge.

Nelson, R.R. and Winter, S.G. (1977). "In Search of A Useful Theory of Innovation", *Research Policy*, 6, 1, pp. 36-76.

Orléan, A. (ed) (1994). *Analyse économique des conventions*. Paris: Presses Universitaires de France.

Orr, J. (1990). *Talking about Machines: A Ethnography of a Modern Job.* Unpublished Ph.D. Thesis, Cornell University.

Paquet, G. (1998a). "Evolutionary Cognitive Economics" *Information Economics and Policy,* Elsevier.

Paquet, G. (1998b). "Lamberton's Road to Cognitive Economics" in S.Macdonald and J. Nightingale (eds) *Information and Organization: A Tribute to the Work of Don Lamberton,* Amsterdam: Elsevier.

Paquet, G. (1998c). "Canada as a Disconcerted Learning Economy: A Governance Challenge" *Transactions of the Royal Society of Canada.*

Paquet, G. and Pigeon, L. (1998). "In Search of a New Covenant" in Evert Lindquist (ed.) *Government Restructuring and the Future of Career Public Service in Canada,* Toronto: Institute of Public Administration of Canada.

Pecqueur, B. (ed) (1996). *Dynamiques territoriales et mutations économiques.* Paris: L'Harmattan.

Pecqueur, B. (1997). "Processus cognitifs et construction des territoires économiques" in B. Guilhon *et al,* pp. 154-176.

Porter, M.E. (1990). *The Competitive Advantage of Nations,* Free Press, New York.

Putnam, R.D. (1995). "Bowling Alone: America's Declining Social Capital" *Journal of Democracy,* 6, 1, pp.65-78.

Richardson, G.B. (1972). "The Organization of Industry" *Economic Journal,* 82, pp. 883-896.

Salais, R. (1996). Préface in B. Pecqueur (ed.) *Dynamiques territoriales et mutations économiques,* Paris: L'Harmattan.

Saxenian, A. (1994). *Regional Advantage: Culture and Competition in Silicon Valley and Route 128,* Cambridge: Harvard University Press.

Schelling, T.C. (1960). *The Strategy of Conflict.* Oxford: Oxford University Press.

Simon, H.A. (1993). "Altruism and Economics", *American Economic Review 83,* 2, pp. 156-161.

Spinosa, C. *et al* (1997). *Disclosing New Worlds.* Cambridge: The MIT Press.

Storper, M. and Salais, R. (1997). *Worlds of Production.* Cambridge: Harvard University Press.

Storper, M. (1992). "The Limits to Globalization: Technology Districts and International Trade," *Economic Geography 68.*

Wikstrom, S. and Normann, R. (1994). *Knowledge and Value.* London: Routledge.

Ziman, J. (1991). "A Neural Net Model of Innovation" *Science and Public Policy, 18,* 1, pp. 65-75.

PART B

CONCEPTUAL PERSPECTIVES

2 CALIBRATING THE LEARNING REGION

Richard Florida
Carnegie Mellon University

2.1 INTRODUCTION

Regions are becoming focal points for knowledge creation and learning in the new global, knowledge-intensive, capitalism. In effect, they are becoming learning regions. These learning regions function as collectors and repositories of knowledge and ideas, and provide the underlying environment or infrastructure which facilitates the flow of knowledge, ideas and learning. Quite simply, regions are becoming more important modes of economic repeated and technological organization in a global economy.

In Silicon Valley, a center for new technology has emerged where entrepreneurs and technologists with global venture capital invent new technologies of software, personalized information and biotechnology that will shape our future. In the financial centers of Tokyo, New York and London, computerized financial markets provide instantaneous capital and credit to companies and entrepreneurs across the vast reaches of the world. In the film studios of Los Angeles, computer technicians work alongside actors and film directors to produce the software that will run on new generations of home electronics products produced by television and semiconductor companies throughout Asia. Computer scientists and software engineers in Seattle work with computer game makers in Kyoto, Osaka and Tokyo to turn out new generations of high-technology computer games. In Milan, highly computerized factories produce designer fashion goods tailored to the needs of consumers in Paris, New York and Tokyo almost instantaneously. Teams of automotive designers in Los Angeles, Tokyo and Milan create designs for new generations of cars, while workers in Kyushu work to the rhythm of classical music in the world's most advanced automotive assembly factories to produce these cars for consumers across the globe. Throughout Japan, a new generation of knowledge workers operate the controls of mammoth automated factory complexes to produce the most basic of industrial products-steel. A new industrial revolution sweeps through Taiwan, Singapore, Korea, Malaysia, Thailand, Indonesia, and extends its reach to formerly undeveloped nations such as Mexico and China. And, once-written-off regions, like the former rustbelt of the USA are being revived through international investment and the creative destruction of traditional industries.

But, even though there have been numerous excellent studies of the dynamics of individual regions, the role of regions in the new knowledge-based, global capitalism remains rather poorly understood this book goes some distance to rectifying the situation. And, while several outstanding studies have chronicled the rise of knowledge-based capitalism, outlined the contours of learning organizations, and described the knowledge-creating company, virtually no one has developed a comparable theory of what such changes portend for regions and regional organization.

This chapter suggests that regions are a key element of the new age of global, knowledge-based capitalism. Its central argument - in harmony with the other contributors to this volume - is that regions are themselves becoming focal points for knowledge-creation and learning in the new age of capitalism, as they take on the characteristics of *learning regions*. Learning regions function as collectors and repositories of knowledge and ideas, and provide an underlying environment or infrastructure which facilitates the flow of knowledge, ideas and learning. Learning regions are increasingly important sources of innovation and economic growth, and are vehicles for globalization. In elaborating this thesis, the following sections provide brief descriptions of the new era of knowledge-based capitalism and its global scope, before turning to our discussion of the dynamics of learning regions.

2.2   THE KNOWLEDGE REVOLUTION

Capitalism, as writers as diverse as Peter Drucker and Ikujiro Nonaka point out, is entering into a new age of knowledge creation and continuous learning. This new system of knowledge-intensive capitalism is based on a synthesis of intellectual and physical labour - a melding of *innovation and production*. In fact, the main source of value and economic growth in knowledge-intensive capitalism is not the firm, but the human mind. Knowledge-intensive capitalism represents a major advance over previous systems of Taylorist scientific management or Fordist assembly-line systems, where the principal source of value and productivity growth was repeated physical labor. The shift to knowledge-based capitalism represents an epochal transition in the nature of advanced economies and societies. Ever since the transition from feudalism to capitalism, the basic source of productivity, value and economic growth has been physical labor and manual skill. In the knowledge-intensive organization, intelligence, creativity and intellectual labor replace physical labor as the fundamental sources of value and profit.

The new age of capitalism makes use of the entirety of human intellectual and creative capabilities. Both R&D scientists and workers on the factory floor are the sources of ideas and continuous innovation. Workers on the factory floor use their deep and intimate knowledge of machines and production processes to devise new, more efficient production processes. This new system of economic organization harnesses the knowledge and intelligence of the team - a sharp break with the conception of individual knowledge embodied in the heroic lone inventor or great scientist. Teams of R&D scientists, engineers and factory workers become collective agents of innovation. The lines between the factory and the laboratory have blurred.

The factory is itself becoming more like a laboratory - a place where new ideas and concepts are generated, tested and implemented. Like a laboratory, the knowledge-intensive factory is an increasingly clean, technologically advanced and information-rich environment. In an increasing number of factories, workers perform their tasks in clean room environments, alongside robots and machines which conduct the physical aspects of the work. In some knowledge-intensive factories, laboratory-like spaces are available for workers, which may include sophisticated laboratory-like equipment-computerized measuring equipment, advanced monitoring devices, and test equipment. Workers use these laboratory-like spaces together with R&D scientists and engineers to analyses fine-tune, and improve products and production processes.

2.3 THE GLOBAL SHIFT

This new age of capitalism is taking the form of an increasingly integrated economic system, with globe-straddling networks of transnational corporations and high levels of foreign direct investment between and among nations. Such investment is a vehicle for diffusing advanced technologies and state-of-the-art management practices and is a powerful contributor to the global flow of knowledge. Indeed, international investment has surpassed global trade as the defining feature of the new global economy. A United Nations report shows that today transnational corporations operate some 170,000 factories and branches throughout the globe. In 1992, this worldwide network of foreign affiliates generated $5.5 trillion in sales, exceeding world exports of $4 trillion, one-third of which took the form of intra-firm trade.

Globalization is increasingly taking place through *transplant* companies and in some instances through integrated complexes of transplant factories and surrounding supplier and product development activities. The best examples of such complexes include Toyota and Honda's massive production complexes in the USA. In fact, Japanese automotive production in North America takes the form of an integrated transplant complex comprising seven major automotive assembly complexes and more than 400 suppliers located in and around the traditional industrial heartland region of the USA.

Transplant investment is the source of important productivity improvement and economic growth. According to a recent study by the McKinsey Global Institute, transplants increase productivity by accelerating the adoption and diffusion of best-practice organization and management, and placing pressure on domestic industries to adopt those best practices. The McKinsey study notes that:

Transplants from leading-edge producers: (1) directly contribute to higher levels of domestic productivity, (2) prove that leading-edge productivity can be achieved with local inputs, (3) put competitive pressure on other domestic producers, and (4) transfer knowledge of best-practices to other domestic producers through natural movement of personnel. Moreover, foreign direct investment has provoked less political opposition than trade because it creates jobs instead of destroying them. Thus, it is likely to grow faster in years to come.

A recent OECD study provides additional empirical evidence of the link between foreign direct investment, productivity improvement and economic growth. Comparing investment and productivity patterns in 15 advanced industrial nations, the OECD study found that foreign-owned companies are typically more efficient than domestic firms in both absolute levels and in rates of productivity growth. The study found that these productivity gains resulted from more advanced technology than domestic industries, or from adding capacity. By contrast, productivity increases at locally owned companies more often resulted from downsizing and lay-offs. The study also found that international investment has been a key source of employment growth across the advanced industrial nations. In 10 of 15 countries studied, foreign-owned companies created new employment more rapidly than did their domestically owned counterparts, sometimes expanding their operations while domestic firms were contracting. In three others, they eliminated jobs, but they did so more slowly than domestically owned enterprises. The study found that the largest employment declines occurred in Japan and Germany, where soaring costs during the 1980s caused international investors to cut a significant number of jobs. Furthermore, the OECD study points to a link between investment and trade, as foreign subsidiaries tended to export and import more than domestic firms, with most of the imports taking the form of intra-firm trade.

Foreign direct investment has played a key role in the economic revival of the USA. For example, productivity grew more rapidly in foreign-owned transplant manufacturing companies in the US than for the manufacturing sector as a whole during the 1980s. The real output of transplant manufacturers rose nearly four times as fast as all manufacturing establishments between 1980 and 1987. Transplant companies generated productivity increases and value-added which outdistance US-owned companies. From 1987 to 1990, for example, the rate of increase in plant and equipment expenditures for transplant industrial enterprises (eg non-bank, nonagricultural business) was five times greater than that for US-owned business. As of 1989, value-added per employee was substantially higher in transplants than for US-owned manufacturers. And, transplant companies have played an important role in the economic resurgence of the US industrial midwest-a region which produced more than $350 billion in manufacturing output, making it the third largest manufacturing economy in the world.

Technology and innovative activity are also undergoing considerable globalization. For most of the Cold War, the USA was the world's overwhelming generator of research and technology. However, by the early 1990s, the combined R&D expenditures of the EC and Japan exceeded those of the USA, and their R&D efforts were much more focussed on commercial technology. Furthermore, the share of patents to non-US inventors has increased dramatically, with non-US inventors accounting for nearly half of all US patents in 1992.

As the pace of innovation has accelerated and the global sources of technology have grown, corporations have expanded their global innovative activities and cross-border alliances. A global survey of companies in the USA, Europe and Japan found that corporations are substantially increasing their reliance on external sources of research and technology for both basic research and product and development. Furthermore, a growing number of corporations are establishing R&D facilities abroad. US companies conducted roughly 12% of their total R&D activities abroad in 1991, the most recent year for which reliable data are available. Japanese companies have established a global network of more than 200 research, development and design facilities.

The past decade has seen the progressive globalization of the US technology base, as the USA has become the hub in the global science and technology system. Since 1980, foreign companies have invested tens of billions of dollars in roughly 400 research, development and design centers in the USA. The annual R&D outlays of these facilities has risen from $4.5 billion in 1982 to $10.7 billion in 1992, and the share of total industrial R&D they comprise has grown from 9% to nearly 17% over the same period, roughly one out of every six dollars of industrial R&D spending in the USA. R&D spending by foreign companies is highly concentrated in sectors where foreign industries are highly competitive-European companies in chemicals and pharmaceuticals and Japanese and German companies in automotive-related technologies and electronics. The globalization of innovation is required to tap into the sources of knowledge and ideas, and scientific and technical talent which are embedded in cutting edge regional innovation complexes such as Silicon Valley in the USA, Tokyo or Osaka in Japan, Stuttgart in Germany, and many others.

2.4   TOWARD THE LEARNING REGION

The shift to knowledge-intensive capitalism goes beyond the particular business and management strategies of individual firms. It involves the development of new inputs and a broader infrastructure at the regional level on which individual firms and production complexes of firms can draw. The nature of this economic transformation makes regions key economic units in the global economy. In essence, globalism and regionalism are part of the same process of economic transformation. In an important and provocative essay in *Foreign Affairs*, Kenichi Ohmae suggests that regions, or what he calls *region-states*, are coming to replace the nation state as the centerpiece of economic activity.

> The nation state has become an unnatural, even dysfunctional unit for organizing human activity and managing economic endeavor in a borderless world. It represents no genuine, shared community of economic interests; it defines no meaningful flows of economic activity. On the global economic map the lines that now matter are those defining what may be called region states. Region states are natural economic zones. They may or may not fall within the geographic limits of a particular nation-whether they do is an accident of history. Sometimes these distinct economic units are formed by parts of states. At other times, they may be formed by economic patterns that overlap existing national boundaries, such as those between San Diego and Tiajuana. In today's borderless world, these are natural economic zones and what

matters is that each possesses, in one or another combination, the key ingredients for successful participation in the global economy.

Region-states, Ohmae points out, are fundamentally tied to the global economy through mechanisms such as trade, export, and both inward and outward foreign investment-the most competitive region-states are home not only to domestic or indigenous companies, but are attractive to the best companies from around the world. Region-states can be distinguished by the level and extent of their insertion in the international economy and by their willingness to participate in global trade.

The primary linkages of region states tend to be with the global economy, and not with host nations. Region states make such effective points of entry into the global economy because the very characteristics that define them are shaped by the demands of that economy. Region states tend to have between five million and 20 million people. A region state must be small enough for its citizens to share certain economic and consumer interests but of adequate size to justify the infrastructure-communications and transportation links and quality professional services-necessary to participate economically on a global scale. It must for example, have at least one international airport and, more than likely, one good harbor with international-class freight-handling facilities. A region state must also be large enough to provide an attractive market for the broad development of leading consumer products. In other words, region states are not defined by their economies of scale in production (which, after all, can be leveraged from a base of any size through exports to the rest of the world) but rather by having reached efficient economies of scale in their consumption, infrastructure and professional services.

For most of the 20th century, successful regional as well as national economies grew by extracting natural resources such as coal and iron ore, making materials such as steel and chemicals, and manufacturing durable goods such as automobiles, appliances and industrial machinery. The wealth of regions and of nations in turn stemmed from their abilities to leverage so-called natural comparative advantages that allowed them to be mass producers of commodities competing largely on the basis of relatively low production costs. However, the new age of capitalism has shifted the nexus of competition to ideas. In this new economic environment, regions build economic advantage through their ability to mobilize and to harness knowledge and ideas. In fact, regionally based complexes of innovation and production are increasingly the preferred vehicle used to harness knowledge and intelligence across the globe.

The new age of capitalism requires a new kind of region. In effect, regions are increasingly defined by the same criteria and elements which comprise a knowledge-intensive firm-continuous improvement, new ideas, knowledge creation and organizational learning. Regions must adopt the principles of knowledge creation and continuous learning; they must in effect become learning regions. Learning regions provide a series of related infrastructures which can facilitate the flow of knowledge, ideas and learning.

Regions possess a basic set of ingredients that constitute a production system (see Table 1). They all have a *manufacturing infrastructures* - a network of firms that

Table 1 - From mass production to learning regions

	Mass Production Region	**Learning Region**
Basis of competitiveness	Comparative advantage based on: • Natural resources • Physical labor	Sustainable advantage based on: • Knowledge creation • Continuous improvement
Production system	Mass production • Physical labor as source of value • Separation of innovation and production	Knowledge-based production • Continuous creation • Knowledge as source of value • Synthesis of innovation and production
Manufacturing infrastructure	Arm's length supplier relations	Firm networks and supplier systems as sources of innovation
Human Infrastructure	• Low-skill low-cost labor • Taylorist work force • Taylorist education and training	• Knowledge workers • Continuous improvement of human resources • Continuous education and training • Globally oriented physical and communication infrastructure • Electronic data exchange
Physical and communication infrastructure	Domestically oriented physical infrastructure	• Mutually dependent relationships • Network organization • Flexible regulatory framework
Industrial governance system	• Adversarial relationships • Command and control regulatory framework	

produce goods and services. Mass production organization was defined by a high degree of vertical integration and internalization of capabilities. External supplies tended to involve ancillary or non-essential elements, were generally purchased largely on price, and stored in huge inventories in the plant. Knowledge-intensive economic organization is characterized by a much higher degree of reliance on outside suppliers and the development of co-dependent complexes of end-users and suppliers. In heavy industries, such as automobile manufacturing, large assembly facilities play the role of hub, surrounding themselves with a spoke network of customers and suppliers in order to harness innovative capabilities of the complex, enhance quality and continuously reduce costs.

Regions have a *human infrastructure* – labor market from which firms draw knowledge workers. Mass production industrial organization was characterized by a schism between physical and intellectual labor–a large mass of relatively unskilled workers who could perform physical tasks but had little formal involvement in managerial, technical or intellectual activities, and a relatively small group of managers and executives responsible for planning and technological development. The human infrastructure system of mass production–the system of public schools, vocational training, and college and university professional programs in business and engineering–evolved over time to meet the needs of this mass production system turning out a large mass of 'cogs-in-the-machine' and a smaller technocratic elite of engineers and managers. The human infrastructure required for a learning region is quite different. As its name implies, a learning region requires a human infrastructure of knowledge workers who can apply their intelligence in production. The education and training system must be a learning system that can facilitate life-long learning and provide the high levels of group orientation and teaming required for knowledge-intensive economic organization.

Regions possess a *physical and communications infrastructure* upon which organizations deliver their goods and services and communicate with one another. The physical infrastructure of mass production facilitated the flow of raw materials to factory complexes and the movement of goods and services to largely domestic markets. Knowledge-intensive firms are global players. Thus, the physical infrastructure of the new economy must develop links to and facilitate the movement of people, information, goods and services on a global basis. Furthermore, knowledge-intensive organization draws a great portion of its power from the rapid and constant sharing of information and increasingly electronic exchange of key data between customers, end-users and their suppliers. For example, seat suppliers for Toyota receive a computer broadcast of what seats to build as Toyota cars start down the assembly line. A learning region requires a physical and communication infrastructure which facilitates the movement of goods, people and information on a just-in-time basis.

To ensure growth of existing firms and the birth of new ones, regions have a capital allocation system and financial market which channel credit and capital to firms. Existing financial systems create impediments to the adoption of new management practices. For example, interviews with executives and surveys of knowledge-intensive firms in the USA indicate that banks and financial institutions often require inventory to be held as collateral, creating a sizeable barrier to the just-in-time inventory and supply practices which define knowledge-intensive economic organization. The capital allocation system of a learning region must create incentives for knowledge-based economic organization, for example, by collateralizing knowledge assets rather than physical assets.

Regions also establish mechanisms for *industrial governance* – formal rules, regulations and standards, and informal patterns of behavior between and among firms, and between firms and government organizations. Mass production regions were characterized by top-down relationships, vertical hierarchy, high degrees of functional or task specialization, and command-and-control modes of regulation. Learning regions must develop governance structures which reflect and mimic those of knowledge-intensive firms, that is co-dependent relations, network organization, decentralized decision making, flexibility, and a focus on customer needs and requirements.

Learning regions provide the crucial inputs required for knowledge-intensive economic organization to flourish: a manufacturing infrastructure of interconnected vendors and suppliers; a human infrastructure that can produce knowledge workers, facilitates the development of a team orientation, and which is organized around life-long learning; a physical and communication infrastructure which facilitates and supports constant sharing of information, electronic exchange of data and information, just-in-time delivery of goods and services, and integration into the global economy; and capital allocation and industrial governance systems attuned to the needs of knowledge-intensive organizations.

2.5 BUILDING THE FUTURE

For most of the past two decades, experts predicted a shift from manufacturing to a post-industrial service economy, or from basic industries to high technology. In the wake of the predictions, efforts were undertaken to invest in new critical technologies and industries. But, the change under way is not one of old sectors giving way to new, but a more fundamental change in the way goods are produced and the economy itself is organized-from mass production to a knowledge-based economy. The implications of the epochal economic transformation are indeed sweeping.

For firms and organizations, the challenge will be to shift towards the principles of knowledge-based organization, and to adopt new organizational and management systems which harness knowledge and intelligence at all points of the organization from the R&D laboratory to the factory floor. Maintaining a balance between cutting-edge innovation and high-quality and efficient production will be a critical issue. To do so, organizations will increasingly adopt best-practice techniques throughout the world, creating new and more powerful forms of knowledge-intensive organizations. Such organizational mechanisms are likely to blend the ability of 'Silicon Valley' style high-technology companies to spur individual genius and creativity, with strategies and techniques for continuous improvement and the collective mobilization of knowledge. Knowledge-intensive firms and organizations will be called on to build integrated and dense global webs of innovation and production. And these firms will increasingly be forced to build and maintain new regional infrastructures which can support knowledge-based production systems.

The new age of capitalism holds even greater challenges for regions. The very fabric of regional organization will change, as regions gradually adopt the principles of knowledge creation and learning. Learning regions will be called on to supply the requisite human, manufacturing and technological infrastructures required to support knowledge-intensive forms of innovation and production. Rather than ushering in the 'end of geography,' globalization is likely to occur increasingly through complex systems of regional interdependence and integration. And, as the nation-state is squeezed between the poles of accelerating globalization and rising regional economic organization, regions will become focal points for economic, technological, political and social organization.

At a broader level, there is likely to be a shift from strategies and policies which emphasize national competitiveness to ones which revolve around the concept of sustainable advantage at the regional as well as national scale. *Sustainable advantage* means that organizations, regions and nations shift their focus from short-run economic performance to re-creating, maintaining and sustaining the conditions required to be world-class performers through continuous improvement of technology, continuous development of human resources, the use of clean production technology, elimination of waste, and a commitment to continuous environmental improvement. Indeed, the concept of sustainable advantage has the potential to become central organizing principles for economic and political governance at the international, national and regional scales. In this sense, there is some possibility that over time it may come to replace the increasingly dysfunctional Fordist model of nationally based political-economic regulation.

The industrial and innovation systems of the 21st century will be remarkably different from those which have operated for most of the 20th. Knowledge and human intelligence will replace physical labor as the main source of value. Technological change will accelerate at a pace heretofore unknown: innovation will be perpetual and continuous. Knowledge-intensive organizations based on networks and teams will replace vertical bureaucracy, the cornerstone of the 20th century. The intersection of relentless globalization and the emergence of learning regions are likely to erode the power and authority of the nation-state-the paragon of 19th and 20th century political economy. Whole new institutions for international trade, investment, environment and security will doubtless be created. While the new century holds out great hope, it will require tremendous energy and effort to set in motion the necessary changes, and an unparalleled collective effort to bring them about.

3 REGIONAL SYSTEMS OF INNOVATION AND THE BLURRED FIRM

AnnaLee Saxenian
University of California at Berkeley

3.1   INTRODUCTION

This chapter argues that it is the region and its relationships, rather than the firm, that defines opportunities for individual and collective advance in Silicon Valley. It suggests that open labor markets offer important competitive advantages over traditional corporate job ladders in a volatile economic environment. The essential advantage of regional, rather than firm-based, labor markets lies in the multiple opportunities they provide for learning.

Learning occurs in Silicon Valley as individuals move between firms and industries acquire new skills, experiences, and know-how. It occurs as they exchange technical and market information in both formal and informal foras, and as shifting teams of entrepreneurs regroup to experiment with new technologies and applications. Learning occurs as firms of different sizes and specializations jointly solve shared problems. Above all, learning occurs through failure, which is more common than success. In short, learning in Silicon Valley is a collective process that is rarely confined within the boundaries of individual firms and ultimately draws of the resources of the region as a whole.

The case of Silicon Valley offers important lessons. The region provides a clear example of "boundaryless careers" in action. And while much scholarship on work and employment concerns the often wrenching transition from traditional organizational job ladders toward less bounded labor markets, this transition has been avoided in Silicon Valley. From the earliest days, careers in Silicon Valley had few boundaries. As a result, the Silicon Valley experience provides important insights into both the social and institutional prerequisites and the learning advantages of "boundaryless careers."

In contrast with standard economic models of labor market behavior that rely on atomistic assumptions concerning individual job search and human capital, this essay emphasizes the social embeddedness of labor markets (Granovetter, 1995, 1988). It demonstrates the extent to which career mobility in Silicon Valley depends centrally on participation in local networks of social relations. These networks not only transcend company and industry lines, but also blur the boundaries between the economy and social life. In the words of one theorist of open labor markets:

> ... individuals secure their long-term employability through participation in neighborhood groups, hobby clubs, or other professional and social networks outside the firm. Only those who participate in such multiple, loosely connected networks are likely to know when their current jobs are in danger, where new opportunities lie, and what skills are required to seize them. The more open corporate labor markets become, the greater the economic compulsion to participate in the social activities they organize. (Sabel, 1991)

These words were not written to describe Silicon Valley, although they certainly could have been.

In the following section I describe the origins of the technical community in Silicon Valley--one that supports open labor markets. Subsequent sections demonstrate how the region's social and professional networks support high rates of job mobility and contribute to ongoing experimentation, entrepreneurship, and collective learning at the regional level. In a final section I contrast Silicon Valley's labor markets with those of the region's leading technology competitor, Boston's Route 128, where firms have, until recently, maintained more traditional organizational boundaries and career paths. While the comparison cannot be fully developed here, research suggests that these differences help account for the divergent fortunes of these two regional economies (Saxenian, 1994).

3.2   THE ORIGINS OF A TECHNICAL COMMUNITY IN SILICON VALLEY

Silicon Valley's early engineers and scientists saw themselves as the pioneers of a new industry in a new region. They were at once forging a new settlement in the West, and advancing the development of a revolutionary technology, semiconductor electronics. As newcomers to a region that lacked prior industrial traditions, Silicon Valley's pioneers had the freedom to experiment with institutions and organizational forms, as well as technologies. These young engineers, having left behind families, friends and established communities, were unusually open to risk-taking and experimentation.

The shared experience of working at the Fairchild Semiconductor Corporation also served as a powerful bond for many of the region's early semiconductor engineers. During the 1960s, it seemed as if every engineer in Silicon Valley had worked there. Even today, many of the region's entrepreneurs and managers still speak of Fairchild as an important managerial training ground and applaud the education they got at "Fairchild University." Similar shared professional experiences continued to reinforce the sense of community in the region even after individuals had moved on to different, often competing, firms.

A poster of the Fairchild family tree, which traces the genealogy of the scores of Fairchild spin-offs, hangs on the walls of many Silicon Valley firms. This family tree symbolizes the complex mix of social solidarity and individualistic competition that characterizes Silicon Valley. The tree graphically illustrates the common ancestry of the region's semiconductor industry and reminds engineers of the personal ties which enabled people, technology and money to frequently recombine into new ventures. The importance of these overlapping, quasi-familial ties is reflected in continuing references, more than three decades later, to the "fathers" (or "grandfathers") of Silicon Valley and their offspring, the "Fairchildren."

At the same time, the family tree glorifies the risk-taking and competitive individualism which distinguish the region's business culture. Silicon Valley's heroes are the successful entrepreneurs who have taken aggressive professional and technical risks: the garage tinkerers who created successful companies. These entrepreneurial heroes are celebrated for their technical achievements and for the often considerable wealth that success has brought them.

Habits of informal cooperation among Silicon Valley engineers predate the semiconductor industry. Stanford Dean of Engineering Frederick Terman actively pursued a vision of building a "community of technical scholars" around Stanford during the 1940s and 1950s in order to enhance the region's industrial base. He promoted collaboration between the university and local technology firms in a variety of ways ranging from providing start-ups with financial support and low-rent space in the newly created Stanford Industrial Park, offering continuing technical education for employees of local companies, and encouraging ongoing interaction between businesses and university faculty and students. His support of engineering students such as William Hewlett and David Packard and the Varian brothers thus far exceeded traditional professorial encouragement of promising graduate students.

These students in turn extended this tradition of assistance in their relations with other firms in the region, providing encouragement, advice, computer time, space, and even financing to new entrepreneurs. A San Jose journalist later noted that:

As their company grew, both Hewlett and Packard became very involved in the formation and growth of other companies. They encouraged entrepreneurs, went out of their way to share what they learned, and were instrumental in getting electronics companies to work together on common problems... Largely because of them, there's an unusual spirit of cooperation in the local electronics industry.

The shared identities which grow out of these social and professional networks underlie the common practices of information exchange among the region's producers. The Wagon Wheel bar in Sunnyvale, a popular watering hole where engineers met during the 1960s and 1970s to exchange ideas and gossip, has been termed "the fountainhead of the semiconductor industry." As Tom Wolfe (1983: 362) described it:

Every year there was some place, the Wagon Wheel, Chez Yvonne, Rickey's, the Roundhouse, where members of this esoteric fraternity, the young men and women of the semiconductor industry, would head after work to have a drink and gossip and brag and trade war stories about phase jitters, phantom circuits, bubble memories, pulse trains, bounceless contacts, burst modes, leapfrog tests, p-n junctions, sleeping sickness modes, slow-death episodes, RAMs, NAKs, MOSes, PCMs, PROMs, PROM blowers, PROM blasters, and teramagnitudes, meaning multiples of a million millions.

By all accounts, these informal conversations were pervasive, and served as an ~~important~~ source of up-to-date information about competitors, customers, and changes ~~in markets~~ and technologies. Local entrepreneurs came to see social relationships and even gossip as a crucial aspect of their businesses. In an industry characterized by vigorous technological change and intense competition, informal communication was often of more value than formal, but less timely forums, such as industry journals.

Informal information exchange was not limited to after hours, but continued on the job. Competitors consulted each other on technical matters with a frequency unheard of in other areas of the country. According to one executive: "I have people call me quite frequently and say, 'Hey, did you ever run into this one?' and you say 'Yeh, seven or eight years ago. Why don't you try this, that or the other thing?' We all get calls like that." (Braun and Macdonald, 1978: 130)

Information exchange also occurred through frequent gatherings organized by the Western Electronics Manufacturers Association (WEMA) and at other industry conferences and trade shows. According to the president of WEMA: "Easterners tell me that people there don't talk to their competitors. Here they will not only sit down with you, but they will share the problems and experiences they have had." (Bylinsky, 1974)

A variety of more and less formal gatherings of specialists also served as forums for exchange. The Homebrew Computer Club, for example, was founded in 1975 by a group of local microcomputer enthusiasts who had been influenced by the counter-culture ethic of the sixties. They placed a notice on bulletin boards inviting those interested in computers to: "come to a gathering of people with like-minded interests. Exchange information, swap ideas, help work on a project, whatever..." (Levy, 1984: 194) Within months, the club's membership had reached about 500 regular members, mostly young "hackers," who came to meetings to trade, sell or give away computer hardware and software and to get advice. The club became the center of an informal network of microcomputer experts in the region which survived even after the group itself folded. Eventually, more than twenty computer companies, including Apple Computer, were started by Homebrew members.

While Homebrew is a particularly well known example, it was not exceptional. Such groups continue to be pervasive in Silicon Valley. The CEO of a semiconductor materials firm noted that: "There are people gathered together once or once every two months to discuss every area of common scientific interest in the Valley. Around every technological subject, or every engineering concern, you have meeting groups that tend to foster new ideas and innovate. People rub shoulders and share ideas." (Delbecq and Weiss, 1988)

3.3   SOCIAL NETWORKS

The region's dense social and professional networks are not simply conduits for the dissemination of technical and market information. They also function as highly efficient job search networks, contributing to the unusually high rates of inter-firm mobility in the region. Gathering places like the Wagon Wheel served as informal recruiting centers as well as listening posts; job information flowed freely along with shop talk. As one engineer reported: "In this business there's really a network. You just don't hire people out of the blue. In general, its people you know, or you know someone who knows them." (Gregory, 1984: 445)

Engineers in Silicon Valley shifted so frequently between firms that mobility not only became socially acceptable, it became the norm. The preferred career option in Silicon Valley was to join a small company or a start-up, rather than an established company with a good reputation. In fact, the superiority of innovative, small firms over large corporations became an article of faith among the region's engineers.

By the 1970s, Silicon Valley was distinguished by the highest levels of job-hopping in the nation. Average annual employee turnover in local electronics firms exceeded 35% and was as high as 59% in the region's small firms (American Electronics Association, 1981). It was almost unheard of for a technical professional in Silicon Valley to have a career in a single company. An anthropologist studying the career paths of the region's computer professionals concluded that job tenures in Silicon Valley averaged two years. One engineer explained to her:

> Two or three years is about max [at a job] for the Valley because there's always something more interesting across the street. You don't see someone staying twenty years at a job here. If they've been in a small company with 200 to 300 people for 10 or 11 years you tend to wonder about them. We see those types coming in from the East coast. (Gregory, 1984: 216)

The region's engineers developed loyalties to each other and to advancing technology, rather than to individual firms or even industries. In the words of the co-founder of LSI Logic: "Here in Silicon Valley there's far greater loyalty to one's craft than to one's company. A company is just a vehicle which allows you to work. If you're a circuit designer its most important for you to do excellent work. If you can't in one firm, you'll move on to another one."

When John Sculley was recruited from the East coast in 1985 to become the CEO of Apple Computer, he was amazed by the extent to which mobility had become a norm in Silicon Valley:

> The mobility among people strikes me as radically different than the world I came from out East. There is far more mobility and there is far less real risk in people's careers. When someone is fired or leaves on the East coast, it's a real trauma in their lives. If they are fired or leave here it doesn't mean very much. They just go off and do something else... (Delbecq and Weiss, 1988: 37)

The geographic proximity of firms in Silicon Valley facilitated these high levels of mobility. Moving from job to job in Silicon Valley was not as disruptive of personal, social or professional ties as it could be elsewhere in the country. According to one engineer:

> If you left Texas Instruments for another job, it was a major psychological move, all the way to one coast or the other, or at least as far as Phoenix. Out here, it wasn't that big a catastrophe to quit your job on Friday and have another job on Monday. This was just as true for company executives. You didn't necessarily have to tell your wife. You just drove

off in another direction on Monday morning. You didn't have to sell your house, and your kids didn't have to change schools." (Hanson, 1982)

As another Silicon Valley executive put it, "People change jobs out here without changing car pools." Ironically, many Silicon Valley of these "job-hoppers" may well have led more stable lives than the upwardly mobile "organization men" of the 1950s who were transferred from place to place by the same employer.

Local technology companies in turn competed intensely for experienced engineering talent. They offered bonuses, stock options, high salaries, pleasant work conditions, interesting projects, and other incentives to attract good people away from their competitors. Early efforts to take legal action against employees who left to join new companies proved inconclusive or protracted, and most firms came to accept high turnover as a cost of business in the region. In fact, employees often left for new opportunities with the "blessings" of top management, and the understanding that if it didn't work out, they could return. Such understandings created an important safety net for local engineers.

As individuals move from firm to firm in Silicon Valley, their paths overlap repeatedly: a colleague might become a customer or a competitor, today's boss could be tomorrow's subordinate. Individuals move both within and between industry sectors: from semiconductors to personal computers or from semiconductor equipment to software. They move from established firms to start-ups, and vice versa. And they move from electronics producers to service providers such as venture capital or consulting firms--and back again.

Professional loyalties and friendships generally survive this turmoil. Few presume that the long-term relationships needed for professional success will be found within the four walls of any particular company. Many rely on trade shows, technical conferences, and informal social gatherings to maintain and extend their social and professional networks. In the words of the CEO of a local semiconductor firm:

> The network in Silicon Valley transcends company loyalties. We treat people fairly and they are loyal to us, but there is an even higher level of loyalty--to their network. I have senior engineers who are constantly on the phone and sharing information with our competitors. I know what my competitors say in their speeches and they know what I say in private conversations.

Informal exchange and collaboration thus coexist with fierce inter-firm competition in Silicon Valley to support careers that are rarely confined to individual companies.

In the words of anthropologist Kathleen Gregory: "...negotiating a career in Silicon Valley is best viewed as an intricate free form dance between employees and employers that rewards continuous monitoring, but cannot be fully choreographed. Careers in computing do not take place by design, but are emergent and negotiated between ever changing individuals and employers." (Gregory, 1984: 205)

3.4   LEARNING THROUGH EXPERIMENTATION (AND FAILURE)

Starting a new firm became increasingly legitimate in Silicon Valley with the emergence of successive generations of successful role models. The generations of entrepreneurial start-ups depicted by the Fairchild family tree were replicated during the 1960s, 1970s and 1980s in sector after sector--from computers and disk drives to networking, software and multimedia. This proliferation of successful new ventures depends upon and in turn reproduces a regional culture of risk-taking and open exchange.

The culture of Silicon Valley accords the highest status to those who start firms. Not only is risk-taking glorified in the region, but failure is socially acceptable. And the successes of generation after generation of start-ups reinforces the belief that anyone can be a successful entrepreneur; there are no boundaries of age, status or social stratum that preclude the possibility of a new beginning. Unlike elsewhere, there is little embarrassment or shame associated with business failure. In fact, the list of individuals who have failed, even repeatedly, only to later succeed, is well known in the region.

As Apple CEO John Sculley put it: "In Silicon Valley, if someone fails, we know they're in all likelihood going to reappear in some other company in a matter of months." (McKenna, 1989: 85) New ventures in Silicon Valley are typically started by engineers who have acquired operating experience and technical skills working in other firms in the region. The archetypical Silicon Valley start-up is formed by a group of friends and/or former colleagues with an idea for a new product or application. Some are frustrated by the difficulties of pursuing their ideas within an established firm, others simply seek new challenges. They seek funding and advice from local venture capitalists--often former entrepreneurs and engineers themselves--and rely on an ever expanding circle of specialized suppliers, consultants, university researchers, and market researchers for additional assistance starting the new enterprise. As they grow, they often recruit employees from their networks of professional friends and acquaintances.

In short, the region's social and technical networks operate as a kind of super-organization through which individuals, in shifting combinations, organize a decentralized process of experimentation and entrepreneurship. Individuals move between firms and projects without the alienation that might be expected with such a high degree of mobility because their social and professional relationships remain intact (Gregory, 1984). In Silicon Valley, the region and its networks, rather than individual firms, is the engine of technological advance.

Widespread job mobility and open information exchange accelerates the diffusion of technological capabilities and understandings in the region. Departing employees are typically required to sign non-disclosure statements which prevent them from revealing company secrets. However much of the useful knowledge in the industry grows out of the experience of developing technology. When engineers move between companies, they take with them the know-how and skills acquired at their previous jobs. What distinguishes Silicon Valley is the extent to which the region's networks ensure the rapid spread of knowledge and skill within a localized industrial community.

Experimentation--and even failure--are recognized in Silicon Valley as critical opportunities for learning. One executive recruiter notes that: "Everybody knows that some of the best presidents in the valley are people that have stumbled." In the words of another commentator:

> The value of failure is its role in the learning process; unless failure is possible, no learning is possible. . . in the realm of ideas, unless falsification is possible, learning isn't possible. As a matter of fact, in information theory, no information is transmitted unless negation is possible, and so the tolerance of failure is absolutely critical to the success of Silicon Valley. If you don't tolerate failure, you can't permit success. The successful people have a lot more failures than the failures do.

Vice President of Marketing at semiconductor producer Integrated Devices Technology, Larry Jordan, described how the learning that occurs through continuous recombinations strengthens the region's industrial fabric:

> There is a unique atmosphere here that continually revitalizes itself by virtue of the fact that today's collective understandings are informed by yesterday's frustrations and modified by tomorrow's recombinations. . . Learning occurs through these recombinations. No other geographic area creates recombinations so effectively with so little disruption. The entire industrial fabric is strengthened by this process.

This localized process of recombination encourages both technological and organizational innovation. Many more technical paths are pursued in Silicon Valley than would occur either in the traditional large firm or in a region with less fluid social and industrial structures, because of the relative ease of new firm formation. Within most large firms (or stable regions) a single technical option is selected and pursued--typically leaving multiple viable alternatives untapped. Over time, the organization becomes increasingly committed, or locked-in, to a particular trajectory. A more flexible regional economy continues to generate and pursue technological alternatives.

While it was no longer true by the 1980s and 1990s that "everyone knows everyone" in Silicon Valley, local executives still regard the density and openness of the region's social and professional networks, and the corresponding frequency of informal exchange, as a crucial advantage. According to one local manager:

> Over a lunch conversation or a beer, you'll learn that company A or company B has a technology that you want, and you'll find out about it. If it fits your needs, you'll build it into your next product. Even if you don't, you get a lot of pointers from networking: you learn about the latest start-ups and you learn about people, about who's good and who's not. You also learn about technology. You learn about things that are possible. You sometimes learn that what you thought wasn't possible can in fact be done. So you just call the people up and learn more. You might not call your competitor directly, but you can call people in the supply base, and they'll share anything.

Technical information thus continues to diffuse rapidly in the region, along with the tacit technological capabilities and understandings that move as engineers move, paving the way for new opportunities and enterprises.

This environment of open communication also facilitates collaboration with customers. An executive at a 1980s semiconductor firm claims a sophisticated local customer base is absolutely critical to his firm:

When we come out with the specs for a new product, we take them to a series of companies that we have relations with and that have good technical horsepower, and they'll give us feedback on the features they like and don't like. Its an iterative process: we define a product, we get feedback and improve it, we refine it and develop associated products. The process feeds on itself. And the fact that these customers are nearby means that the iterations are faster; rapid communication is absolutely critical to insuring fast time-to-market.

Another executive similarly comments on the importance of sophisticated customers in close proximity for a process of mutual learning and adaptation:

Its not necessary that our customers are geographically close, but the fact that the valley has some of the leaders in systems and computers is a vital part of the cross-pollination process. We use others' existence to create our own existence: our form is vitally effected by their presence. We change and so do they.

These interactive relationships between customers and vendors foster cross-fertilization, learning and innovation along the entire production chain in Silicon Valley, from semiconductors to disk drives and software to final systems.

This is not to suggest that conflicts are absent in Silicon Valley. The very intensity of competition among local producers is a continuing spur to imitation and technological innovation. Competitive rivalries are often highly personalized, as status is defined as much by technical excellence as by market share. And by the 1980s, lawsuits and conflicts over intellectual property were commonplace in the region.

Yet even as competitive pressures intensify, an underlying loyalty and shared commitment to technological excellence unifies the members of this industrial community. Local firms compete intensely for market share and technical leadership, while simultaneously relying on the collaborative practices which distinguish the region. The paradox of Silicon Valley is that competition demands continuous innovation, but innovation requires inter-firm cooperation. Nothing is prized more than individual initiative and technological advance, which is impossible without access to the information, ideas, and experience that reside in the Valley's dense social and professional networks.

Thus while Silicon Valley's high rates of job mobility and new firm formation may lead to losses for individual firms, they also foster a dynamic process of industrial adaptation in the region. Knowledge of the latest techniques in design, production, and marketing diffuse rapidly throughout the area. And the ease of recombining new ideas with existing skill and know-how ensures that firms in the region pursue a multiplicity of technological opportunities, many of which would have been bypassed under a more stable industrial regime.

၁.၁ A PASSING PHASE?

Observers often conclude that Silicon Valley's open labor markets and collaborative practices are appropriate only to the early stages of an industry life cycle, when firms are small and technologies fluid. Once the industry matures, the argument goes, and companies gain scale and self-sufficiency, they will abandon local relationships (Harrison, 1994; Markusen, 1985). The obvious implication is that over time, Silicon Valley's largest firms will establish internal labor markets and more traditional organizational boundaries.

A comparison with the Route 128 region in Massachusetts-- where firms compete in the same technology sectors as Silicon Valley yet have failed to adapt successfully to changing markets--suggests that open labor markets are not simply a passing phase. Rather they are a component of a flexible and enduring alternative to the traditional organizational model, one that under conditions of market and technological volatility such as the present, offers important competitive advantages.

As late as the 1980s, the most desirable career path on Route 128 was to move up the corporate ladder of a large company with a good reputation. Whereas inter-firm mobility was a way of life in Silicon Valley, Route 128 executives were more likely to consider "job hopping" as unacceptable and express a preference for professionals who were "in it for the long term." The employees of the region's technology firms in turn were very loyal, generally expecting to stay with a company for the long term. What little mobility there was in the region tended to be between established, relatively safe, companies.

This difference in career patterns was reflected in the regional compensation practices. In Silicon Valley, the early firms like HP and Intel offered their employees profit sharing plans and generous stock options in order to attract and motivate talented individuals. Today, while there is little interest in retirement plans in because company tenures are so short, stock options are typically available to all employees--from the rank and file to top managers. One venture capitalist described stock as "the mothers milk of Silicon Valley" and suggested that local firms would lose their competitive edge if people didn't have equity ownership in their firms. In Route 128, by contrast, even recent graduates worry primarily about vesting a comfortable pension and it is rare for firms to offer stock options to employees, except perhaps to top executives.

The practice of leaving a large company to join a small firm or a promising start-up, so common in Silicon Valley, was virtually unheard of in Route 128. As one veteran employee of Honeywell put it:

> There is tremendous loyalty to the company and tremendous will to make things succeed within the company [on Route 128]. There were pockets of brilliance at Honeywell, but these individuals never took the leap to go off on their own or join another company. I stayed at Honeywell for more than twenty years. I had lots of opportunities to leave, but I never took them seriously because I had too many personal commitments and business ties. When I finally left it was like an 8.5 on the Richter scale. Everyone was shocked, they just couldn't believe it!

Indeed while there were a handful of high profile entrepreneurs in the region, starting a company in Route 128 was still regarded as extremely risky. A former DEC executive who is now based in Silicon Valley similarly reports that: "We never talked about start-ups back East. Out here we're always talking about who's doing what, what's succeeded. As a result, everyone in Silicon Valley is motivated to do start-ups, while in the East coast nobody is." Not surprisingly, local engineers have a hard time identifying gathering spots or opportunities in Route 128 for socializing or informal information exchange. Indeed, most report that work and social life remain largely separate in the region.

The comparison of the leading computer firms in the two regions--Silicon Valley's Hewlett-Packard (HP) and Route 128's Digital Equipment Corporation (DEC)--demonstrates the competitive advantages of open labor markets. DEC, which relies almost exclusively on internal labor markets and long term employment, has failed to keep pace with its California counterpart, with its open boundaries and integration into regional networks. Both DEC and HP were comparable sized companies in 1990 and the largest and oldest civilian employers in their respective regions. Yet by 1995, HP had decisively surpassed DEC technologically and had established itself as the second largest computer company in the nation, after IBM.

Corporate performance has multiple causes, to be sure, but the differences in the firms' organization--and in particular the openness of their boundaries--help explain these differences in performance. While HP actively participates in local labor markets and the associational life of the region, and in so doing helps to reproduce the open labor markets of Silicon Valley, DEC has historically maintained clear boundaries between itself and the regional economy. This comparison suggests the limits of bounded organizations in the current environment.

For close to fifty years, Hewlett and Packard actively encouraged their employees to involve themselves in the region and its rich professional and social networks. Following the model of its founders, successive generations of HP executives have maintained congenial relations with individuals who leave the firm (at times even assisting them with start-ups), they have rewarded the creation of outside relationships, and they remain active participants in local civic and political affairs.

DEC, by contrast, detached itself almost completely from the region. By the late 1960s, DEC was the largest employer in the town of Maynard, yet none of the firm's senior management belonged to the local Chamber of Commerce and CEO Ken Olsen discouraged his managers from participating in community affairs. One observer has described DEC as "a sociological unit, a world unto itself." (Rifkin and Harrar, 1990: 106) Indeed executives at the company are quick to point out that DEC does not see itself as part of Route 128 or even New England; but rather as part of the national and global economies.

Not surprisingly, this insularity is reflected in labor markets and career paths. Promotions at DEC have historically come almost exclusively from building strong internal relations, not from success in dealing with the external world. According to one long-time employee:

> Getting hired into DEC . . . is like getting married: you meet your wife's mother and father and her aunts, uncles and cousins. It is a bonding process to an extended group of peers, as well as executives higher up and workers lower down. (Rifkin and Harrar, 1990: 119)

Strong bonds of mutual support developed between DEC employees, creating intense loyalty and dedication to the organization. As a result, employee turnover at DEC was among the lowest in the computer industry.

While engineers leaving HP--or most other Silicon Valley firms--typically stayed in touch with former colleagues for the rest of their careers, those who left DEC were ostracized and cut off from the DEC community. They were often labeled as failures rather than seen as potential resources. And once having left, there was no option of return. According to one executive who did leave: "If you're stupid enough to cut yourself off from the mother church, Digital's attitude is, 'Don't bother to come back.'" (Rifkin and Harrar, 1990: 121) The closed nature of the DEC corporate culture thus insured that all of the social and professional relationships that mattered were internal to the firm.

The blurring of organizational boundaries does not simply benefit individual firms, but also contributes directly to the health of the regional economy. While Hewlett Packard served as prolific training ground for Silicon Valley entrepreneurs, DEC's inward-looking organization was less conducive to developing entrepreneurial skill. HP's decentralized and quasi-autonomous divisions provided the autonomy and general management responsibilities that are essential for a start-up. As a result, former HP executives were responsible for starting more than 18 firms in Silicon Valley between 1974 and 1984, including such notable successes such as Rolm, Tandem, and Pyramid Technology.

In contrast, DEC's matrix organization--which represented only a partial break from traditional functional corporate hierarchies--stifled the development of managerial skill and initiative in the Route 128 region. The matrix demanded continuous negotiations to reach consensus, and despite the addition of cross-functional relations among product groups, final authority remained highly centralized in the firm (Schein, 1985). With the exception of its acrimonious split with Data General, it is difficult to identify successful spin-offs from DEC. As one local venture capitalist noted:

Its well known among the venture capital community here that it is a significant disadvantage to hire anyone from DEC onto a startup management team. The matrix management and decision by consensus means that a manager who's been at DEC for a long time is going to be indecisive.

HP and DEC are not exceptions in their regions: rather they reflect widespread differences in corporate organization between Silicon Valley and Route 128. Management scholars Delbecq and Weiss (1988) conclude that Silicon Valley firms tend to be organized as highly decentralized confederations of autonomous work teams, linked by intense informal communications. They argue that the critical unit in the region is not the firm, but the "loosely coupled engineering team," which they define as a:

set of individuals with a strong sense of entrepreneurship, joined around a project mission associated with technology-driven change, who remain in contact frequently and informally with multiple levels and functions within the company through intense informal communications.

This characterization--loosely linked engineering teams embedded in dense informal networks -- applies to life outside as well as inside of Silicon Valley companies; indeed it is an apt characterization of the organization of the regional economy.

Route 128's technology firms, by contrast, are more stable, formal and orderly organizations. While a handful of firms, including DEC, consciously experimented with decentralized forms of organization, Weiss and Delbecq (1987) conclude that most Route 128 firms rely on a "formal, vertical structure" and that their management styles are "conservative and top-down oriented" relative to their Silicon Valley counterparts.

The experience of a DEC research lab in Palo Alto during the mid-1980s illustrates some of these differences. DEC employees who had previously worked on the east coast were amazed as they began to participate in Silicon Valley's social and technical networks. One engineer concluded that:

> DEC definitely relates differently to the regional economy in Silicon Valley than in Route 128. DEC is the largest employer in Route 128 and you come to think that the center of the universe is North of the Mass Pike and East of Route 128. The thinking is totally DEC-centric: all the adversaries are within the company. Even the non-DEC guys compete only with DEC. DEC Palo Alto is a completely different world. DEC is just another face in the crowd in Silicon Valley.

He described his years with the DEC group in Palo Alto:

> We had an immense amount of autonomy, and we cherished the distance from home base . . . and all the endless meetings. It was an idyllic situation, a group of exceptionally talented people who were well connected to Stanford and to the Silicon Valley networks. People would come out from Maynard and say 'this feels like a different company.' The longer they stayed, the more astounded they were.

DEC was ultimately unable to assimilate the lessons of its geographically distant Palo Alto research group, in spite of their technical advances, and in 1992 transferred it back to Maynard headquarters. Not surprisingly, many of the key employees of the lab left DEC to work for Silicon Valley companies.

In short, Route 128 firms like DEC strictly defended their corporate boundaries at a time when job mobility and open information sharing were widely accepted practices in Silicon Valley. Networking and informal exchange on Route 128 occurred almost exclusively within the large firms, not between them. As a result, information about labor and product markets remained trapped within the boundaries of individual firms, rather than diffusing rapidly through local social and professional networks as in Silicon Valley. This insularity deprived the Route 128 region of many of the opportunities for collective learning that distinguish Silicon Valley.

The ultimate irony is that when the Route 128 economy faltered in the late 1980s and engineers began to leave DEC and the region's other large firms (often involuntarily), many chose to go to Silicon Valley rather than staying in the region. As one local venture capitalist noted: "We always worked on the theory that there were good people in the old traditional companies and when they were ready to come out if we had the money. Now we wake up to find that half of them moved to California to start their companies."

3.6   CONCLUSIONS

The blurring of the boundaries between firms provides a regional advantage to Silicon Valley. Open labor markets allow individuals and firms to experiment and to learn by continually recombining local knowledge, skills, and technology. And the resilience of the regional economy suggests that the learning advantages of open labor markets--when embedded in a rich fabric of social relationships--outweigh their costs. Silicon Valley's professional networks minimize the search and switching costs incurred by high rates of interfirm mobility. In other words, they provide both the social capital (Raider and Burt, 1995) and the informational signals (Rosenbaum and Miller, 1995) needed insure career success outside of internal labor markets.

As individuals move between project teams and companies in an open labor market like Silicon Valley they also blur the boundaries between firms and the institutions of the surrounding community--undermining the distinction between work and social life and between the economy and civil society. Individuals in Silicon Valley rely for career success on bars, health clubs and other informal social gatherings, hobbyists clubs, technical and professional associations, training programs, community colleges, and universities, and a variety of other networks that cross company lines and reach from the economy into social and even family life. As a result, the success of local firms is linked to the reinvigoration of the local community. The nature of the relationship is nicely captured by Best and Forrant's term (1995) "community-based careers."

Silicon Valley firms continue to outperform competitors with traditional corporate boundaries located in Route 128 and elsewhere. The question remains, however, of the extent to which this model of open labor markets embedded in social networks will diffuse. Will the more traditional bounded organization continue to dominate and absorb civil society (Perrow, 1995) or will the boundaries between firms and between the economy and society increasingly blur with the dismantling of internal labor markets? While the Silicon Valley model is still far from widespread, businesses and policy-makers in regions from Seattle to Texas are attempting to foster the relationships that distinguish Silicon Valley. They may well have more success than older industrial regions that have deeply entrenched organizational boundaries to overcome. The regions that succeed will be those that create social and institutional fabrics that can support both companies and careers that know no boundaries.

REFERENCES

Best, M. and Forrant, R. (1995). "Community-Based Careers and Economic Virtue: Arming, Disarming, and Rearming the Armory" in M. Arthur and D. Rousseau, (eds.) *Boundaryless Careers: Work, Mobility and Learning in the New Organizational Era,* Oxford University Press, New York.

Braun, E. and Macdonald, S. (1978). *Revolution in Miniature: The History and Impact of Semiconductor Electronics,* Cambridge: Cambridge University Press.

Bylinsky, G. (1974). "California's Great Breeding Ground for Industry", *Fortune,* June.

Delbecq, A. and Weiss, J. (1988). "The Business Culture of Silicon Valley: Is it a Model for the Future?" in J. Weiss (ed.) *Regional Cultures, Managerial Behavior and Entrepreneurship*, New York: Quorum Books.

Granovetter, M. (1995). *Getting a Job: A Study of Contacts and Careers* (second edition) Chicago: University of Chicago Press.

Granovetter, M. (1988). "The Sociological and Economic Approaches to Labor Market Analysis" in G. Farkas and P. England, (ed.), *Industries, Firms, and Jobs,* New York: Plenum Press.

Gregory, K. (1984). *Signing Up: The Culture and Careers of Silicon Valley Computer People*, Unpublished PhD Dissertation, Department of Anthropology, Northwestern University.

Hanson, D. (1982). *The New Alchemists: Silicon Valley and the Microelectronics Revolution,* Boston, MA: Little Brown and Co.

Harrison, B. (1994). *Lean and Mean: The Changing Landscape of Corporate Power in the Age of Flexibility*. Basic Books. New York.

Lévy, S. (1984). *Hackers: Heroes of the Computer Revolution,* Garden City, NY: Anchor Press/Doubleday

Markusen, A. (1985). *Profit Cycles, Oligopoly, and Regional Development,* Cambridge: MIT Press.

McKenna, R. (1989). *Who's Afraid of Big Blue?,* Reading, MA: Addison-Wesley.

Perrow, C. (1995). "The Bounded Career and the Demise of Civil Society" in M. Arthur and D. Rousseau, (eds.) *Boundaryless Careers: Work, Mobility and Learning in the New Organizational Era,* Oxford University Press, New York.

Rifkin, G. and Harrar, G. (1990). *The Ultimate Entrepreneur: The Story of Ken Olsen and Digital Equipment Corporation,* Rocklin, CA: Prima Publishing.

Sabel, C. (1991). "Moebius-Strip Organizations and Open Labor Markets: Some Consequences of the Reintegration of Conception and Execution in a Volatile Economy" in P. Bourdieu and J. Coleman, (eds.) *Social Theory for a Changing Society,* Bolder, CO: Westview.

Saxenian, A. (1994). *Regional Advantage: Culture and Competition in Silicon Valley and Route 128,* Cambridge: Harvard University Press.

Schein, E. (1985). *Organizational Culture and Leadership,* San Francisco: Jossey-Bass.

Weiss, J. and Delbecq, A. (1987). "High Technology Cultures and Management: Silicon Valley and Route 128", *Group and Organization Studies,* 12, 1: 39-54.

Wolfe, T. (1983). "The Tinkerings of Robert Noyce: How the Sun Rose on Silicon Valley", *Esquire,* December.

4 MODELING REGIONAL INNOVATION AND COMPETITIVENESS

Tim Padmore
University of British Columbia
Hervey Gibson
Glasgow Caledonian University

4.1 INTRODUCTION

The importance of science and technology to industrial development is unchallenged. What has occupied many researchers in the last half of the 20th century is trying to understand the *process* of technological progress and industrial development in order to better manage it. An early notion was that progress proceeded in a linear fashion, starting with scientific research producing fundamental knowledge, which then stimulated applied research and development (R&D). This R&D then led to marketable products, newer and better than the old ones. This simple idea has been set aside in favor of much more complicated systems with multiple feedback loops and interactions between technological, economic, social, and management systems. Firms are no longer regarded as independent or isolated actors, but are more properly seen as being parts of a system, linked together (technologically, economically, socially, and managerially) in groups or clusters.

Science and technology as an activity is therefore embedded in a very complex system. In coming to grips with this, researchers and policy makers have enlarged the context for discussion from science and technology to "innovation", and now talk about "systems of innovation." The earliest work was around the way whole countries organize for innovation. However, as this book shows, much of the industrial dynamic takes place at a *regional* scale and, recently, the concept of *regional* systems of innovation has become important (Acs, 1999). Firms are elements of this system, as are labour markets, public institutions and government programs, and other entities that support or connect innovative activity. In this context, innovation refers to more than just technological innovation but also encompasses marketing, finance, management and other issues. Much effort has gone into designing policy to strengthen systems of innovation, primarily at the national and lately at the regional level.

Increasingly, nations and regions see themselves operating in a world context, and economic development policy has framed as its basic challenge to create and enhance industrial competitiveness. Policy makers need to know what makes a region competitive and which of those factors can be influenced by regional governments and industries. Cogent policy design requires first an understanding of the systems to be managed and, second, measurements of system processes to optimize policy design and monitor the impacts.

The InnoCom BC project[1] was established to pursue this goal. The project will look at the regional innovation system and competitiveness of industrial clusters in British Columbia, which is a convenient laboratory for a regional study (a relatively simple economy, well-defined geography, cohesive culture), and selected sub-regions.

This chapter outlines a theoretical framework developed to serve as a context for InnoCom BC studies. The genesis of the work was a model developed for a comparative study on the evolution of science and technology policies in European and Canadian regions. The study, known as Ec/Tech,[2] aimed both to describe the policies and evaluate their impact, and the model evolved into the Cycle model of innovation systems presented below. The GEM model, for assessing competitiveness of industrial clusters, was evolved from consulting projects (CSRD, 1996, and Western Economic Diversification, 1996) as a working tool to assist economic development professionals to visualize the strengths and weaknesses of industry sectors. We have since tried to strengthen the underpinnings (Padmore, et al, 1997, Padmore and Gibson, 1997) and refine the application of the GEM model. The work presented here is largely conceptual rather than empirical, although we do discuss some recent quantitative applications of the GEM model and comment on some of the implications of the Cycle model for existing work on measurement of innovation.

4.2   RESEARCH, INNOVATION AND COMPETITIVENESS

This chapter attempts to bring together a variety of perspectives on how firms are organized and how they interact with their environment. The models presented below attempt to bridge these perspectives to provide a more systematic framework for reflection and measurement.

For the scientist studying a complex system, it is often useful to specialize and focus on identifiable parts of the system that are at once interesting, understandable and significant. The manager of a firm, on the other hand, must take into account the whole range of factors significant to the business and its prospects. The policy maker, too, should understand the broad range of factors that affect a firm, an industry, or an economy, and also needs to understand the component parts of the system.

The evident importance of technology and innovation in modern industry and society drew the attention of social scientists to the research enterprise, which we show at the top of Figure 1 Views of scientific development current since the 1950s trace the development of knowledge from "pure" or basic science through applied science to engineering and application. This linear model, indicated schematically by the chain of boxes in the upper oval, is still common in political debate -- which is often conducted in terms of near-market versus basic research. The linear model suggests that everything originates with the discovery, where the science is completed and packaged before becoming available for an invention. The invention is then perfected, incorporated in a product, and introduced to the marketplace.

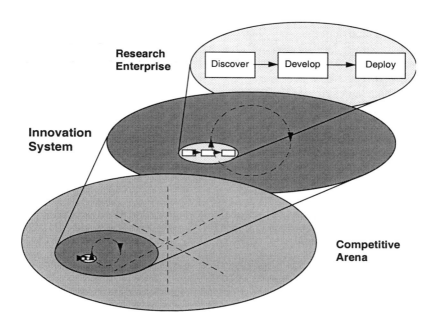

Figure 1 - Three system perspectives

In the basic model there are no arrows from right to left, which might represent a need calling forth a particular area of research or discovery, or for trial and error in market innovation. However, the three boxes do reflect the way that knowledge is organized, with enabling knowledge supporting more complex or specific applications. Real innovation involves complex re-iterations of the basic pattern. It also involves interactions tangential to the technology development (e.g. financing or labour relations) and, indeed, sometimes may not involve technology at all.

The milieu for all these interactions is the "system of innovation". Originally, the concept of systems of innovation evolved in a national context (Freeman, 1988; Niosi, 1991; OECD, 1991, Lundvall 1992; Nelson, 1993). The idea has been generalized to regional systems of innovation (Acs, 1996; Bergman, 1991; Saxenian, 1994; Wolfe, 1993; Ernste and Meier, 1992; Higgins and Savoie, 1988; Miller and Côté, 1987; Scott and Storper, 1992; Hilpert 1991). These discussions (which deal mainly with technological innovation) elaborate the types and roles of different players: firms and their various functional units, public institutions, suppliers and customers. The research enterprise is only a small part of this complex system.

The system of innovation is the object of the first of our models (section), indicated schematically by the dotted circle in the middle oval.

What do we mean by the word innovation? To us, an innovation is any change in inputs, methods, or outputs which improves the commercial position of a firm and that is new to the firm's operating market. In principle, we accord comparable importance to innovation in areas such as management, marketing and finance as we do to technological innovation, reflecting the reality of the business world which can not afford to draw a line between different types of innovation. This is more than the observation that management (Fairtlough, 1994, Martin, 1994), marketing (Littler, 1994) or financing (Tylecote, 1994) play a critical enabling role in innovation. We think that these areas of endeavor represent distinct arenas of innovation or what Lamberton, 1992, calls "exploratory behavior." Our definition is crisper, but not much different in substance, from that offered by Schumpeter in 1934.[3]

We have attempted to draw our models generally enough to accommodate the breadth of this definition. However, the importance of non-technological innovation, while recognized, has been barely explored (OECD, 1996a) except in the management literature. Consequently, most of the issues raised here and the evidence cited will relate to the better-studied and better-understood subject of technological innovation.

The impetus for considering *regional* systems of innovation was recognition that regional conditions, including regional policy environments, were important to economic development outcomes. National endowments and policies were a part, but only a part of the regional picture. What is the region? Governments are most interested in administrative regions. Policies are generally uniform across an administrative unit, so this makes sense. We have constructed our models so they can be readily applied to a political region, but we prefer to think of regions being defined by the pattern of industrial structures and linkages.

At the other end of the spectrum of specialization is the arena of competitiveness. Research in this area tends to be found in the management literature (Porter, 1990, Ohmae, 1995). Competitiveness depends intimately on innovation, but it also involves many other factors: availability of financial capital, natural resource endowments, knowledgeable customers and so on. A natural object of competitiveness analysis is the industrial "cluster", and our second main model (section) provides a way of assessing regional clusters. The model is based on ratings of six determinants of competitiveness, suggested by the six pointed star in the bottom oval.

4.3   SYSTEMS OF INNOVATION: CYCLE MODEL

Probably the best known depiction of the system of innovation is the so-called chain-link model developed by Kline and Rosenberg (Figure 2).

The chain link model represented an important conceptual and analytic advance because it recognized the multi-dimensional nature of the innovation process, and the many feedback links among the steps in product development and sources of knowledge outside the firm. It has provided a context for policy measures aimed at strengthening the system at points where it interacts with commercial activities and is cited explicitly by the OECD as the basis of the new "Oslo Manual" on measuring innovation (OECD, 1996a).

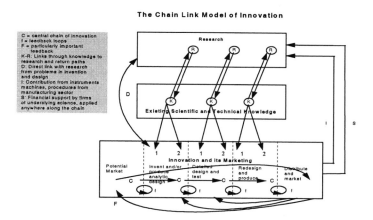

Figure 2 - The chain link model on innovation system

The chain link model is high level in that it does not show the relationships between different kinds of institutions. Even so, in our work with policy makers, we have found that the chain link model is seen as being rather complicated. We have decided to attempt a symmetrization of the model that would make it more readily understandable and more easily manipulated. The structure of the model is explained in detail and a number of examples of its application are given in Padmore and Gibson (1997).

The system is shown from the perspective of the firm. The *circle* of linked arrows represents all the activities in the firm, organized according to the product-development process. The activities depend on a two-way flow of knowledge between units of the firm and the "innovation system," represented by the *central disc*. The innovation system includes the firm itself and some of the flows represent communication with other parts of the firm (which measurement must be careful not to double count).

The model is based on a continuous product-development cycle,[4] divided into admittedly rather arbitrary stages but which can be fit to a broad spectrum of enterprises producing either goods or services. At any one time, an enterprise may be involved several products, at various stages.

The circle of activities can be divided up in many ways. We have chosen divisions that correspond as closely as possible to the categories used in the Oslo manual to classify "expenditure on innovation." We have, however, separated out Prototype ("an original model or test situation which includes all the technical characteristics and performances of the new product or process")[5] and Production, which includes tooling up and capital costs. The Prototype stage would include testing, but not ongoing quality control. We have added two categories which may be less important from a technology perspective but are important commercially, account for a large fraction of expenditures and certainly involve innovation. They are Process/system development, which means adjustments to reduce the manufacturing costs or improve the quality of a good, or increase the cost-effectiveness of a service; and, Distribution/Sales. Distribution is the physical distribution of a good, and Selling, which applies to both goods and services, means the ongoing promotion and sales activities that connect the production process to the market.

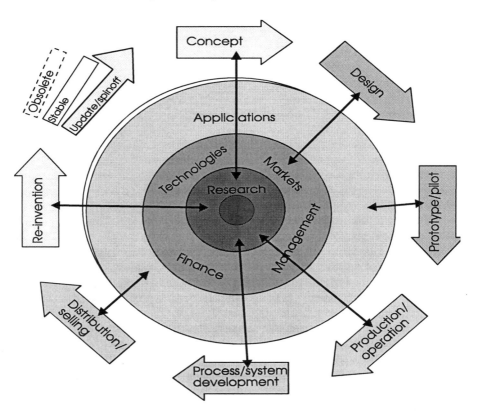

Figure 3 - The cycle model of an innovation system

The Re-invention/Concept stage comprises all the work done beginning with the reconsideration of existing products, investigation of new technologies and markets and the potential means for producing them, and estimation of feasibility. This really is a single stage but we spread it out in the picture to show the relationship of the cycle to different market conditions. (Grossman and Helpman, 1991, and Coriat, 1992, provide complementary dissections of this step.)

The stages are shown as being sequential but, as Rothwell (1994) has pointed out, they may not be. In his "fourth generation" model, the activities of R&D, marketing and production proceed in parallel, but they are still separable in terms of the organization of the firm. Similarly, the history of a particular product development may involve a return to an earlier stage, for example a failed prototype may take the product "back to the drawing board". These are important distinctions, but do not matter too much for the Cycle model, which takes the "subject approach" to innovation (OECD, 1991 and 1996). In the subject approach, all of a firm's innovation activities in a specific time frame are considered, while the "object approach" follows the path of specific innovations.

But what about the central disc - the "system" that the firm interacts with? The first organizing principle is the readiness-for-use of knowledge. At the outside of the disc are applications that can be used "off-the-shelf" to support the product-development cycle, for example an established packaging system, a packaged software application, or provisions of an existing labour contract. The next circle comprises the supporting body of component technologies, marketing knowledge, management expertise, etc. The inner ring is the discovery enterprise, including both social and natural sciences research, much applied research and even some experimental development.

In the diagram the double-headed arrows indicate information flows between the product-development cycle and the innovation system. It is not always true, of course, that the Design step can make do only with off-the-shelf technologies, but this is often the case. It is also possible to estimate the *strength* of the interactions by a measurement of the flow of knowledge at each step, which is the subject of the next section.

The old linear model can be recovered by taking a radial slice through the central disc. The closest correspondence to the classical model is a slice at the 12 o'clock position, then following the circle around to the market place. Every radial flow can be interpreted as a partial or complete turn of the linear-model crank.

Much of the structure of the Chain Link Model can be recovered by cutting the Cycle model apart at the Re-invention stage and spreading it out flat. Where the Cycle model most differs is in its symmetrical structure, which has two advantages. First, it invites various transformations, for example the mappings in section . Second, it is more useful as a communication tool: it is easy for people to "find themselves" in the picture and see their actual or potential contribution.[6] Third, it emphasizes the continuity (in good practice) between perception of the market and conceptualization of the product; the model promotes this link above a mere feedback loop.

4.4 MEASURING INNOVATION

To be useful a model of reality must emphasize some aspects and subordinate others. The emphasis in the Cycle model is on flows of information. These are shown pictorially and, of course, we would like to measure them.

What is the unit of measurement of innovation? (This is a question that de la Mothe and Paquet have raised, in this volume and elsewhere). The Cycle model represents a system of innovation by diagraming flows of knowledge. We think the most useful way to think of these flows is in terms of economic value. This creates a framework for understanding how firms allocate their effort and where they look for innovation (von Hippel, 1988). An accounting framework allows checking for consistency and makes it easier to estimate missing data. However, assigning monetary values to the flows is, practically, very difficult. The value of intellectual property transferred by a licensing agreement is reasonably reflected by its price, but it is much more uncertain what is the value of information obtained by know-how trading with a competitor, what part of the value of new equipment represents a purchase of innovation, or how much of the work of a given employee is innovative. In practice, it is necessary and even desirable to accept other measures that may relate more directly to a given policy question and this has been the pattern with most innovation surveys to date (for examples see European Commission, 1994a, OECD 1996b).

To the extent that monetary estimation is feasible for the Cycle model, we believe the appropriate basis is the selling price of the product, which we take to represent the true value of the product. This includes the costs of inputs and the value added by the firm itself. Valuation is discussed further in section .

To document and understand patterns of innovation in the context of the cycle model, we need data on the distribution of the selling price across various categories.

For example, a reasonably complete decomposition by the channels of transfer can be organized under the following headings:

Embodied in acquired goods: This covers the technological innovation obtained by firms when they purchase sophisticated plant, equipment or materials (Evangelista, 1996). We would also include product-specific knowledge obtained from suppliers: manuals, advice on production methods, and staff training. This heading would also cover goods purchased from competitors, for example for reverse engineering.

Embodied in acquired services: Examples would be testing services and specialized surface treatments. Most of the ready examples of innovation embodied in *goods* are technological, but in the case of *services* the innovation is often of the management, financing, or marketing types, e.g. a firm may pay an advisor to design a creative financing instrument or ownership structure new to the sector.

Acquisition of intellectual property: The category covers not only the conventional forms of packaged intellectual property such as patents, licenses and copyrights but also knowledge acquired through know-how trading with competitors or peers or at low or zero cost from journals, conferences, and public sources.

Acquisition of human capital: Arguably the most important category, human capital, can be increased by hiring skilled people, through internal training and through informal learning processes. Pricing can be tricky. For most employees, wages are a good proxy for value but a star employee can be worth more than the market price.

There are still other ways to slice the cake. Geographical mapping for example, would be useful for studying the extent of innovating regions and industrial clusters. One of the most important decompositions for policy purposes is by the *source* of innovation, the subject of the next section.

4.4.1 Sources of Innovation

Understanding where knowledge comes from is basic to public policy and private strategies. Knowledge to sustain innovation includes not only the "bright idea" that is sometimes the trigger for innovation, but all the other expertise needed to bring products to the market. Who is involved, and to what degree? We have slightly reorganized what the OECD (OECD, 1996a) refers to as "sources of information for innovation" into five broad categories

In-house: This is innovation that is managed by the firm, using the firm's own resources. Innovation in-house involves interaction between different departments and links different parts of the product-development cycle. In-house innovation can be quite complex, particularly in large firms with many divisions, multiple sites, and internal pricing arrangements that may be designed for other purposes than efficient resource use.

Suppliers: Firms obtain advantage (Sako, 1994) by procuring innovative inputs from suppliers embodied in purchased goods or services. The innovation content may be small (in a purchased near-commodity, for example) or large (in an exclusive, custom product that confers a decisive edge through a technical breakthrough). Increasingly, suppliers interact directly with firms in key-supplier relationships or through strategic alliances (Rose, 1995).

Peers: 'Peers' in this case are similar firms, typically competitors but not necessarily. Collaboration is increasingly an important way for firms to gain information for innovation (Dodgson, 1994). Firms learn from competitors by copying or improving their products and practices and through know-how trading (von Hippel, 1988). This occurs even in highly secretive sectors. Some information is transferred through formal licensing or partnering arrangements.

Customers: In the post-Fordist economy with its emphasis on meeting more and more specialized needs and tastes, obtaining information from customers (marketing) has become a major activity for many firms (Shaw, 1994) highly correlated with competitiveness. Much information about customer appetites is tacit or obtained informally. A customer can be a point of final demand, or it can be another firm.

Public Sector: In much traditional policy thinking, the public sector is a fundamental source of innovation, standing at the root of a technology-transfer process that sees basic discoveries in research institutions moving out to applied departments or to industry where they become commercial products. In practice, the relationship is much more complex. The means of transfer of knowledge from the public sector to the firm are highly varied: including institutions like technical libraries or research laboratories, through public "events" like conferences, through government policy bodies like regulatory agencies, and so on. As views of the role of public institutions have changed, there has been a more deliberate attempt to develop more varied and flexible relationships with industry.

The classification is a fairly simple one, but it is meant to be comprehensive, that is, it should be possible to classify any specific source under one and only one of the five headings.

Roles can vary. A government department that buys a product is a customer for that transaction, but on another occasion it may be a public sector body providing information as part of its public service mandate.   There are fine points of classification, only some of which we have worked out. A research lab that provides services on a subsidized basis is a public sector source; if the services are sold on a fully commercial basis (normal return on capital) it would be logical to classify it instead as a supplier.   Key employees hired away from a competitor are not a peer source but are rather an in-house source whose value is approximated by the wages and other inducements paid to obtain their services.

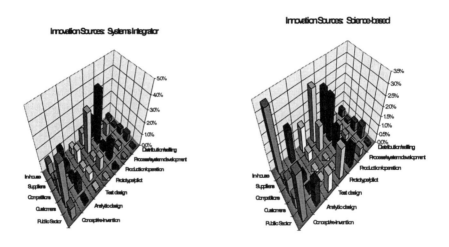

Figure 4 - Innovation sources for two innovative firms

Existing innovation surveys have collected a great deal of information on sources of innovation (European Commission, 1994a and 1995).   For example, Lhuillery, 1996, contrasts the sourcing patterns of small and large firms in France, and the diversity of sources used by different industry sectors. This kind of analysis (see also Evangelista, 1996) would be facilitated by the more comprehensive accounting implicit in the Cycle model.

We have used some limited data available on the distribution of inputs to the selling price of high tech firms (MRED, 1989) and the sourcing patterns of different types of highly innovative firms (National Research Council of Canada, 1994), together with some reasonable guesses by the authors, to generate innovation sourcing profiles for different types of innovative firms. Two examples are shown below. The chart distributes innovation effort, measured as a percentage of the selling price, by source and by stage in the product-development cycle.

In the NRC model, science-based firms (e.g. bio-pharmaceuticals) are highly secretive, participate in R&D races, are highly dependent on in-house innovation, often drawing on a close relationship with a public sector research institution, and are often dependent on strategic alliances with larger firms for marketing and distribution. Systems integrators (e.g. telecommunications) are very dependent for innovation on suppliers of components and subsystems, and also on customers (products are one-of-a-kind or highly customized). Governments and large firms are key customers.

The patterns are distinctive. For example, in the early stages of the product-development cycle, the skyline of the science-based firm is hollowed out in the middle while the systems integrator shows the opposite pattern. The charts capture some of the distinctive character of each category, suggest further elaborations of the taxonomy, spotlight possible avenues for government policy, and, of course, invite quantitative verification. The pictures are also an avenue for investigating product *life* cycles, which is another important organizing framework for innovation (Thomas, 1985). A life cycle will involve many development cycles, occurring in many firms. In the course of a product life cycle, we would expect characteristic shifts in the patterns of innovation for a given industry (Hobday, 1994; Rothwell and Dodgson, 1994).

4.4.2 Valuation Questions

Knowledge transfers take many forms and are not easy to value. We would like a measurement in monetary terms and have said that we believe the right framework for this is to identify the values as a percentage of sales.[7] Sales include the cost of material inputs, wages and salaries, interest costs, taxes and profits. Expenditures on goods and services from suppliers, expenditures on wages for in-house R&D, and market research fit easily into this framework and the Oslo manual provides reasonable guidelines for what should and should not be included in the important case of technological innovation.

It is harder to deal with the value of information obtained from analysis of a competitor's products, or the value of information obtained free or at a low cost from a public source. We see the taxes paid by the firm as a form of payment for public services of all kinds. Normal profits reflect a normal return on investment, including investments that are specifically innovative. However, a firm that benefits inordinately from free public information will enjoy, at least for a time, extra profits, and we would expect managers to value the public information in proportion to that. Excess profits may also reflect a temporary advantage in the trading of knowledge with competitors or an excellent relationship with customers. Some innovation activities have a low out-of-pocket cost, for example IBM making its PC technology available to others, but a high value, related in this example to the greatly expanded market penetration of PC technology, which will be reflected in profits. So in principle, there is a home in the framework for these transfers.

Of course, there are serious practical problems of inter-firm, inter-industry and temporal aggregation in making associations between innovation values and taxes or profits, e.g. an unprofitable firm may consume more public services than a profitable one.   In practice, the notion can be no more than approximately useful.   Senior managers, however, have a great deal of information -- not always precise, but operationally significant -- about the value of knowledge obtained from various sources and in various forms. We do not yet know if asking managers to dissect out innovation values from total sales in this way will yield reliable results, but the question is a fair one.

In principle (again) this accounting captures "spill-over" phenomena, since the values of all innovations are ultimately reflected in sales somewhere.   However, "ultimately," "somewhere" and "sales" involve vast complications, e.g. what about public sector outputs?   Looking, as we do, from the perspective of a single firm, we will see some "spill-ins" clearly (purchases of innovations embodied in imported technology) and some only vaguely (value attributable to knowledge about competitors' products).

4.4.3 Innovation Surveys

Much of the information collected in innovation surveys can be organized within the framework developed in this chapter. In another publication (Padmore *et al*, 1997) we examined in detail how the Cycle model maps onto the model questionnaire developed by the European Commission for use by members (European Commission, 1994a). The Community Innovation Survey (CIS) is a fairly long survey, done from the "subject perspective" adopted here, that countries can shorten or adapt to their specific needs.   The CIS is based on the so-called Oslo manual and its revisions (OECD, 1991 and 1996), which captures much of the recent work on defining and describing innovation and distills out some consistent perspectives.

One difference with the approach advocated here is that the Oslo manual deals almost exclusively with technological innovation (TPP, or technological product and process innovation).   Marketing innovation is included, so long as there is a technological element. Management or organizational innovation is distinguished but the framework is not developed and it is excluded from TPP. As a practical matter, this makes sense, given the history of innovation research, but in the long run the broader view will need to be developed.

Both approaches divide up the innovation enterprise along various dimensions (sources, channels, etc.) and we have attempted to use similar terminology and categories. The unit of measurement varies. For example, sources of information a measured by "importance" on a five-point scale, channels by a simple "yes/no". The Oslo category lists are often very specific, reflecting sources of particular interest for policy. Items of less interest must go into the slots marked "Other." This is most evident in the public sector categories. We have attempted to define broad, mutually exclusive, and exhaustive, categories.

Overall, we believe that innovation surveys could be strengthened by:

- adopting more comprehensive categories, while retaining the option for fine-grained measurement within the categories
- moving toward a common unit of measurement

Given the degree of correspondence that already exists, the first of these would not be difficult. Establishing a common unit of measurement will be a more lengthy and problematic exercise. If both objectives are achieved, all the advantages of an accounting framework follow, including the ability to cross-check sums and estimate missing data.

We do want innovation surveys to become onerous accounting exercises. The simple regional innovation survey described in Chapter 9 incorporates several modifications suggested by this Cycle model. A combination of detailed case studies and surveys, conducted in a common framework, can provide credible system data.

4.5   ASSESSING COMPETITIVENESS

This section briefly reviews a new model of cluster competitiveness that is described in more detail in Padmore and Gibson (1997). Our main objective was to develop a description that would be useful to economic development professionals, which meant that it should be simple and comprehensive, that it could be quantified to facilitate comparative studies, that it could be readily applied in a *regional* as opposed to national context, and that it should use the kind of information that can be elicited from firms that wish to understand their competitive environment or and from public sector agencies involved in economic development.

4.5.1 The GEM Model

A cluster is a concentration of firms that prosper in part because of their interaction, whether that is through competition or cooperation, or by serving as suppliers or customers in the value-chain. Clustering is a real phenomenon, which occurs beyond what is needed to be near final consumers or basic inputs (Ellison and Glaeser, 1994; Head *et al*, 1994). Clustering is related to the ease and availability of linkages that allow people to form personal relationships of trust, cooperation, and competition (Lorenz, 1992). In conceptual terms, linkages can be either vertical, moving from supplier to client, or horizontal, among similar firms or between firms and various forms of economic infrastructure (Hanson, 1994). The linkages all involve information exchange, which relates cluster dynamics closely to the information flows of the Cycle model.

Many regions have some degree of industrial concentration. But not every concentration is a cluster (Acs and de la Mothe, 1999). Clustering depends on effective linkages. Beyond that, not every cluster is a successful cluster. For example, subsistence agriculture is an industrial concentration in many parts of the world.[8] In many of these areas, enterprises may have good linkages with each other, with local suppliers, and with local customers. But lack of technology, financing or other factors keep productivity and profits low. Cluster success depends on more than innovation and more than linkages.

We developed the GEM model to get at these factors. There are six determinants covered in the GEM analysis, grouped in three pairs whose initials provide the acronym: Groundings, Enterprises, and Markets. We have tried to define the determinant groups in a logical and consistent way but, inevitably, there are ambiguities on the edges of the definitions. The sequence of the six determinants is deliberately chosen to concentrate these ambiguities between adjacent items, which greatly simplifies the assessment and increases the robustness of the scoring or assay described below. The six determinants therefore form a rough continuum into which, we believe, can be fitted most of the factors relevant to cluster success or failure.

Groundings (Supply Determinants)

The supply determinants are the inputs to the productive process that originate outside the cluster.

1. **Resources:** Resources are natural, inherited or developed endowments available within the region. These include natural resources like forests, mineral deposits and fish stocks as well as land, a labour supply that is skilled, flexible and reasonably priced, strategic geographical location, financial capital and technology.
2. **Infrastructure**: Infrastructure consists of physical structures and institutional arrangements that facilitate access to resources and support other business functions. It includes physical infrastructure like roads, ports, pipelines and communications as well as intangible infrastructure like business associations, research laboratories, training systems, tax and regulatory regime, national monetary policy, financial markets, business and labor climate, quality of life (housing, crime, etc.).

Abundant and effective infrastructure can compensate to some extent for resource shortages, and vice versa. This is particularly true in a regional or local context where resources may be controlled at a higher level of government but access to them greatly affected by local infrastructure.

Enterprises (Structural Determinants)

Structural issues determine the efficiency of production in the cluster. The concepts and the terminology closely echo Porter, 1990.

3. **Supplier and related industries**: Success factors include diversity, quality, cost and proficiency of *local* suppliers, as well as the quality of the buyer-supplier relationships.[9] *Related* firms use similar technology, transferable human resources, similar specialized infrastructure or serve common markets. Success factors include the number and quality of these related firms, and the existence of formal and informal linkages between them and the cluster firms.

4. **Firm structure and strategies**: The firms in this case are the firms in the cluster itself, generally meaning enterprises directly in the value-chain for the product line that names the cluster. Considerations include the number and size of firms, birth and death rates, ownership and financial strength. Concentration is important because it is correlated with growth of market share (Henderson et al, 1993). The assessment also includes competitive and growth strategies.[10] Note that appropriate strategies are a strong function of the stage of technological development of a particular cluster (Utterback and Suarez, 1993), cultural and historical issues.

The distinction between "cluster" firms and supplier and related firms can be a fuzzy one, but the classification is not critical as long as the analysis is inclusive of the important relationships.

Markets (Demand Determinants)

Markets include both final and intermediate demand, including intermediate demand from the cluster itself.

5. **Local markets:** By local markets, we mean markets *within the region.* Important are the size of the market, market share, growth and prospects, extent of local sourcing by purchasers, standards and quality expected, distinctiveness of local demand, and willingness of buyers to work with the local cluster

6. **Access to external markets**: In principle, regions face a more or less common set of external markets. What differentiates among regions therefore is *accessibility*. Issues include closeness of markets, their size and growth rates, global market share for the cluster, characteristics of end users, existing market relationships, barriers to entry, trade and export barriers. Note: "external" here means external to the region, not necessarily international. However, in this as for the other determinants, the *standards* used to judge the cluster are global.[11]

The relative importance of internal and external markets varies according to the cluster. Porter, 1990, emphasizes the importance of discriminating internal markets even for clusters that are normally thought of as export oriented (e.g. Japanese consumer electronics). For other clusters, access to external markets is critical from the outset (Sharp and Holmes, 1989).

It is important to establish the size[12] of the target region and the scope of the cluster (is the cluster based on "information technology," "communications," "multi-media," or "animation"?). The GEM framework is fairly accommodating. For example, a market that might have been "local" if the region were larger, simply becomes an accessible part of the "external" market. Similarly, some of the suppliers to a narrowly defined cluster become part of a cluster defined more broadly. The overall GEM score is insensitive to these shifts, because of the way the six determinant scores are defined and combined. This is important in practice, because the unit of analysis (industry or region) is often defined by a client. However, the fit between cluster and region and the choice of scope should be reasonable, or it will be hard to make comparisons with competing and comparable clusters elsewhere.

4.5.2 Relationship to Earlier Work

There is no unique way of analyzing a complex system like an industrial cluster, nor even a uniquely good way. Building on the insights of earlier workers, we have tried to reduce the arbitrariness and increase the usefulness of the analytic framework.

The most influential model of cluster competitiveness is the Porter "diamond" of four factors that he found in a series of national case studies to be closely related to competitive success (Porter, 1990). The diamond model has four determinants: physical and human resources, supporting and related industries, domestic markets, and firm structure and strategies. Two additional factors, "chance" and "government" were seen as elements exogenous to business but sometimes important to competitive outcomes. The Porter model has been critiqued[13] and a number of alternatives proposed, but it remains very influential. Dunning, 1991, was principally concerned with the impact of a country's involvement in international business activity and proposed an additional exogenous factor to account for this. In our model, Dunning's international business activity factor is included in the determinant Access to External Markets. Narula (Narula,1991 and 1993) changes the diamond to better incorporate technology-driven innovation. Narula argues that innovation is important in all four of Porter's determinants, which we would support, and indeed we find that innovation is important in all six of the GEM determinants (see section for details). Rugman and D'Cruz, 1993, wanted to better account for the activities of multi-national enterprises and developed for North America a "double-diamond" formulation linking the diamonds for Canada and the United States. At the *regional* level, the issue of multi-nationality becomes an issue of "multi-regionality" and is ubiquitous. In the GEM formulation, branches of firms operating outside the target region are important as sources of technology and skills (Resources), strategies (Structure and strategies -- often a negative in this case), and for opening doors to external markets. They may also act as suppliers or customers. Cartwright (1992 and 1993), and Daly (1993), offer a generalized multiple-diamond linking any number of off-shore centres. Cartwright also introduces scales to quantify the four diamond determinants.

In the GEM, the four points of the Porter diamond are found, slightly modified, as determinants #1, #3, #4, and #5. The main modifications are in the scope of the Resources determinant and the specialization to a regional perspective. Porter, however, downplays the importance of government infrastructure, and generally treats external markets as an externality. In our view, in modern economies, infrastructure is equally as important as -- and to some extent a substitute for -- for resources. (The development of the Japanese economy in the 60s and the 70s is a good example.) Infrastructure is of signal importance for public policy, as it is one area where governments have a great deal of leverage, as Porter recognizes. As for external markets, it is true that as the value-per-kilogram of exports climbs, trade barriers fall, cultures converge and financial markets become more efficient, then will issues of location become less important. However the world is still far from the limiting case, and Access to External Markets remains a critical issue.

There is a second main thread in the analysis of industrial competitiveness, which sees industrial systems in more conventional production-system terms. We recognized that this framework (Supply $\Rightarrow$ Structure $\Rightarrow$ Demand) could be used to organize our six determinants into pairs, as explained above. This approach, developed in the consulting community,[14] emphasizes infrastructure or "economic foundations," which captures not only our definition of infrastructure but also some elements, e.g. financial institutions, that we would classify in the adjacent categories, namely Resources, and Supplier/Related firms.

4.5.3 Scoring the GEM

The first step is to score each of the determinants, on a world standard, using a scale from 1 to 10 (highest). At this stage of development, the score is subjective, based on information from a variety of sources, including the experience base of the economic development authorities, industrial consultants, and regional industry representatives, as well as objective data.

The process is calibrated conceptually by reference to a notion of competitiveness based on market share and profitability. We say a cluster is competitive if we can see it maintaining (or increasing) market share with average (or above-average) return on investment for its product line. We adjust the scoring so that a cluster with middle ratings of 5 on each determinant would be barely competitive. Note that the rating is a relative one. For example, consider a cluster with a product for which world demand is growing rapidly. The cluster may be growing but still be uncompetitive, if it is falling behind its competitors.

To summarize the overall competitiveness or wealth generation potential of a region we use an heuristic function based on the GEM determinant scores. We argue that the members of each determinant *pair* as grouped above are rough substitutes, and we model this by combining the scores within a pair additively and with equal weights:

$$Pair\ Score = (D_{2i-1} + D_{2i})\,/\,2$$

where the D_i, $i = 1, 6$, are the individual determinant scores. On the other hand, each of the pairs is (at least roughly) complementary to the others. For example, previously dominant clusters that lose their position often do so by failing to innovate structurally, weakening determinants #3 and #4 together, so the cluster fails even if markets, resources and infrastructure remain strong.

We model this by combining the pair scores as a geometric mean with equal exponent weights:

$$Linear\ Cluster\ Score = \prod_{i=1,3} (Pair\ Score_i)^{1/3}$$

Finally, we do two scaling transformations. One is substantive, converting the score from a linear to a quadratic form. We want a score that reflects the wealth-building potential of the cluster. The quadratic choice reflects our guess that markets confer increasing returns on successful clusters: added to the impact of increased market share will be the impact of increased profit margins.[15] The second transformation is cosmetic, scaling the overall rating so that the ideal maximum is 1,000. The final form for the GEM assay is:

$$GEM = 2.5 \; (\prod_{i=1,3}(D_{2i-1} + D_{2i}))^{2/3}$$

A neutral, marginally competitive cluster would have *GEM* = 250 on this scale.

We have found the model useful as a background for expert discussions of cluster strengths and weaknesses. Although the rating scale is subjective, we also find a reasonable degree of inter-rater consistency and the process of developing consensus on scoring is often very formative, especially in a group setting. The ratings have also been used to develop action plans to fill in GEM gaps (see section), which are now being implemented in several jurisdictions (Western Economic Diversification, 1996).

4.6   GEM MEASUREMENT AND INTERPRETATION

The GEM is supposed to measure competitiveness and we have described in a general way what we mean by competitiveness in terms of market share and profitability, and we have outlined what are the main determinants of competitiveness. This conceptual framework appears quite sufficient for subjective assessments of cluster strengths and weaknesses and the GEM score provides a sensible way of comparing overall competitiveness of different clusters.

From a research perspective, however, we believe it is important to try to make this framework more quantitative. For example, we would like to be able to associate GEM scores with statistical indicators. We would also like to test the heuristic structure and possibly improve it. This is an ambitious project. For example, different industries have very different structures -- there will be no single set of indicators and weightings. Also, it can be very hard to find numerical indicators for many important issues that raters consider in scoring the GEM.

In the this section we present some new work on the measurement and data issues and discuss a more formal mathematical framework for the GEM function.

4.6.1 Measuring the Determinants

In section we sketched some of the issues that are important to each of the GEM determinants. Not all of these lend themselves to quantitative measurement, and some are more amenable than others, or more significant.

With respect to *Groundings* we would, with Porter, place emphasis on the specializations developed in application of resources. In a quantitative assessment much employment and some investment data can be presented under this heading, along with data on innovation and research.

Enterprises means both supporting and supplying activities and the entrepreneurial climate in the industry itself. In a quantitative assessment we would include data on linkages derived from input-output analysis, cost structures, concentration ratios, firm and unit size, foreign ownership and inward investment.

Data on *markets* will include input-output based analysis of local markets (including the degree of import penetration), existing sales by market at home and abroad, and growth trends in external markets.

Angenendt (1994), lists about 50 competitiveness indicators, which he calls "cause parameters," classified under the four Porter determinants. Some of these are quantifiable and some not and not all are equally important.

What can be done in practice depends on available data. Competition being what it is, much important information for policy will not be found in public sources. But an essential screening device, and one which can direct the investigator towards likely sources of critical information, is analysis from available desk resources including the following:

Census of Production

The structure of sales, purchases, value added and primary inputs of capital and labour are in most countries recorded on an establishment basis, usually on an annual or four-yearly cycle. Issues of disclosure can be a serious hindrance to non-governmental investigators or those outside the statistical agency. It is usually important to obtain the data in as much detail as possible. Recent international harmonization has been of some help in making the classifications in different countries compatible through the ISIC, but there are still many problems. Although it is not yet operational, we have found that reference to the draft Central Product Classification, which combines production and trade classifications, is helpful in resolving conceptual issues.

Census of Employment

Often employment data is more available, and can be used as a (poor) substitute for value-added data. For example, value added data available for a wide region or industry group may be subdivided using detailed employment ratios.

Input-Output Tables

Input-output tables contain the principal linkages in a cluster, because they show the sales from every industry in an economy to every other industry. If the tables are presented in sufficient detail then it is possible to trace linkages between sectors, identifying clusters around the row and column maxim of a table. They also provide a starting point for models of costs structures within an industry, and in principle can provide a large number of measures of market share.

Global Economic Activity Metric

We have developed, using a wide variety of data sources but principally OECD and various United Nations agencies, a matrix setting out the value-added structure of the entire global economy, broken down by more than 130 countries and some 70 industries. The output of each industry has been adjusted so that the country's GDP corresponds to its GDP measured at Purchasing Power Parity, with the adjustment falling mainly on industries. The model allows us not only to measure overall competitiveness, but also to assess the strength of key linkages.

National Accounts and Detailed Local Economic Measurement

The framework embodied in the UN System of National accounts has now been adopted by almost every nation of the world, and we have found it valuable to tie cluster measures back to this standard scale. This has also meant the development of an expertise in interpolating national accounts to a detailed local level, typically in 125 or so industrial sectors. Apart from its statistical utility, national accounts have the advantages of providing a common language linking industrialists and policy makers, and providing enough detail on individual clusters for structures to be discerned..

An example of indicator development for the GEM determinants of a specific industry is sketched in section .

4.6.2 Measuring Competitiveness

Competitiveness is sometimes measured as narrow cost comparisons – for example conventionally international bodies measure changes in international cost competitiveness by changes in relative unit labour costs or relative export prices. However, quite apart from difficult issues in determining a baseline for these scales, to treat competitiveness in this way is to restrict the view of competition to price alone.

An approach we have found more fruitful is to measure competitiveness through revealed preference – in effect by market share. Over time, for example, by measuring price elasticities and adjustment models econometrically we have been able to separate price effects out of changes in competitiveness and thus to gauge non-price competitiveness for national industries.

On the cross-section basis appropriate to global comparison, Porter, 1990, measures competitiveness as revealed by shares of internationally traded business. This measure is problematic at the regional level, because for most regions, statistics for international trade do not exist, or are sparse, unreliable, not detailed and not promptly published. In addition, with the majority of global economic activity now concentrated in services, there are no reliable comprehensive detailed statistics for the growing trade in services.

For these reasons we often measure competitiveness in terms of share of global value added (output). This basis is both more practical and in our view theoretically better because of its comprehensive coverage. Statistics for service industries are not perfect, but in principle the GDP statistics are comprehensive of economies and of all industries within them. Competitiveness as revealed by shares of output can be extended, with caution, to goods and services which are not always readily traded, such as military equipment or health services.

Revealed competitive advantage is measured and summarized graphically in a salience chart.

Salience of industry i in country j is defined as

$$\sigma = \ln(x_{ij}.x_{IJ}/(x_{Ij}.x_{iJ}))$$

where x_{ij} is output of industry i in country j, capital letter J denotes summation across all countries in the world and I across all industries globally.

The logarithmic format shows strength (positive salience) and weakness (negative salience) symmetrically and is additive: the salience of region A relative to nation B plus the salience of nation B relative to the world, is the salience of A relative to the world.

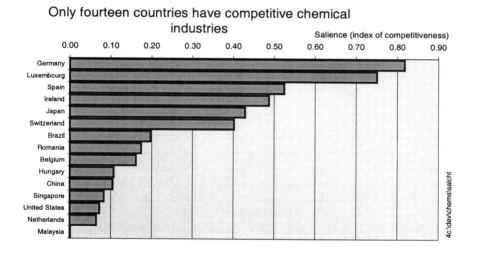

Figure 5 - Saliences of national chemical industries

However, we do not want to restrict the measurement of competitiveness to market share alone. We would argue that the aim of economic development is to enable industries to thrive, benefitting their region. "Thriving" and "benefits" can be difficult concepts to tie down, but clearly include notions not only of market share, but of general prosperity deriving from solid profits and good wages. Growth could be added to the list, but has to be considered relative to growth of the world industry. These can all be measured to some degree. Angenendt (1994), lists more than a dozen candidate competitiveness "result parameters" including industry balance of trade, reputation and product quality (but not wages). In our conceptual backgrounding to set up the GEM scoring, we ask raters to consider both market share and profitability as normative factors for competitiveness.

In practical terms, our measurement priority would be to develop comparable profitability measures (recognizing that reporting requirements vary between nations) representing returns to capital, and later to add wage measures representing returns to labour. Together, these would represent the "prosperity" derived from a unit of output in the industry.

4.6.3 Relating the GEM to Competitiveness Measures

The GEM is an assessment of the current structure and therefore the competitive potential of an industrial cluster. The situation now is not necessarily the same as the situation in the past or as it will be in future, for example, and major policy change may materially improve the prospects of the cluster. However, profitability and market share will take some time to catch up. The approach to equilibrium will proceed at a pace specific to the industry(ies) involved in the cluster. To reflect the dynamic relation between competitiveness and outcomes, we relate the GEM score to *equilibrium* values for profitability and share, which are not directly observable. The equilibrium values P_{eq} and S_{eq} can, however, are not directly measurable, but can be derived from time series for the observed variables p and S through the relationships,

$$dp / dt = (p_{eq} - p) / \tau$$
$$ds / dt = (S_{eq} - S) / \tau$$

where τ is the adjustment time for the industry, p_{eq} is the equilibrium profitability (or prosperity, which includes returns to labour as well as capital), and S_{eq} is the equilibrium share. This formulation describes an equilibrating process, where the actual profitability and share move towards the equilibrium values at a rate proportional to the distance from equilibrium. If the equilibrium values are constant in time, then the relaxation is a simple exponential relationship: $1 - \exp(- t / \tau)$.

The notion of an adjustment time is approximate, of course, and τ would have to be estimated for each industry. Capacity adjustment may require replication, replacement (or depreciation) of expensive physical plant. We believe that the adjustment time will be related to the average length of a product development cycle (see section) which typically involves some renewal of both physical and human capital. For example, the adjustment time would be a few months for the software industry, but many years for pharmaceuticals. Note that changes in technology or industry strategy can change the adjustment time. Formulating the model in this way, cluster *growth* is a derived quantity, determined by the adjustment time and the equilibrium share.

We would define competitiveness Γ as follows:

$$\Gamma = \Gamma_0 \, (p_{eq} / p_n)(S_{eq} / S_n)$$

where p_n is the normal profitability[16] for the industry and S_n is the normal share.[17] In section we argued a functional form for a competitiveness GEM that is quadratic in six determinants: Resources, Infrastructure, Supplier and Related Firms, Firm Structure and Strategies, Local Markets, and Access to External Markets. If we identify the predictive GEM function and the competitiveness outcome measure Γ, we can test this structure.

In our work to date, we have relied on subjective assessments of the GEM determinants, on a scale of 1 to 10, with a constant of proportionality that makes the maximum GEM equal to 1000. In contrast, the competitiveness Γ can range, in principal, from 0 to infinity. Logically, if the GEM is identified with Γ, it would make more sense to do the ratings on a scale from 0 to infinity as well, but raters find it difficult to consistently apply a scale with no upper limit, and likewise policy makers find it easier to interpret scores with a finite maximum. However, it is a simple matter to scale the ratings to the desired range, for example to define determinants

$$D_i = D_i / (10 - D_i).$$

The normal value of each d_i is 1. Supposing Γ to be determined by the d_i in the same way as the GEM by the D_i, we would hope to predict Γ by:

$$\Gamma = \Gamma_0 \left(\prod_{i=1,3} (d2i - 1 + d2i) \right)^{2/3} / 4$$

with a normal value equal to the chosen Γ_0.

This formulation invites construction of more objective indicators that could be combined with appropriate weights to calculate the six determinants. Statistical indicators can be scaled to have the right range and mean. Where an indicator reflects measurements of capacity, e.g. the size of a mineral resource or number of specially qualified personnel it should generally be linear in these measures. An example is described in section 7.2.

4.6.4 Generalizations

As well as searching for appropriate indicators, it would make sense to try to optimize the functional form of the GEM function. We have done our best to define determinants that are of approximately equal importance for a broad range of industries, and practitioners seem to be comfortable using the framework, but the structure should be checked. For example, while we argue that the GEM should be quadratic in its determinants it would be reasonable to try different weights on the Groundings, Enterprises and Markets pairs, that is to test a function of the form:

$$\Gamma \propto (g^\alpha \bullet e^\beta \bullet m^\gamma)$$

with the constraint $\alpha + \beta + \gamma = 2$ Here g, e and m are the scores for each of the determinant pairs, that is,

$$g = (d_1 + d_2) / 2, \text{etc.}$$

This formulation allows one or other of the g, e and m to be more critical than others. For example, in the export/commodity oriented forest industry, we might want competitiveness to depend more sensitively on g (especially Resources) than on m.

Also, it would be interesting to check the weighting within the pairs. For example, a training system (Infrastructure) may not be a perfect substitute for skilled human resources (Resources) because the former involves a lag that could be dangerous in a new industry, such as Internet software, where firms are competing vigorously to establish technological dominance. Thus another generalization is .

$$g = \phi \, d_1 + (\tfrac{1}{2} - \phi) \, d_2), \, 0 \le \phi \le \tfrac{1}{2}.$$

Currently, raters make this kind of adjustment on an intuitive basis.

Finally, the GEM function uses geometrical and arithmetic means to reflect the relative complementarity and substitutability of the six GEM determinants. Our basic formulation assumes that Groundings, Enterprises, and Markets are pure complements, and that the determinants going into each pair are pure substitutes. This is not quite so, of course, and for some industries the differences may be important. In the forest industry example, transportation infrastructure is a critical link, which accounts for the weakness of the industry in Siberia, where there is abundant fibre but poor transport.

4.7   APPLICATIONS IN COMPETITIVENESS ANALYSIS

In this section, we present some recent applications of the model described in section in client-defined projects.

4.7.1 Technology Policy -- Western Canada Diversification

What regional development managers need is an idea of which current industries are poised for success, and what they can do to accelerate that success. The GEM approach was constructed with this in mind: the GEM function measures the potential of the *current* industrial structure, and inspection of its components reveals current opportunities for increasing the potential, and a rough gauge of their impact. The GEM model was developed as a tool for economic development, to help agencies and their industry partners to understand the strengths and weaknesses of their regional clusters and to help them devise strategies to improve them. The best targets for intervention[18] are parts of the value chain that mesh with industry and infrastructure strengths in the region, offering clear and sustainable advantages in terms of the flow of information and product that regional concentration makes possible (Brezis and Krugman, 1993). The region can look at its particular set of resources (human, social, economic, geographic) and identify packages that match the attributes of multinational or other corporate activities. The region can ask what new infrastructure would support the cluster, what industrial strengths could be encouraged, and what firm strategies should be considered.

We give an example from an actual cluster study in Western Canada (Western Economic Diversification, 1996). In this project, a study of cluster structure and dynamics based on desk analysis and interviews was followed by a series of expert workshops to identify specific projects that might be undertaken by government or government and industry together to strengthen each of three clusters identified for analysis.

Following is an abbreviated discussion of the GEM competitiveness determinants for the region's information technology (IT) cluster, against world standards, with the subjective scores shown after each heading.

Resources   6

The principal resource issue is human resources.   There is a good supply of domestically produced skills of generally high standard but high cost compared to some competitors.   Workshops identified several critical shortages (software, radio frequency engineering), and firms commonly recruit internationally for highly qualified personnel.   There is a large pool of manpower next door in the USA, but recruiting is sometimes difficult.   Experienced management is a critical deficit.   Technology sources in universities are good, but there is less breadth than in USA benchmarks.

Infrastructure   5

This is the most developed advanced-technology sector in western Canada and has for some time been an object for strategic attention by government. As a result there is a considerable amount of supporting public sector research and some industry networking organizations.   There is an average tax and regulatory environment (improving with the North American Free Trade Agreement), and overall average perceived quality of life. Conventional communications infrastructure is cheap and reliable, but high speed networks are in their infancy.   The benchmark is a high one as many competing jurisdictions have worked hard at improving infrastructure for IT; there is no successful IT cluster anywhere that has not benefitted from massive government intervention.

Supplier and Related Firms   3

Firms can't count on local sources for most technology needs, although this is less important for established systems integrators with international networks.   There is a shortage of sophisticated financial and marketing services.

Firms Structure and Strategy   **3.5**

There is a mix (variable among provinces) of medium, small and very small firms, many of them very young. Many 'firms' producing IT services are still part of their client organizations. The lack of large established firms is a weakness, but the small/medium-sized enterprises pattern is a common one in all but the strongest half dozen competing jurisdictions.

Local Markets   **4**

Local markets are underdeveloped. Public sector procurement is not systematically organized to stimulate the industry. Except for oil and gas in Alberta and some forest applications, the resource sectors make light use of innovative information technologies. The business services sector is an eager local market, but is not salient itself in western Canada; a good part of the business-service demand is for packaged software, not a western specialty.

Access to External Markets   **6**

External markets are large and growing rapidly. Growth is less rapid for the USA, the closest major market. Access to US and other external markets is stronger in BC than in the Prairies, but even there falls short of many competing regions.

These ratings give an overall GEM score of 200, implying that, as things now stand, the cluster is likely to continue strong growth, but the growth will be slightly slower than world averages.

The workshops developed business plans, in varying levels of detail, for seven initiatives. The consultants estimated the likely impact, cost and time-frame of these initiatives, and re-rated the GEM determinants assuming that all initiatives were successfully implemented. The following chart shows the original GEM (heavy outline) and the potential GEM (light outline) that could reasonably be expected if all the strategies were implemented.

The analysis indicates that the initiatives from the workshops have the potential to move the cluster from marginally uncompetitive to marginally competitive on a world scale. However the impact will not be as strong as it might be with a more strategic approach.

The key weaknesses: the industry itself is below critical mass and lacks large firms, the framework of supplier firms is incomplete, and local markets (originally the driving clusters) are insufficiently demanding.

The following were suggested as strategic considerations:

*   The priority initiatives do not do much to address the cross-cutting structural weaknesses.

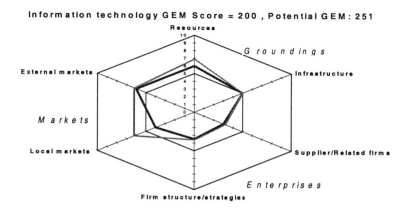

Figure 6 - GEM analysis of a technology cluster

Addressing these would amplify the impact of all the other changes. Planners could go back to the "long lists" of initiatives developed early in the workshops, which included, for example, a comprehensive set of incentives to increase availability of venture capital, a Related Firms issue. However, the most decisive intervention in this area, which would strengthen both structural dimensions, would be to attract large computer, component or software/systems makers to the region.

• The largest number of initiatives address an area which is already fairly strong, namely human resources.

The question is one of balancing government resources and the effect of diminishing returns. Planners may want to focus on the initiatives with the highest impact/cost ratio.

• The initiative with the most impact potentially is also expensive and very long term.

From the workshop discussion, what was envisaged here was very ambitious compared to alternative "public relations" approaches. The project involved putting a great deal of information technology into the education system, which we believe would ultimately have a large impact but would also involve large capital costs.

• There are no infrastructure initiatives.

While this is a dimension that is already marginally competitive (reflecting past efforts to establish appropriate infrastructure) we believe it is an area of some opportunity, since many of the desirable infrastructure investments could have broad social benefits as well. The most urgent weakness we see is coordination of *institutional* infrastructure.

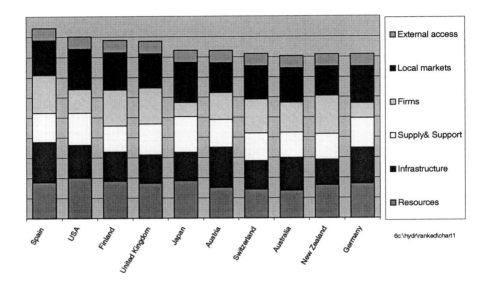

Figure 7 - GEM determinants scores for electricity generation

4.7.2 Measuring GEMs -- Electricity Generation

Most of our GEM assessments have relied on qualitative judgments, but we have done some preliminary work on quantitative indicators. For a consulting project, we were asked to measure the "health" of national electrical industries. A revealed-preference analysis based on market share was not helpful. By its nature electricity is not readily traded, and beyond this the history of state ownership or close regulation of production and distribution systems means that the world industry is far from competitive equilibrium. The GEM model was therefore adopted to assess the determinants of competitiveness, even though competitiveness itself could not be measured directly.

The industry shows variations worldwide. The amount of electricity generated and used varies enormously, from barely 50 kilowatt-hours per head per year in warm poor countries with few resources like Bangladesh and Nepal, to 500 times that amount in cold, rich ones like Norway and Canada. Growth is far from evenly spread, and different parts of the world start off from radically different positions.

We have used the available evidence to rate each of 88 countries on five of the GEM factors, omitting Access to External Markets.

The score for each country is made up in different ways, so that for example in Japan the market is growing and demanding, but the company structure is conservative. The chart compares the determinant scores for the 10 countries with the highest GEM scores.

The USA, for example, owes its standing to resources and affluent local markets, while for the UK the strengths are an active commercial structure and resources, including fuels, technology and people.

The sources of data were:

- Electronic databases of government-supported international organizations such as the International Monetary Fund
- Economic and energy data assembled electronically by the authors
- Statistical reports published by government-supported national and international organizations (e.g. International Energy Agency, Central Intelligence Agency) and by British Petroleum
- Regional and sectoral electricity industry market research reports (e.g. Financial Times Energy Publishing, McGraw Hill)
- Individual country reports (e.g. Economist Intelligence Unit, the UK Department of Trade and Industry)

For some countries the information was incomplete and for others different publications contained contradictory information. Some judgment was therefore needed in forming views about individual countries. However, extensive supporting information is available, broadening the basis of the different country assessments.

Resources

The main physical resources needed for an electricity industry are accessible, properly-priced, diverse primary energy sources. We measured the availability of indigenous primary fuels as a percentage of GDP, based on BP and World Bank or OECD data. Where a country has more than one indigenous fuel this is a strength, so we calculated an index of the available diversity of indigenous primary fuels. Human resources are summarized by the level of participation in secondary and tertiary education in a country, as recorded by the World Bank. We have placed more weight on secondary.

Availability of internal financial resources and economic stability were gauged by the overall balance of payments (the full account, not just the current account)[19] and by low interest rates in both money terms and real terms. This data was taken from the IMF database International Financial Statistics. We appraised the diversity of technology strengths by a measure of the diversity of generating technologies and of thermal generating technologies. This was collected from industry market research sources and from UNIPEDE.

Infrastructure

Key infrastructure includes: A political, legal and financial system to secure and remunerate large long term investments; physical access for construction, operation and maintenance; a research apparatus and education/training system for staff; environmental framework; and power distribution system.

Collecting appropriate data to assess these was not straightforward. We used assessments of political, commercial and financial risks and regarded an absence of government regulation as positive. Physical access was estimated by population density, but the research/education infrastructure could only be measured globally by tertiary education penetration. No comparable transmission network information data with global coverage was presented.

Supplier and Related Firms

The businesses required to support a thriving electricity industry include construction and engineering industries to create and maintain plant and networks, financial and business services industries, transport and fuel supply. For most of these we used the salience of the appropriate industry as an indicator of strength.

Firm Structure and Strategies

The requirements include: freedom to act, flexible degree of integration (regional distribution companies, grids, generators); efficient resource and margin management; focus on service and adding value; ease of entry, exit and market growth.

It was not possible to analyze the strategies and rivalry of so many companies in so many countries. The estimation focused on an analysis of commercial and political openings.

Local Markets

The criteria were that the local market should be rich enough and growing in terms of output and number of generating sets, diverse, changing and flexible, and fit well with generating and transmission capabilities (capacity mix, fuels, location, grid). GDP per head, paying particular attention to differences among the less developed countries, was the key variable used for ranking. This was followed by percentage and absolute growth in electricity production and consumption, and high price levels. Weight was given to the presence of power-using industries.

Access to External Markets

We would hope to measure: information/expertise flow (technical and commercial); foreign ownership; interconnection; and other issues. We did not have adequate data sources for this and so attributed the normal score of 5 for each country.

Indicators were constructed for the first five determinants, scaled to range from 0 to 10 with a mean of 5 and a standard deviation of 2. Determinant scores were combinations of indicators using subjective weights. On the basis of the calculated GEMs, the ten most vibrant national electricity industries turn out to be primarily developed countries with growth potential remaining, namely:

Spain, USA, Finland, United Kingdom, Japan, Austria, Switzerland, Australia, New Zealand, Germany

Spain had the highest GEM score, 594. A total of 27 out of the 88 countries examined had GEM scores above the marginal competitiveness threshold of 250.

4.8   LINKING THE TWO MODELS

Clearly, innovation has a great deal to do with competitiveness. What, then, is the relationship between the Cycle model of a regional innovation system and the GEM model of competitiveness? There is a fairly straightforward mapping of the sources of innovation to the GEM determinants, as illustrated in the table below. The table gives the GEM determinants principally influenced by the pace and quality of innovation from each source. In the third column is a single illustrative example of a specific mechanism for transfer from the indicated source and affecting the indicated determinant.

It is no surprise to find that innovation is important in every sector of the GEM, but equally important to note that innovation is not the whole story. Natural endowments of physical and human resources, flexible and well run legal and financial systems, political constraints and opportunities, and cultural links to important markets or sources, are examples of other factors, discussed more completely in Padmore and Gibson, 1997, that are important to competitiveness. [20]

Innovation source	*GEM determinant*	*Example*
In-house	Resources	Skilled workforce
	Firms structure, strategy, rivalry	R&D intensity
Suppliers	Supplier and related firms	Embodied technology
Peers	Firms structure, strategy, rivalry	Know-how trading
Customers	Local markets	Government technology procurement
	Access to external markets	Favoured- nation status
Public sector	Infrastructure	University-industry liaison
	Resources	University researchers

Table 1 - Mapping of the cycle model onto GEM model

ENDNOTES

1.   The principal investigators are based at the Centre for Policy Studies on Science and Technology at Simon Fraser University and the Centre for Policy Studies in Education at the University of BC, with active associates at the University of Ottawa and Glasgow Caledonian University.

2.   Considerable detail on the Ec/Tech study and related work can be found in papers presented at an international workshop on *The Role and Impact of Science and Technology in Innovation and Regional Economic Development: Models, Approaches, Policies and Their Effectiveness* that was held at the University of British Columbia in August, 1995. Proceedings will be forthcoming and copies of abstracts and papers are available from author Padmore.

3.   Schumpeter, 1934 distinguishes five types of innovations: i) introduction of a new product or a qualitative change in a existing product; ii) process innovation new to an industry; iii) the opening of a new market; iv) development of new sources of supply for raw materials or other inputs; v) changes in industrial organization. The most significant difference is that our definition would qualify as innovation an action "opening a new market" only if that action involved a change in inputs, methods or outputs novel to that market.

4.   Not to be confused with a product *life cycle*, which will involve many development cycles as the market for the product and its successor versions develops, matures and decays. See section.

5.   OECD, 1996, p. 43.

6.   We use the model routinely in discussions with Canadian and UK policy makers.

7.   The Oslo manual (OECD, 1996) and the European Community innovation surveys both use selling price/sales as a reference for comparing costs of technological innovation expenditures. A difficulty with this approach is that because of lags, current sales may not be relevant to current innovation activities. However, serial measurements would allow testing of this idea.

8.   Together, agriculture, forestry and fishing represent approximately 7 per cent of global GDP. Out of 135 countries measured, 85 have a concentration and in 68 of them it is greater than 10 per cent of GDP.

9.   See Scherer, 1984, especially Ch. 15, for examples of buyer-supplier ties relating to technology flows and productivity growth.)

10.   See Ernste and Meier, 1992, Ch. 3, and Scherer, 1984, Section III, especially on firm size.

11.   For a study expected to command significant policy or development resources the appropriate focus of analysis must be *global* competitiveness. We see only limited value in comparing with the generality of the local trading bloc. The bloc may not be a particularly competitive one in the industry, and the concern of cluster analysis is not to refine the local division of labour, but to point to ways to create competitive advantage. The local industry should be looked at in the context of overall global totals.

12.   A cluster may crowd into one corner of a province, span several cities and suburbs, or straddle an international border. See Scott and Storper, 1992b, p. 6, for a diagram that usefully captures some of these complexities. The commonest cluster span is a metropolitan area comprising a major city and related communities (Ellison and Glaeser, 1994).

13.   A good summary of the literature on the Porter diamond can be found in Penttinen, 1994.

14.   DRI/McGraw Hill have used this approach in a number of projects, but there are no publications in the open literature. (Steven Waldhorn, private communication.)

15.   Sorenson, 1995, argues that this is the case at least for markets where firms compete on quality. European Commission, 1994b, presents other evidence at the firm level linking market share and return on invested capital.

16.   By normal profitability we mean the average return on capital for the selected industries world-wide. Normal "prosperity" would be the weighted average returns to capital and labour. Because profitability is highly variable, and sometimes negative, it would be appropriate to consider long term expected profits or expected present value of future profits, both of which can be assumed to be positive.

17.   Normal share is the total size of the regional economy (GDP) divided by the size of the world economy. The industry salience for the region is

18.   For evidence of effectiveness see Head et al, 1994b, or Rauch, 1993. For case studies see Sharp and Holmes, 1989.

19.   The full account, not just the current account. It may sometimes be argued that under perfectly flexible exchange rates, the balance of payments is an uninteresting residual. In this application, where countries have varying degrees of flexibility and openness, we propose that variations in the balance of payments reflect variations in availability of capital resources.

20.   Paul Foulsham did much of the work digging out data on individual electrical industries from various sources. We thank the Social Sciences and Humanities Research Council of Canada, the National Research Council of Canada and Statistics Canada for partial support of this work. We would also like to acknowledge the contribution of Humphrey Stead for a particularly careful review of and commentary on the Cycle model of an innovation system, and John Helliwell for useful criticism of the GEM model.

REFERENCES

Acs, Z. (ed.) (1999). *Regional Innovation, Knowledge and Global Change*, London, Pinter.

Acs, Z. and de la Mothe, J. (1999). "Cities, Information and 'Smart Holes'", in Acs, Z. (ed.).

Acs, Z. (1996). US High Technology Clusters, in J. de la Mothe and G. Paquet (eds.), *Evolutionary Economics and the New International Political Economy*, London: Pinter, 183-219.

Angenendt, G., (1994). "Identification and Discussion of Parameters That Can Be Used To Analyze Industries With Michael E. Porter's System of Determinants that Influence the Competitive Position of Nation's Industries" (Research Institute of the Finnish Economy, Helsinki), discussion paper #482.

Bergman, E., Maier, G., and Tödtling, F. (eds.), (1991). *Regions Reconsidered - Economic Networks, Innovation, and Local Development in Industrialized Countries*, Mansell, London.

Brezis, E. and Krugman, P. (1993). "Technology and the Life Cycle of Cities", NBER Working Paper 4561.

Cartwright, W., (1993). Multiple Linked "Diamonds" and the International Competitiveness of Export-Dependent Industries: The New Zealand Experience, *Management International Review*, special issue 33:2, 55-70.

Cartwright, W., (1992). "Canada at the Crossroads Dialogue", *Business Quarterly* 57:2, 10-12.

Coriat, B. (1992). The Revitalization of Mass Production in the Computer Age, in Storper and A.J. Scott (Eds), *Pathways to Industrialization and Regional Development*, Routledge, London, pp. 137-156.

Daly, D.J. (1993). "Porter's Diamond and Exchange Rates", *Management International Review*, 33:2, 119-134.

Dodgson, M. and Rothwell, R. (eds). (1994). *The Handbook of Industrial Innovation*, Brookfield, Edward Elgar.

Ellison, G. and Glaeser, E.L, (1994). "Geographic Concentration in U.S. Manufacturing Industries: A Dartboard Approach", NBER Working Paper WP4840.

Ernste, H., and Meier, V. (eds), (1992). *Regional Development and Contemporary Industrial Response: Extending Flexible Specialization,* Bellhaven Press, London.

European Commission (1995). *Green Paper on Innovation*, Volume 1 and Volume 2 (annexes).

European Commission (1994a). *The Community Innovation Survey: Status and Perspectives.*

Evangelista, R. (1996). "Embodied and Disembodied Innovative Activities: Evidence from the Italian Innovation Survey", in Innovation, Patents and Technological Strategies, OECD, Paris, pp 139-162.

Fairtlough, (1994), "Innovation and Organization", in Dodgson, M., Rothwell, R. (eds), *The Handbook of Industrial Innovation* Aldershot Edward Elgar.

Grossman, G., and Helpman, E. (1991). "Quality Ladders in the Theory of Growth", *Review of Economic Studies, 58*, 43-61.

Hanson, G.H. (1994). *Localization Economies, Vertical Integration and Trade*, NBER Working Paper 4744.

Head, C.K., Ries, J. and Swenson, D. (1994). *Agglomeration Benefits and Location Choice: Evidence from Japanese Manufacturing Investment in the U.S.*, NBER Working Paper 4767.

Henderson, J.V., Kuncoro, A. and Turner, M. (1993). *Industrial Development in Cities*, NBER Working Paper 4178.

Higgins, B. and Savoie, D.J. (1988). *Regional Economic Development: Essays in Honour of Francois Perroux,* Unwin Hyman, Boston.

Hilpert, U. (1991). *Regional Innovation and Decentralization: High Tech Industry and Government Policy,* Routledge, London and New York.

Hobday, M. (1994). "Innovation in Semiconductor Technology: The Limits of the Silicon Valley Network Model", in Dodgson, M., and Rothwell, R. (eds), *The Handbook of Industrial Innovation*, Edward Elgar, Aldershot, pp. 154-168.

Lhuillery, S. (1996). *Innovation in French Manufacturing Industry: A Review of the Findings of the Community Innovation Survey, in Innovation, Patents and Technological Strategies,* OECD, Paris.

Lorenz, E.H. (1992). "Trust, Community, and Cooperation: Toward a Theory of Industrial Districts", in Michael Storper and Allen J. Scott (eds), *Pathways to Industrialization and Regional Development,* Routledge, London and New York.

Lundvall, B.-Å. (ed.) (1992). *National Innovation Systems: Towards a Theory of Innovation and Interactive Learning*, Pinter, London.

Miller, R. and Cote, R. (1987). *Growing the Next Silicon Valley*, Lexington, Mass.: D.C. Heath.

Nelson, R. (ed.) (1993). *National Innovation Systems. A Comparative Analysis*, New York. Oxford University Press.

Niosi, J. (1991). Canada's National System of Innovation, *Science and Public Policy, 18.*

OECD (1996). *Technology, Productivity and Job Creation.* Paris: OECD.

OECD (1996b). *Innovation, Patents and Technological Strategies,* Paris, OECD.

Ohmae, K. (1995). *The End of the Nation State,* New York, Harper Collins.

Porter, M.E. (1990). *The Competitive Advantage of Nations,* Free Press, New York.

Rugman, A. and D'Cruz, J. "The "Double Diamond" Model of International Competitiveness: The Canadian Experience", *Management International Review,* volume 33:2.

Saxenian, A. (1994). *Regional Advantage: Culture and Competition in Silicon Valley and Route 128,* Cambridge: Harvard University Press.

Scott, A.J. and Storper, M. (1992). "Industrialization and Regional Development", in Michael Storper and Allen J. Scott (eds), *Pathways to Industrialization and Regional Development,* London: Routledge.

Sharp, M. and Holmes, P. (1989). *Strategies for New Technology: Case Studies from Britain and France,* New York: Philip Allan.

Thomas, M.D. (1985). "Regional Economic Development and the Role of Innovation and Technological Change", in A.T. Thwaites and R.P. Oakey (eds), *The Regional Impact of Technological Change,* London: Pinter.

Tylecote, A. (1994). "Financial Systems and Innovation", in Dodgson, M., and Rothwell, R., *The Handbook of Industrial Innovation,* Edward Elgar, Aldershot, Hants., England.

Utterback, J.M. and Suarez, F.F. (1993). "Innovation, Competition and Industry Structure", *Research Policy,* 22.

von Hippel, E. (1988). *Sources of Innovation,* Oxford, Oxford University Press.

Western Economic Diversification (1996). *Building Technology Bridges: Cluster-Based Economic Development for Western Canada,* Ottawa, National Research Council.

PART C

INTERNATIONAL AND
INTER-REGIONAL PERSPECTIVES

5   KNOWLEDGE-BASED INDUSTRIAL CLUSTERING: INTERNATIONAL COMPARISONS

Roger Voyer

Coopers and Lybrand Consulting, Ottawa

"Economic life develops by grace of innovating; it expands by grace of import-replacing. These two master economic processes are closely related, both being functions of city economies".

Jane Jacobs

5.1   INTRODUCTION - WHAT IS INDUSTRIAL CLUSTERING?

Knowledge-based industrial clusters are regional or urban concentrations of firms including manufacturers, suppliers and service providers, in one or more industrial sectors. These firms are supported by an infrastructure made up of universities and colleges, research institutes, financing institutions, incubators, business services and advanced communications/transportation systems. The concept of industrial clustering[1] fits the notion of systems of innovation[2] well since both deal with capabilities and relationships.

With the shift to the "new economy", sub-national regions and municipalities around the world are setting in place the infrastructure and mechanisms needed to support technology-intensive industrial development. I have estimated that there are about 200 or so sub-national regions and municipalities active in developing strategies to attract knowledge/technology-based investment. The global economy is driven by a mosaic of regional/local clusters. Some are described in the following sections.

The idea of industrial clustering has a long history and is well anchored in the study of economic geography.[3] Benefits can accrue to an area from the activities of firms in that area. These benefits typically arise from the fact that a firm cannot capture *all* the economic benefits from its innovation process (i.e. bringing its products to market). There are spillovers out from the firm that can benefit the community at large if there are suitable structures and receptors in place to take advantage of them.

For example, people with expertise leave firms to work for other firms or to set up their own firms. Capturing these spillovers leads to the establishment of new capabilities and more growth in the community.

With globalization and the shift to a knowledge-based world economy, time-to-market and just-in-time delivery become more critical. This encourages the clustering of capabilities in regional/local centres to support the innovation process and thus to minimize the "leakage" of external benefits outside the community. Firms are attracted to communities that can provide the key functions needed to bring their products or services to market rapidly.

Very few regions/localities around the world have industrial clusters with more than 100,000 people working in industry. After some 50 years of development, Silicon Valley, California is such a cluster concentrated in the information technology and related microelectronics area, with about 1 million people in more than 6,000 firms generating more than $200 billion in sales. At that level the cluster is usually self-sufficient or complete, in that it has in place all the essential technical, business, financial, legal, etc. capabilities needed to sustain industrial activity in the cluster. The more firms and the more people working in industry in the cluster, the more it tends to be self-sufficient, i.e. fewer outside resources (i.e. imports) are needed. The "leakage" outside the cluster increases with fewer and fewer firms and employees in the cluster.

In summary then, the key characteristics of industrial clustering are:
- strong linkages among firms and the supporting technological and business infrastructure in a region stimulate the innovation process and the growth of the cluster;
- geographic proximity of firms, educational and research institutions, financial and other business institutions enhances the effectiveness of the innovation process;
- the larger the cluster (e.g. large number of firms and workers) the higher the level of self-sufficiency; i.e. less need to get key functions (e.g. supplies, financing) supplied from outside; that is there is less "leakage" outside the cluster.

Selected knowledge-based clusters in regions, metropolises, technopoles and remote regions around the world are described in the following sections to illustrate the key characteristics of industrial clustering.

5.2   KNOWLEDGE-BASED REGIONS

Knowledge-based regions are sub-national areas which are usually within political jurisdictions such as states, provinces, landers, etc. However, they can extend beyond such jurisdictions. For example, in Canada, the Ottawa area cluster, which is in the Province of Ontario, stretches across the border of the Province of Québec into the neighboring Outaouais region. But, because geographical proximity is important to the innovation process, clusters are concentrated in areas which provide the primary linkages between business and technical capabilities.

5.2.1 The Four Motors for Europe

Four major regions, Lombardy, Baden-Wuerttemburg, Rhônes-Alpes and Catalonia have grouped together as a loosely tied association to market themselves internationally as "The Four Motors for Europe".

5.2.1.1 Lombardy (Italy)

Lombardy, whose capital city is Milan, is the economic engine of Italy. This state, with its population of 9 million, is responsible for about 21% of Italy's GDP. Its per capita income is 30% above the Italian average. Lombardy is a major trading centre and accounts for some 30% of Italy's exports.

The region accounts for one-third of the active corporations in Italy. While there are many large conglomerates, such as Montedison, most of the 200,000 industrial firms, which represent about 45% of industrial employment in Italy, are small and medium-sized enterprises (SMEs). It has major concentrations in the telecommunications informatics and chemical sectors as well in services such as engineering consulting.

Lombardy accounts for 32% of Italy's government R&D and 40% of its private sector R&D expenditures. Milan itself has four universities, one of which, the Polytechnico is said to be the MIT of Italy. Lombardy is more R&D intensive than any other region in Italy.

To ensure that the SMEs in the region gain access to the latest technology, the regional government has entered into a 50/50 partnership with a number of industry associations to establish an innovation centre, CESTEC. The mandate of CESTEC is to provide the business and technical support for the development of technology within SMEs. CESTEC focuses its activities on microelectronics, robotics, CAD/CAM software, office automation, lasers, new materials, biotechnology and environmental technologies. CESTEC also is empowered to allocate dedicated regional funds to new research activities and technological innovation initiatives.

Lombardy, is not only Italy's economic engine, it is also its technology engine. These favourable characteristics help to attract investment.

5.2.1.2 Baden-Wuerttemberg (Germany)

Baden-Wuerttemberg, whose capital is Stuttgart, has a population of 9.4 million and produces about 17% of Germany's exports. Many of these exports are "high-end" automotive products since Mercedes-Benz, Porsche and Bosch are located in this state. Zeiss also adds to the high value-added exports of the state. Overall, there are some 12,000 manufacturing firms in the state, 95% of which are SMEs. Baden-Wuerttemberg accounts for about one-quarter of Germany's automotive and electronics/electrical production.

Given its industrial mix, Baden-Wuerttemberg has highly developed expertise in electronics and electrical goods, machinery, automotive engineering and precision engineering. The state has the highest concentration of research institutes in Europe and accounts for 30% of Germany's R&D capability. There are nine universities, 23 polytechnics, 11 Max Planck Institutes and 14 Fraunhofer institutes and research centres.

A key element of Baden-Wuerttemberg's technical infrastructure is the Steinbeis Foundation for Economic Promotion, a not-for-profit corporation, whose mandate is to support the development of industry in the state, especially SMEs, through the provision of R&D, technical advice and financial support. The Foundation, whose budget is about DM 80 million, has 114 technology centres spread throughout the state. More than 70% of its revenue comes from consulting and R&D services. The Chairman of Board is also the Baden-Wuerttemberg Government Commissioner for Technology Transfer which gives an indication of the importance that the state attaches to technology development.

The State Commissioner for Technology Transfer co-ordinates the activities that encourage firms to strengthen their technological position. In this way, Baden-Wuerttemberg implements its strategy of ensuring that it remains the leading research state in Germany.

5.2.1.3   Rhône-Alpes (France)

The Rhône-Alpes region, with a population of 5.2 million and about 10% of the national economic activity, has the largest concentration of S&T resources and personnel in France, after Paris. Some 20,000 people (10% of French researchers) work in research centres in the region. These centres are situated mainly in Lyon, Grenoble, St. Etienne and Annecy, with the first two municipalities each accounting for 45% of S&T activities.

The region has nine universities. The Centre National de Recherches Scientifiques (CNRS), has 140 laboratories and 2,000 researchers in the region. There are also other major research centres, such as the Commissariat de l'Energie Atomique (CEA) with its 2,600 researchers. The major industrial research centres are situated in French companies such as l'Institut Mérieux, Merlin-Gérin, Rhône-Poulenc, ELF, Péchiney, Air Liquide and Thomson. As well, there are major multi-national firms in the region such as Hewlett-Packard, Sun Microsystems and Caterpillar Tractors, and several hundred small and medium-sized enterprises (SMEs).

The level of S&T activities in the region is significant on a European level, as well as on a regional level. The Université de Grenoble, for example, was recently named one of two "European" universities and will grant "supra" national degrees. As well, the European Synchroton (ESRC) is located in Grenoble.

The emergence of Grenoble was due to a few individuals who attracted major laboratories to the city due early on in the post-war period. One was Louis Néel, the Nobel Prize physicist. Another one was Mr. G. Merlin who founded the Merlin-Gérin electrical equipment group. These individuals were influential in attracting the CNRS and the nuclear research centre, CENG in the 1950s. In the 1960s, the Laue-Langevin Franco-German neutron laboratory was established. Another pioneer was Mr. H.

Dubedout, a nuclear scientist, who as the Mayor of Grenoble for 18 years, developed the city's scientific infrastructure.

One of Mr. Dubedout's projects was the business park called the "Zone pour l'innovation et les réalisations scientifiques et techniques" (ZIRST) set up in 1972 which today has more than 200 high-technology firms. The business plans of tenants are vetted by a committee. In return, the financing for a firm's buildings are guaranteed by the municipal government. An impetus for the development of the ZIRST came in 1975 when engineers and researchers from the SEMS (a Honeywell-Bell, Télémécanique and Thomson joint venture) set up their own firm in the park.

The development of advanced technology sectors such as electronics, informatics and biotechnology has been given priority by the regional government which spends about 50% of its budget (approximately FF 4 Billion) on education, training and research. As in other regions of Europe, the local Chambers of Commerce play a particularly aggressive role in encouraging high technology industrial development and international investment in the region.

Because the Rhône-Alpes region has a significant S&T budget, it is able to fund specific S&T programs and lever funds from national and European programs.

5.2.1.4 Catalonia (Spain)

Catalonia, whose capital is Barcelona, is Spain's industrial centre. It has a population of 6 million and generates nearly 20% of Spain's GDP and accounts for 27% of Spain's industrial output. Some 40% of its industrial activity is in manufacturing. Catalonia attracts industry because it can offer a skilled labour force at relatively lower wage rates than the rest industrialized Europe.

The 1992 Barcelona Summer Olympics spurred the development of new transportation, communications and municipal infrastructure (e.g., sewers).

In 1987, the Valles Technological Park (PTV) was established to encourage the development high-technology development. The PTV is co-located with the National Microelectronics Centre and the University of Barcelona. The five areas of interest of the PTV are microelectronics, telecommunications, advanced automation, biotechnology and new materials.

Catalonia sees itself in a catch-up situation vis-à-vis the other three Motors and has an aggressive policy to attract foreign investment. For example, it receives about 70% of all Japanese investment in Spain involving firms such as Sony, Nissan, Toshiba and Sanyo.

5.2.2 Silicon Valley, California

Silicon Valley, the 300 square mile region stretching from Palo Alto to San Jose, has more than 1 million people working in 6,000 high-technology firms, most of which centre on microelectronics development and computers.

The origins of Silicon Valley can be traced back to the work of Lee de Forest in Palo Alto, in 1912 (see Exhibit 5-1). Two key individuals responsible for the development of the region were Frederick Terman, Professor of Electrical Engineering at Stanford University, and William Shockley the co-inventor of the transistor at Bell Labs – who moved to Palo Alto in 1955 to establish Shockley Semiconductor Laboratory.

In 1938, Terman gave a loan of $538 to his two bright students, William Hewlett and David Packard to develop Hewlett's variable-frequency oscillator. Hewlett and Packard started production in a garage behind their rooming house in Palo Alto. An early customer was Disney Productions which bought eight oscillators to use for special effects in the film Fantasia. Terman was also instrumental in the establishment of the Stanford Research Park in 1951 to stimulate university – industry linkages.

Shockley, on the other hand, was able to attract brilliant engineers and physicists to his company. Eight of these employees left to launch Fairchild Semiconductor. This company was the spawning ground for other spin-offs as indicated in Exhibit 5-1.

The development and growth of firms in Silicon Valley was stimulated by the dramatic expansion of military and aerospace demand for electronic devices in the late 1950s and 1960s. Opportunities were seized by entrepreneurial scientists and engineers who had a shared culture of innovation. Social networks developed and played a decisive role in the development of Silicon Valley.[4]

Walker's Wagon Wheel Bar and Grill in Mountain View became a focal point for discussions of technical problems and the planning of spin-offs among the engineers from various firms. These discussions and the mobility of personnel between firms made it virtually impossible to protect intellectual property. The only way out was to keep ahead by accelerating the innovation process.

The explosive growth of Silicon Valley leveled off in the late 1980s due to a number of reasons, including the following:

- competition from Japan and other Asian countries;
- high costs led to firms looking for cheaper prices, labour costs, cheaper power, etc. in other jurisdictions;
- semiconductor firms began seeking locations closer to major customers;
- the push for a global presence led to the establishment of activities around the world; and
- a sense that the quality of life in the region had deteriorated also played a factor in companies' diversification plans.

Business leaders who became concerned with the economy of Silicon Valley set up the "Joint Venture: Silicon Valley" initiative to overcome the growing gap between industry, government and the local communities. Following a year of consultations and consensus development the "JV:SV Network" was formed as a non-profit organization to act on 13 actions that were proposed. These actions are:

1. Defense/Space Consortium, Inc.
2. Smart Valley, Inc.
3. Environmental Partnership, Inc.
4. Software Industry Coalition
5. New Business Enterprise Clusters
6. Enterprise Clusters

Exhibit 5.1 - A Chronology of the Important Inventions, Events, and People in the Microelectronics High Technology Industry in Silicon Valley

Year	Event
1912	Lee de Forest discovers the amplification qualities of the vacuum tube in Palo Alto, California, thus making possible radio, television, film, and other communication technologies.
1938	Hewlett-Packard is founded in a garage in Palo Alto by William Hewlett and David Packard, two of the first entrepreneurs in Silicon Valley.
1946	ENIAC, the first mainframe computer, with 18,000 vacuum tubes, is invented at the University of Pennsylvania.
1947	William Shockley, John Bardeen, and Walter Brattain invent the transistor at Bell Labs in Murray Hill, New Jersey. The transistor eventually replaces vacuum tubes.
1955	Shockley leaves Bell Labs to establish Shockley Semiconductor Laboratory in Palo Alto.
1956	Shockley, Bardeen and Brattain win the Nobel prize in physics.
1957	The entrepreneurial spirit of Silicon Valley gets underway when Robert Noyce and seven other brilliant young engineers quit Shockley Semiconductor Laboratory to launch Fairchild Semiconductor. These cofounders later split off to launch over eighty semiconductor firms in Silicon Valley over the next thirty-five years.
1968	Noyce leaves Fairchild to start Intel.
1971	Invention of the microprocessor, a computer control unit on a semiconductor chip, by Ted Hoff of Intel. Silicon Valley is named by the late Don Hoefler, then editor of a local electronics newsletter. Nolan Bushnell designs Pong, launches Atari, and the video game industry is begun.
1976	Steve Jobs and Steve Wozniak build the Apple microcomputer.
1980	Apple goes public: Art Rick, the venture capitalist who had invested $57,000, earns $14 million; Jobs is worth $165 million.
1982	About 3,100 microelectronics firms exist in Silicon Valley; two-thirds have less than 10 employees, and only fifty or so have more than 1,000 workers.
1984	Silicon Valley has 15,000 millionaires and 2 billionaires.

Source: Larsen J.K. and Rogers E.M.; Silicon Valley, in Creating the Technopolis; (Smilor R. et al. ed.) Ballinger, 1990.

7.   Silicon Valley Global Trading Center
8.   Health Care Task Force
9.   Regulatory Forum
10. Council on Tax and Fiscal Policy
11. 21st Century Workforce Initiative
12. Silicon valley Technologies Corporation
13. Economic Development Team

Each action initiative has its own leadership structure, operations mechanisms and volunteer staff activity. Initiative champions who have risen to lead these efforts can rely on the network's staff, director, and board of directors to help the projects find the visibility, research, and fund raising required for success. The Network has become a strong source of regional advocacy and for attracting outside investment. Several Joint Venture/Silicon Valley projects have been successful at harnessing federal dollars from the ongoing Technology Reconstruction Program and the U.S. Department of Commerce.

Silicon Valley appears to experiencing a resurgence of activity along with the rest of California[5]. However, a major challenge is to renew the infrastructure which has deteriorated since the implementation of Proposition 13,[6] severely limiting the taxation power of local government.

5.2.3 Boston's Route 128

Route 128 emerged as the result of industrial restructuring in the 1970s and 1980s. Between 1968 and 1975, Greater Boston lost 252,000 manufacturing jobs. However, between 1975 and 1980, 225,000 new manufacturing jobs were created, mostly in high-technology industries. Most of the new firms located along Highway 128, the suburban beltway of Boston, completed in 1951, which links 20 towns.

The core of Greater Boston's new industrial development is the computer industry, which had its start in 1950s with the establishment of firms such as Digital Equipment Corporation[7]. Most of the firms emerged from MIT, which is credited as being a decisive factor in the ability of the region to reindustrialize. Funds and orders from the Department of Defense catalyzed the early development of the region.

The Boston area pioneered the concept of venture capital. MIT President, Karl Taylor Compton, played a key role in the 1946 formation of American Research and Development, the U.S.'s first modern venture capital fund. ARD was led by General Georges Doriot who taught at the Harvard Business School. He obtained money from Boston insurance companies. ARD's most famous investment is the backing of Ken Olsen of MIT to create Digital Equipment Corp. in 1956.

MIT had a well established and distinguished electrical engineering department by the 1930s and was open to conduct contract research. The expertise built up through wartime government contracts led to an explosion of high-technology start-up in the post-war period. During the 1960s, for example, 175 new local firms were created by former employees of MIT's research laboratories. Many of these entrepreneurs who were creating a new civilian computer industry came out of larger, well established, defense firms such as Raytheon.[8]

Still, the over-dependence on military markets became evident in the late 1980s with cutbacks in military spending. High-technology industries laid off 60,000 workers between 1988-91. Also this region is much less diversified technologically than Silicon Valley, for example. One type of computer, the minicomputer, dominates the computer industry. The shift away from minicomputers to microcomputers had adverse effects on firms such as Wang and DEC.

Moreover, it appears that Route 128 firms lacked the flexibility to adapt to new technologies and the new market environment that was present in Silicon Valley firms due to its social networks. In the words of Saxenian:

> "The Route 128 region, in contrast, is dominated by a small number of relatively integrated corporations. Its industrial system is based on independent firms that internalize a wide range of productive activities. Practices of secrecy and corporate loyalty govern relations between firms and their customers, suppliers, and competitors, reinforcing a regional culture that encourages stability and self-reliance. Corporate hierarchies ensure that authority remains centralized and information tends to flow vertically.[9]

According to Saxenian this difference in socialization patterns between Route 128 and Silicon Valley, was a major cause of the latter outpacing the former by the late 1970s (see Exhibit 5-2)

Exhibit 5.2 - Total High Technology Employment, Silicon Valley and Route 128, 1959-1990. Data from *County Business Patterns*

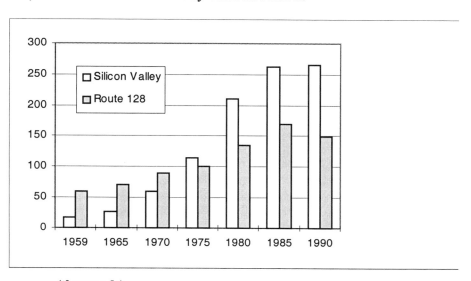

(thousands)

Source: Saxenian, *op. cit.*

Saxenian goes on to diagnose Route 128 current problems as follows:
"The current difficulties of Route 128 are to a great extent the product of its history. The region's technology firms inherited a business model and a social and institutional setting from an earlier industrial era. When technology remained relatively stable over time, vertical integration and corporate centralization offered needed scale economies and market control. In an age of volatile technologies and markets, however, the horizontal coordination provided by interfirm networks enables firms to retain the focus and flexibility needed for continuous innovation.

To be sure, regional institutions and culture are difficult to change. An industrial system is the product of historical processes that are not easily imitated or altered. However, the first step toward the regeneration of the Route 128 economy is self-understanding. The challenge facing the Route 128 region today is to learn from Silicon Valley's success. Managers and policy makers need to overcome their outdated conception of the firm as a separate and self-sufficient entity; they need to recognize that innovation is a collective process as well as an individual one. Adopting a business model that breaks down the institutional and social boundaries that divide firms represents a major challenge for Route 128, but it is decidedly less daunting than the challenges faced by regions with less sophisticated industrial infrastructures.[10]

5.2.4 Western Canada

Western Canada is made up of four provinces; British Columbia, Alberta, Saskatchewan and Manitoba. Collectively, these provinces have a population of about 8 million people.

Western Canada has three clusters as major economic drivers; energy and mining, forest products and agri-food. Over the past ten years, these three clusters have contributed more than 25% to the economic growth of Western Canada. These clusters are driving the development of knowledge-based capabilities in biotechnology, information technology and advanced materials/manufacturing.

A recent study[11] has assessed the development and potential of each of these technical areas in Western Canada. That report summarized the situation as follows:
"Both the driving clusters and the technology industry groups in Western Canada face significant challenges in terms of fierce international competition and the region's economic infrastructure has key gaps in terms of access to technology, available human resources and supportive tax and regulatory policies. In addition, there is a lack of public/private collaboration and generally weak linkages between technology providers and the driving clusters.

Of the three enabling technologies examined, only one subcluster, agricultural biotechnology, has strong enough cluster characteristics to qualify as marginally competitive on world markets. One, advanced materials and manufacturing, is quite weak and even with a fairly vigorous program of government and industry action, it will be some time before the industry can be considered a competitive cluster.

It makes sense to develop strategies for the biotechnology and information technology that aim to strengthen critical weaknesses that currently weaken the cluster, for example, the demand factors affecting healthy care biotechnology and the structural issues in information technology.

Advanced materials, while it may not soon be a self-standing industrial cluster, remains an important supporting component of the driving clusters. Governments could, therefore, consider putting more emphasis on supporting the role of this technology with respect to the driving clusters. Ultimately, of course, this will increase the likelihood that AMM eventually becomes a competitive cluster itself."

This situation has arisen mainly because of the focus on the exploitation of natural resources rather than using these resources to move along the value-added chain to create more value-added products and services.

While the focus of the study was on the four provinces collectively, it concluded that, "most of the important links are found within provinces, or even within single metropolitan areas ..." This supports the notion that the appropriate unit for the study of industrial clustering is at the level of sub-national jurisdictions, not at higher levels where a "bird's eye" view can miss the important fundamentals of the dynamics of industrial clustering.

Western Canada, with its natural resource orientation, its relative remoteness from major markets and its dispersed population is in sharp contrast with the regional clusters described in the previous sections which are well anchored in manufacturing and high-technology development. A major challenge to this region, is to develop a culture that is in resonance with high-technology development. Other remote regions, such as Scotland, have managed to do this.

5.2.5 Summing Up

Regions, especially those that fall within specific political jurisdictions, can be a focus for high-technology cluster development. Since proximity is important to innovation, the local/regional focus is the first level of decision-making for both businesses and governments that support cluster development.

5.3 CLUSTERING AND THE LARGE METROPOLIS

The metropolis is the focus of the markets as well as the major business and financial services that support knowledge-based industries. The metropolis also nurtures the development of knowledge-based industries. The larger metropolis, which has the full array of support services as well as markets, has come to be called a "city-state" because of its self-sufficiency.

5.3.1 Paris

Paris has the largest concentration of high-technology firms and research laboratories, not only in France, but also in Europe. Most of them are found in the southwest suburbs, around the town of Massy. There are more than 10,000 high-tech firms along with major universities and research laboratories.

Government nuclear and defence research centres were established in the 1950s, followed by the decentralization of Paris-based universities. The 1965 "Schéma Directeur de la Région Parisienne" also gave impetus to the development of the region. The outward movement of industry to the suburbs, from the 1960s on, was yet another factor which favoured the development of the Massy area.

However, while the area grew, the links between large and small firms and between research groups and industry remained weak.[12] The elitism of the French "Grandes Ecoles" continues to contribute to the isolation within industry, universities and research institutions.

Nevertheless, the concentration of firms has led to numerous spin-offs from large firms adding to the economic dynamism of the region, giving the region the highest ration of small business firms in the entire Paris metropolitan area.

The Paris Southwest has the industrial and research capabilities of a high-technology cluster. However, the weak linkages among the players could limit the innovative performance of the cluster. Government policy statements, such as the 1990 Livre Blanc are not sufficient by themselves to create the "glue" needed among the players.

5.3.2 London

The high-technology sector is concentrated in London's west side (i.e. the M4 corridor). In true English fashion, the region grew through "laissez-faire", out of the craft workshops in the area. These workshops created the 18th and 19th century machine tool industry. By the end of the 19th century, the British electrical industry was born in the region. It grew rapidly and became the source of electronics/ information technology development after World War II. The defence sector was a very important catalyst in the development of high-technology industries in the region since the major Defence Research Establishments were located there.

As the region grew, it became a focus for the establishment of MNFs in the 1970s. However, since high-technology development was locked into the requirements of the defence sector which are very different from those of the commercial marketplace, the development of a commercially oriented high-technology sector has remained difficult.

5.3.3 Tokyo

Tokyo, situated in the Keihin region, is the largest industrial area in Japan. Its economy has become increasingly based on knowledge-based industries in the post-war period.

The Japanese industrial system is based on the relationship between large corporations and their small subcontractors. The large corporations, which have recently tended to move their factories to the outskirts, depend on the supply of sophisticated parts from SMEs, most of which are still located in Tokyo itself.

R&D functions are in the large firms and remain in the Tokyo area. However, given the links between large and small firms, the increasing sophisticated requirements of the large firms has permitted the smaller firms to upgrade themselves technologically through contracts.

The Tokyo region has historical strengths in the electrical engineering industry which became the basis of the emergence of Japan's electronics industry in the area.

However, the growth pressures on Tokyo have led the government to set in place the Technopolis Program to encourage decentralization of high-technology industries to 26 other municipalities across the country.

5.3.4 Montréal

Montréal has a population of more than 3 million people. There has been a shift to knowledge-based industries in the last 20 years. There are currently some 600 research companies supported by a network of universities and publicly funded research institutions.

There are five high-technology sectors in which Montréal is very competitive; telecommunications, computers and software, aerospace, pharmaceuticals and microelectronics. Much of the activity is in the west side of the municipality.

The province of Québec has set in place generous incentives and programs to stimulate the growth of high-technology industries. For example, a $300 million venture capital fund, Innovatech, has been established to help high-technology start-ups and alliances between universities and the private sector. The provincial government has on-going reviews of the performance of the high-technology sector. The Conseil de la science et technologie, for example, undertook a major review of the state of science and technology in the Montréal area in 1992. As well as making recommendations on improving the competitive position of the five sectors mentioned above, the Conseil also focused on emerging markets (i.e. space, medical technology, software engineering, advanced materials, biotechnology) and established markets (i.e. energy, electrical equipment, printing, food, clothing, financial services and consulting engineering). The Conseil's recommendations are being acted on by groups within the various sectors. This activity gives Montréal an image of being a high-technology metropolis.

5.3.5 Summing Up

The large conurbations described above have foci for knowledge-based industries; they seem to cluster in specific geographic areas within the boundaries of the metropolis.

This geographic proximity should facilitate linkages among firms. However, there can be barriers to establishing networks such as the cultural elitism generated in the French university system. Moreover, knowledge-based industries are but on relatively small element that make-up a vibrant metropolis, and can be "lost' in the bigger scheme of things inhibiting the development of strong linkages.

5.4   TECHNOPOLES

Technopoles are usually municipalities, smaller than metropolises, where a distinct technology focus can be identified.

The word "Technopole" comes from the Greek meaning "a city of technology on which activities pivot".

5.4.1 Cambridge, England

As has been noted in Chapter 4, high-technology development in Cambridge took off in the 1970s. However, Cambridge has had a long tradition of high-technology manufacturing through spin-offs from the University. For example, Cambridge Instruments was set up in 1881 to manufacture scientific equipment for the University. In 1967, the University set up a committee, chaired by Sir Nevill Mott, Head of the Cavendish Laboratory to advise on the relationship between the University and science-based industry. The Committee's report, published in 1969, recommended the establishment of a science park and limited growth of science-based industry. This legitimized the acceptance of science-based industry in Cambridge.

The effect was explosive. By the mid 1980s, high-tech employment was about 15,000 people with an output of £890 million. The key sectors were computers, software, scientific instruments, electronics and increasingly biotechnology. The University was responsible, either directly or indirectly, for firm creation. One analysis proposed three factors to explain the Cambridge phenomenon:

Demand played a role: in some fields (CAD, scanning electron micro-scopy), Cambridge was in a favoured position because it was developing products for which there was a rapidly rising demand.

- *General preconditioning.* The general growth of East Anglia and the favorable business climate for small firms came concurrently with major technical advances in electronics and computer design. The university encouraged research excellence; it enjoyed generous research funding, including major Research Council units. It has combined critical mass with quality: Trinity College alone had won more Nobel Prizes for science than France. It has exploited research in areas where it has done particularly well, and where start-up costs are low. the college-based structure of the

University weakened departmental hierarchies, encouraging individual flair; and the University had an extremely permissive attitude to intellectual property rights, which belong to the researchers. And after 1969 it positively encouraged industrial spin-off. Culturally, the city was small, dominated by the University, and free of old industrial structures; new firms did not feel "lost" here. The Trinity College science park was very visible and was perhaps the most successful university-based science park in Europe; though it played only a minor role in the entire phenomenon, it had become increasingly important because of its policy of offering good premises on short terms at reasonable rents.

• *Special factors.* There were a number of key triggering events: the establishment of Cambridge Consultants in 1960, and of Applied Research of Cambridge in 1969; the Mott Report of 1969; the formation of the Cambridge computer group in 1979; and the lending policies of Barclay's Bank from the late 1970s. There were many young people who could not find university posts, but wanted to stay here.[13]

However, by the late 1980s, activity was leveling off. While there were good university-industry linkages, most new firms remained small and did not network much with each other. They got no help from the larger firms or from government procurement and takeovers were increasing. However, today there are some 1,200 mainly small firms (average size is 12 employees) employing 35,000 people in the area.

5.4.2 Montpellier

Montpellier, a historic city with a population of about 220,000, developed and implemented a strategy to diversify out of tourism, because of the cyclical nature of this sector, through the support of high-technology industries beginning in the early 1980s. The architect of the strategy was the "député-maire" of Montpellier, Mr. Georges Frêche. The Montpellier strategy is based on the development of four "technopoles".

• *Pôle Antenna* – to support the development of telecommunications and broadcasting. As a first step, Mr. Frêche decided to provide the city with an optical fibre cable network as the infrastructure to support the development of advanced videocommunications. This activity is supported by the Grammont Centre International d'Images which has complete facilities for performing arts, television and film production, and the Gutenberg Médiatheque which has audio-visual viewing and teaching facilities.

• *Pôle Informatique* – to support the development of software, artificial intelligence and electronics. The centre-piece of this activity is the IBM plant established in 1965, which manufactures IBM's supercomputers and its network of suppliers. This activity is supported by the University Computer Centre and the National Centre for Supercomputing (CNUSC) the second most important in France. As well, there is the high-tech park (Parc du Millénaire) which has some 150 firms, most of them SMEs.

- *Pôle Euromédicine* – to support the development of biotechnology and pharmaceutical firms. The activity is built around the university's medical faculty, the oldest in Europe (1289) and its hospital complex, the third largest in France. This "pôle" also has its own science park which has about 100 firms, mostly SMEs. Related research institutes such as those of CNRs and INSERM have some 2000 researchers.
- *Pôle Agropolis* – to support the development of agri-business. The related teaching/research complex has 20 institutes and 2000 researchers. This "pôle" also has its own science park which currently has some 20 firms.

The development and promotion of these four "pôles", as well as fifth one on tourism, is the responsibility of Montpellier L.R. Technopole, an economic development agency. Agents are assigned the responsibility for each "pôle" as well as for international and national promotion and for other related activities such as the Centre européen d'entreprise et d'innovation (Cap Alpha) described below.

Cap Alpha is one of the 100 or so business innovation centres set up across Europe under a European community program for regional development. It is a centre which provides business advice and financing to start-ups as well as other commonly accepted incubator support services (e.g., space, secretarial and technical support). Cap Alpha is a mixed enterprise financed by regional agencies, banks and the private sector. It also has its own venture capital firm, Mistral Investissements. To be accepted as a firm in this incubator, the proposed business plan must be approved by the centre. In turn, the centre then places its own staff on the management team for each approved firm, for the duration of occupancy in the centre. There are some 30 or so firms in the centre at present.

What the Montpellier experience illustrates is what a dedicated "champion", in this case, Mr. Frêche, can accomplish through a clear vision, motivation and tenacity.

5.4.3 Austin, Texas

In Austin, the university played a pivotal role in the development of the city's high-technology cluster. Over half the SMEs in the region in 1986 indicated a direct or indirect tie regarding their origin to the University of Texas (UT) at Austin.[14]

The development of the high-technology sector started in the postwar period. An early player was Frank McBee, a UT graduate who founded Tracor Inc., an electronics firm, in 1962. Tracor eventually became a "Fortune 500" company. However, most importantly for the region it became a source of spin-offs.

As the high-technology sector grew, Austin began to attract outside firms. IBM located there in the 1960s. The Microelectronics and Computer Technology Corporation (MCC) chose Austin as its headquarters in 1983, and Sematech settled there in 1988. Sematech officials cited as the main reason for choosing Austin the synergy among business, academic, government and community entities. George Kozmetsky, an academic and an entrepreneur was a champion creating "glue" among the players in Austin's early development.

This synergy is created by the presence of a number of institutions that create linkages. They currently include the following:

- **Austin Technology Incubator** – provides space and management assistance to startups.
- **Texas Capital Network Inc.** – a matching service for investors and entrepreneurs;
- **MOOT Corp.** – a forum for entrepreneurs to present business plans;
- **IC² Institute** – a research centre for the study of innovation;
- **The Know-How Network** – a network of support groups (e.g., legal and accounting firms)that are prepared to offer service at reduced rates to firms during their startup phase; and
- **The Centre for Technology Venturing** – an umbrella group for some of the above.

In their analysis of the development of the high-technology sector in Austin, Smilor et al identified the following points as key to the emergence, growth and maintenance of the region as a technopole;

- The research university has played a pivotal role in the development of the Austin technopolis by 1) achieving scientific preeminence; 2) creating, developing, and maintaining new technologies for emerging industries; 3) educating and training the required workforce and professions for economic development through technology; 4) attracting large technology companies; 5) promoting the development of home-grown technologies; and 6) contributing to improved quality of life and culture.
- Local government has had a significant impact, both positively and negatively, on company formation and relocation, largely from what it has chosen to do or not to do in terms of quality of life, competitive rate structures, and infrastructure.
- State government has had a significant impact, both positively and negatively, on the development of the Austin technopolis through what it has chosen to do or not to do for education, especially in the areas of making and keeping long-term commitments to fund R&D, faculty salaries, student support, and related educational development activities.
- The federal government has played an indirect but supportive role largely through its allocation of research and development moneys, on-site R&D programs, and defense-related activities.
- Continuity in local, state, and federal government policies has an important impact on maintaining the momentum in the growth of a technopolis.
- Large technology companies have played a catalytic role in the expansion of the Austin technopolis by 1) maintaining relationships with major research universities, 2) becoming a source of talent for the development of new companies, and 3) contributing to job creation and an economic base that can support an affordable quality of life.
- Small technology companies in Austin have helped in 1) commercializing technologies, 2) diversifying and broadening the economic base of the area, 3) contributing to job creation, 4) spinning companies out of the university and other research institutes, and 5) providing opportunities for venture capital investment.

- State and local influencers have provided vision, communication, and trust for developing a consensus for economic development and technology diversification, especially through their ability to network with other individuals and institutions in other sectors.
- Consensus among and between the key players is essential for the sustained growth of the technopolis.

5.4.4 Ottawa, Canada

The Ottawa region, with its population of about 1 million has some 700 high-technology firms employing more than 40,000 people and generating $8 billion in revenues. There is a focus on telecommunications and related software with 75% of all Canadian telecommunications R&D being performed in the region.

The origins of the high-technology sector can be traced back to the concentration of government laboratories, especially those of the National Research Council (NRC) and the intense R&D activity carried out during the Second World War. This led to the establishment of the first company, Computing Devices, in the 1950s, to exploit technology developed at NRC for automating aircraft navigation systems. Other companies were subsequently established to exploit opportunities developed in government laboratories. In the 1960s, the Bell-Northern Research Laboratory (now Nortel Technology with more than 5,000 employees) was established, which led to more spin-offs in the region.

In 1984, the Ottawa-Carleton Research Institute (OCRI) was established to create linkages among the firms and with the two universities and the publicly funded research laboratories. OCRI did this by organizing specific research consortia and technology-related events. OCRI became instrument in influencing the development of the high-technology sector. For example, its monthly breakfast seminars attract more than 300 people.

The financing of technology ventures, which started with two individuals who were involved with the local pulp and paper industry, has now developed into a very sophisticated venture capital sector which is involved not only in Ottawa-based investments but also in supporting firms across the country. For example, Capital Alliance Ventures Inc. was established recently to fund high-technology ventures in the Ottawa area.

With the downsizing of government activities, the high-technology sector is now clearly seen as an important cornerstone for the economic future of the region.

5.4.5 Summing Up

The dynamics of cluster development appear sharper at the level of the technopole, possibly because they are relatively small municipalities, which makes interactions and socialization easier, and because they have a more highly focused technological orientation.

5.5   CLUSTERING IN REMOTE AREAS

There have been attempts to create knowledge-based clusters in remote areas, largely as part of regional development policies.

5.5.1 Japan's Technopolis Program

Japan passed its Technopolis Law in 1983 aimed at creating regional high-technology centres to relieve pressure on Tokyo. The project started under a Technopolis Committee, with a tight timetable: to choose the sites by 1984, to complete construction of the physical infrastructure by 1990, and to complete development of each technopolis by the year 2000, with the aim of generating a "Techno-Archipelago" in the twenty-first century. The sites to be chosen had to meet certain rigorous criteria:

- a total area (on one or more sites) of 1,300 square kilometres (500 square miles) or less;
- existing enterprises with potential for high-tech development;
- easily available industrial sites, available water and residential areas;
- an existing city (Mother City) with 150,000 or more people;
- an existing university with high-tech education or research; and
- access to high-speed transportation facilities giving a one-day return trip from Tokyo, Nagoya or Osaka.

Twenty-six sites were chosen. They are listed along with their characteristics in the Appendix. The program did lead to some decentralization of industry out of Tokyo. However, it appears that, to date, these "technopoles" have remained, by and large, "branch plants" of Tokyo.

Key findings of recent evaluations include the following:[15]

- the original vision of creating integrated high-technology complexes linking R&D institutes, educational facilities and firms was not achieved;
- the "branch plant" syndrome dominated with firms mainly making parts for shipment to Tokyo, Osaka and overseas;
- university-industry linkages remain weak;
- there was a lack of "soft" infrastructure (e.g., university research, venture capital funds);
- major businesses have been reluctant to relocate their R&D facilities;
- there was a lack of inter-industry linkages;
- there was a lack of spin-offs;
- there was a failure to attract key knowledge workers;
- the high value of the yen has undermined the technopolis strategy, by forcing firms to move to cheaper offshore jurisdictions; and
- prospects for the sites closer to Tokyo (i.e., within 100 kilometers) appear brighter than those further away.

Despite these negative findings at this time, things could change in the future if linkages among the key players can be developed.

5.5.2 Penang, Malaysia

In 1969, unemployment in Penang was 16%. The government hired a consultant to formulate a plan that would create jobs. The consultant recommended creating an industrial base to spur economic growth. Electronics was targeted because the sector is labour-intensive and export oriented.

In 1970, the government set up an electronics factory and two years later it set up a free industrial zone to lure multinationals that were seeking offshore plants. Some twenty companies, including Hewlett-Packard, Seagate and Quantum set up factories to make disk drivers and components. The infrastructure, government incentives, support industries, quality of skilled people and low labour costs were deciding factors for setting up factories in Penang. Non-electronics firms were also attracted. Toray Industries of Japan began making textiles in 1972 and Mattel Inc. started making toys in 1980.

Growth has been very high and unemployment fell to 2.9% in 1994. A tight labour market has sent wages spiraling upwards leading companies to more automation to stay competitive or to cease production (e.g. Mattel stopped making labour intensive toys such as Barbies).

The government is now planning a second stage of industrialization where multinationals would be encouraged to obtain world product mandates and do R&D in Penang. Intel Corp. is a leader here by setting up an integrated operation which includes microcircuit design. Hewlett-Packard recently established an R&D department to pursue both product and process development.

5.5.3 Scotland

Scotland with its 5 million people, is geographically at the edge of the European Union. It's electronics sector has grown in the post-war period to the point that it now supplies one-third of the European market for PCs and produces 10% of the world's output of PCs and is a major European centre for silicon chip manufacture. In 1994, exports of computer equipment and other electronic goods accounted for half of all Scottish manufactured goods. This compares to less than 9% twenty years earlier.

Scotland's orientation is towards manufacturing and assembly operations which undertake little design and related R&D activities. This limits the potential for innovation. MNFs especially US ones, dominate the IT sector. US firms account for 44% of Scottish electronics jobs.

Scotland has sought foreign investment aggressively and did not reflect much on the truncated industrial structure that resulted until recently. Now there is some disdain for low value-added manufacturing and assembly plants and Scotland has on occasion passed over such opportunities and held out for higher value-added activities with better jobs.

The focus on low-valued manufacturing and assembly jobs did not solve Scotland's brain drain. University graduates continue to migrate to England and elsewhere because of the lack of opportunities requiring higher skills.

5.5.4 Atlantic Canada

Atlantic Canada has a small population of 2.5 million and, analogously of Scotland, is at the periphery of North America.

The strengths, weaknesses, opportunities and threats are shown in Exhibit 5-3.

Atlantic Canada has a commodities based economy (e.g. forestry, fishing, mining, agriculture) and is only now moving towards knowledge-based industries.

Exhibit 5.3 - Strengths, Weaknesses, Opportunities and Threats in the Atlantic Economy

STRENGTHS	OPPORTUNITIES
.Strategic location .Good ports .Strong urban knowledge base .Highly developed centres of natural and applied science .Nationally recognized universities .Growing strength in tradable services .Good communication links .Strong culture and linguistic assets .Good quality of life .Relatively clean environment .Growing regional cooperation among governments	.Transportation centre .Marine industries .Aquaculture .Opportunities for value-added activity .Move to a knowledge-based economy .University-business spin-off .Information industries .Service exports .Adventure tourism .Cultural industries .Home-based and micro business .Increased efficiency of business based on increased competition within the region
WEAKNESSES	**THREATS**
.Resource dependence .Dependence on government .Underdeveloped private sector .Poor access to investment capital .Low level of linkages/weak clusters .Lack of entrepreneurs or aversion to risk .High taxation .Low levels of technology transfer .Road and rail infrastructure .Low levels of productivity .Minimal private sector R&D .Low human capital relative to other regions	.Growth in global competition .Declining federal transfers .Declining role of government .Environmental constraints .Resource supply .Competing transportation networks .Marginalized human resources

Source: DRI Canada; Atlantic Canada; Facing the Challenge of Change (Executive Summary); Prepared for ACOA, Sept. 1994

Like Scotland, Atlantic Canada has a well developed university sector whose graduates tend to migrate out of the region because of an underdeveloped manufacturing and high-technology industrial base. There are only very weak linkages between the university and publicly-funded research base and the existing industries which do not undertake much R&D given the commodities nature of their activities.

Atlantic Canada has the fundamentals to launch a high-technology clustering activity (e.g. highly skilled people, quality of life, good communications links). However, a major effort is needed to shift the culture away from traditional economic activities.[16]

5.5.5 Summing Up

Remote regions aspiring to become high-technology clusters are faced with the problem of attracting industry. Even if there is a well developed publicly funded educational and research infrastructure significant financial incentives appear to be needed to attract industry. And, even then, industry usually only establishes low-value added functions which do not need a sophisticated technical infrastructure.

5.6   CONDITIONS FOR SUCCESS

The 16 clusters described in Sections 2.0 to 5.0 are only a few selected ones among the more than 60 that I have studied over the last 10 years, through various clustering projects that I have undertaken.[17] Through this work I have identified eight ingredients for success. They are:
1. the recognition of the potential of knowledge-based industries by regional/local leaders;
2. the identification and support of regional strengths and assets;
3. the catalytic influence of local champions;
4. the need for entrepreneurial drive and sound business practices;
5. the availability of various sources of investment capital;
6. the cohesion provided by both informal and formal information networks;
7. the need for educational and research institutions; and most importantly,
8. the need to have "staying power" over the long term.

These characteristics are now reviewed in turn.

5.6.1 The Recognition of the Potential of Knowledge-based Industries by Regional/Local Leaders

The conventional wisdom states that clusters develops through "laissez-faire". However, a closer look reveals that either individuals or regional/local authorities have had a hand in shaping the development of clusters.

In the USA where "laissez-faire" is a religion, even Silicon Valley was given an impetus, initially by key individuals such as Professor Frederick Terman of Stanford

University and later by Joint Venture: Silicon Valley, a business initiative set up in the late 1980's when it was perceived that growth was waning. Similarly in Austin, Texas, the business, academic and government communities came together to promote high-technology development, which resulted, for example, in attracting Sematech.

At times, such as in Atlantic Canada or Penang, it requires a crisis to focus the mind on alternatives to traditional industries. In other instances, a vision and strategy lead the way. This was the case of Japan's Technopolis Program and of Grenoble in the Rhône-Alpes region. Even in large metropoles, such as Paris and Montréal, government has shown leadership in identifying directions for high-technology development. This is particularly remarkable, given the competition for attention in large conurbations.

In some instances, the cluster has to reach some level of visibility before it is recognized. This was the case of Lombardy (Milan) and Ottawa where new mechanisms were eventually set in place to co-ordinate various activities and spur further development.

The sense that there can be an opportunity for new economic development is also a reason to focus on knowledge-based cluster development. This was the case in Cambridge in 1967, when the University set up a committee to advise on the relationship between the university and knowledge-based industry. In Montpellier in the 1980's, the Deputy Mayor, Georges Frêche, saw the opportunity of diversifying out of tourism through high-technology development. On the other hand, recognition came only recently in Western Canada, where the provincial governments in co-operation with the federal government commissioned a study to investigate the opportunities related to knowledge-based cluster development.

Therefore, regional/local leaders are seen to become involved at all stages of the development of a cluster; from its initiation to the point where renewal is needed because prospects are flagging.

5.6.2 Identification and Support of Regional Strengths and Assets

The technical strengths in a cluster are usually relatively easy to identify. They can be a group of firms with capabilities in a technological sector, such as telecommunications in Ottawa, the five areas in Montréal and the four "pôles" in Montpellier. They can also be university capabilities as is the case with MIT the University of Texas for example.

There can also be local market opportunities especially in the larger metropolitan areas (e.g. banking, entertainment, government services).

What is often forgotten as strengths or assets is the social infrastructure and quality of life in a locality. Knowledge workers are very much in demand and are "foot-loose". They will migrate to localities with the social and physical amenities that they want. Hence, the migration of scientists and engineers for the socially "stuffy" and congested Paris to Grenoble, with its mountains and well developed social infrastructure. Similarly, in the late 1980's, Silicon Valley was losing skilled people to Austin, Texas and Seattle Washington, not to mention Ottawa, Canada because of a perceived loss of quality of life (e.g. pollution, congestion, crime, housing costs).

The technical, economic, social and physical assets of a locality need to be developed and nurtured if a high-technology trust is to succeed.

5.6.3 The Catalytic Influence of Champions

Influential individuals can drive the development of high-technology clusters. Professor Terman is much considered the "godfather" of Silicon Valley. Georges Kozmetsky was instrumental in the development of Austin, Texas. Georges Frêche's vision drove the high-technology development of Montpellier, while Néel, Merlin, and Dubedout played the same role in the development of Grenoble.

Champions can also be institutions or people with institutional responsibilities. For example, Gerry Turcotte, the President of the Ottawa-Carleton Research Institute (OCRI) for more than 10 years was key to the growth of the Ottawa cluster in the 1985-1996 period. A similar role was played by CESTEC in Lombardy.

Chambers of Commerce have a particularly influential role in Europe. Because they have the ability to levy a tax on firms in their jurisdictions they have sizable budgets and can offer a variety of services (e.g. training, market intelligence, certification) to support local industrial development.

The State Commissioner for Technology Transfer in Balen-Wuerttemberg champions the technological upgrade of firms in that jurisdiction. The same is true of the Vice-President for Research in the Rhône-Alpes Regional Council. In California, the Joint Venture: Silicon Valley consortium has taken over in an institutional manner the task undertaken by Professor Terman earlier on.

A champion, whether it is an individual or an institution can play a key catalyzing role in the development of a cluster.

5.6.4 The Need for Entrepreneurial Drive and Sound Business Practices

Entrepreneurial drive can be found in individuals and in supporting organizations. Silicon Valley is the model for individual entrepreneurship, while European regions depend more on institutions such as the Steinbeis Foundation in Baden-Wuerttemberg and CESTEC in Lombardy along with local level Chambers of Commerce. Canada is in-between, where individual entrepreneurship is supported by various local level organizations (e.g. OCRI in Ottawa).

Even in the USA, while Silicon Valley represents the individualism model, other US centres such as Austin, Texas, did set up institutions early on to support entrepreneurship.

Many of the supporting institutions focus on bringing sound business practices to the entrepreneur by providing training in various aspects related to business (e.g. business plans, marketing, financial planning) that the entrepreneur often neglects. This neglect often leads to needless failure.

Where entrepreneurship is lacking, as in the Japanese Technopolis Program or in Atlantic Canada for example, the consequences are evident - a lack of innovation despite the fact that the technical underpinnings are in place.

Entrepreneurship coupled with sound business practices is essential to the success of firms which in turn will lead to the success of a cluster.

5.6.5 The Availability of Various Sources of Investment Capital

Investment capital is key to the growth of firms and hence to the success of clusters. In all countries, there are a range of financial instruments ranging from venture capital firms to banks and to government programs. High-technology ventures are risky so venture capital comes to play an important role in the start-up phase, since banks and governments are conservative by nature.

The USA has the most well developed venture capital community and much of it developed along with Route 128 and Silicon Valley. Canada's venture capital community is less well developed than that in the USA, but is catching up (e.g. Ottawa venture capital community) and is ahead of that in Europe and Asia where regional governments play an important role. For example, in the Montpellier area, the regional government joined with banks and some individuals to set up Mistral Investissements. In Baden-Wuerttemberg, the Steinbeis Foundation, a Lander supported non-profit organization has funds to invest in ventures. In Lombardy, ventures are funded by CESTEC, a partnership between the regional government and industry.

Pass the start-up stage, firms usually turn to equity markets, banks and various government programs for financing. All of these sources come into play because, as a rule of thumb, a high-technology firm needs $1 of working capital to support $1 of sales as it grows.[18]

Basically, the same array of financing instruments is available in all countries. All of the instruments available are used by firms. However, the mix of utilization will vary from country-to-country depending on culture and tradition.

5.6.6 The Cohesion Provided by Informal and Formal Information Networks

The traditional forms of networking are trade shows and conferences. Most regions have them (e.g. Grenoble's bi-annual TEC series). There are also various organizations, such as Chambers of Commerce and local economic development bodies and professional groups that host events. There are also dedicated groups, like OCRI in Ottawa, that have a mission to create "glue" among the high-technology players in their region.

There are also the informal places where technical and business people meet to discuss ventures. In the early days of Silicon Valley, Walker's Wagon Wheel Bar and Grill in Mountain View was such as place. Now it's the Il Fornaio in Palo Alto. There is also the Boar's Head in Austin and the Space Bar in New York.

High-technology thrives on networking. Saxenian's comparative analysis showed clearly how networking played a key role in Silicon Valley outdistancing more conservative and secretive Route 128, despite the latter's pioneering role in high-technology development. So it is not surprising that governments around the world would like to replicate the Silicon Valley model. But do these cultures lend themselves to the openness needed to make the model work?

5.6.7 The Need for Educational and Research Institutions

Educational institutions provide the skilled people needed. Many stay on in the immediate area. As noted by a Bank of Boston study on the impact of MIT on the region:[19]

> "One of the reasons MIT is so important to the Massachusetts economy is that most of the MIT-related companies never would have been located in Massachusetts if MIT was absent. Only 8.7% of MIT undergraduates grew up on the state, but some 36% of all MIT-related companies are located in Massachusetts."

Research institutions, whether within or independent of universities also play a role in transferring technology to existing firms and to generate spin-offs. Proximity is important in the relationship between industry and local research institutions; the ties wear off with distance. A recent study by Zoltan Acs et al. came to the following conclusion on university research-industry interactions;[20]

> "We found that spillovers of the university research on innovation extended over a range of 75 miles (120 km) from the innovating MSA (Metropolitan Statistical Areas)."

So it is not surprising to see clusters develop in areas where there are significant research activities. Not only are MIT and Stanford seen as the centre of the development of their clusters, but Austin, Cambridge, Montpellier, Grenoble and Ottawa are also examples of clusters which emerged from the local research base.

5.6.8 The Need for "Staying Power"

The evidence shows that it takes a long time to grow a cluster of critical mass. As shown in Exhibit 5-1, the antecedents of Silicon Valley can be traced back to 1912. Some clusters, like Boston's Route 128 and Ottawa, were given some impetus by wartime research. However, most clusters "took off" in the immediate post-war period which means that it can take more than 30 years to build a viable cluster.

This means that a sustained effort is needed to develop clusters. For example, in Grenoble, the early efforts of Néel and Merlin which were then picked up by Dubedout and others illustrate the point. The efforts of Frêche in Montpellier in the 1980s were catalytic but it is questionable if that cluster has yet reached "critical mass". The Ottawa cluster, which goes back to the 1950s only reached "critical mass" in the 1990s, when a sufficient depth of capabilities (e.g. venture capital, networking organizations, business support) became evident.

Therefore, the message for policy makers interested in stimulating knowledge-based cluster development is to set in place mechanisms that transcend the normal 4-5 year political time frame.

5.6.9 Summing Up

The clusters described in Sections 2 to 5 and compared in Section 6 all have some of the ingredients of success. The same is true of other clusters reviewed.[21] However, the extent of the maturity of each characteristic varies from cluster to cluster, and the overall mix of the maturity of the eight characteristics also varies from cluster to cluster, indicating variable levels of maturity.

To illustrate this variability, the eight characteristics for each cluster can be quantified and compared. A template for doing this is shown in Exhibit 5-4. Each characteristic is given a value from 0 to 10 with 10 being the "ideal" situation. These values can be ascribed by an analyst who has studied a cluster or by consensus of a group of knowledgeable individuals.

An illustration of this technique is shown in Exhibit 5-5 for three information technology clusters; Silicon Valley, California; Ottawa, Canada; and Atlantic Canada. In this case Silicon Valley is the most mature and has very well developed characteristics along all eight axes, which makes it the "ideal" model or benchmark. So a value of 10 is assigned to each of the characteristics. Ottawa, Canada, is not as well developed as Silicon Valley overall and there are some special needs in the financing community and within educational institutions, which is reflected in the scoring. Atlantic Canada's information technology cluster is the least mature and faces some major challenges regarding entrepreneurship and financing of ventures. As challenges are met, Ottawa could begin to resemble Silicon Valley and Atlantic Canada could evolve towards the position where Ottawa is today. On the other hand, if challenges are not met all the clusters, including Silicon Valley, could stagnate and regress.

This technique, which illustrates the current status of a cluster and compares it to others, has been found useful to policymakers to indicate where actions should be taken to strengthen one or more ingredients to ensure success of a cluster. As well, corporations are finding the technique helpful in their assessment of possible sites where they could establish a presence.

5.7   CONCLUSION

There are possibly some 200 or so knowledge-based industrial clusters around the world vying for investment. As the shift to a knowledge-based economy accelerates and the globalization process intensifies with the lowering of various trade barriers and through international agreements, sub-national regions and localities will become more important players on the world stage. Which knowledge-based regions/localities succeed in this new environment will depend very much on what they do today to develop the eight ingredients of success.

Exhibit 5-4: Template for Illustrating the Status of Cluster Development

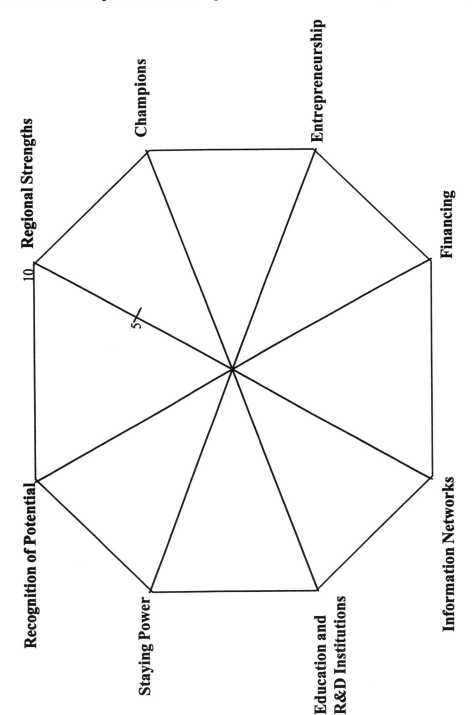

Exhibit 5-5: Development Status of Selected Information Technology Clusters

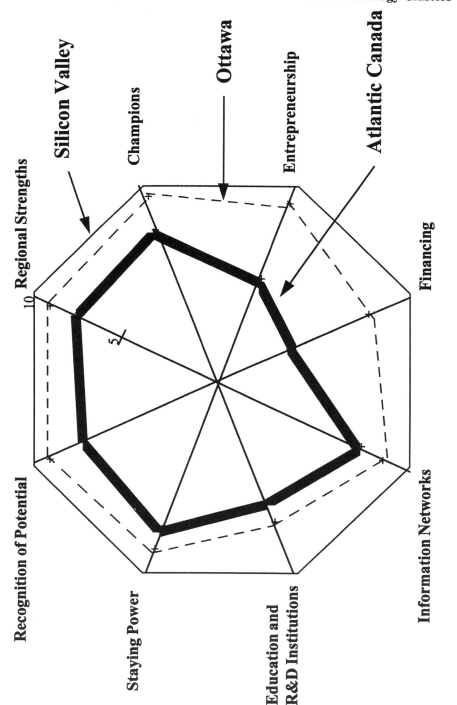

ENDNOTES

1.   Porter M.E.; The Competitive Advantage of Nations; Free Press N.Y. 1990

2.   Lundvall B-A (ed): National Innovation System; Towards a Theory of Innovation and Interactive Learning; Pinter Landon (1992)

3.   Marshall A: Principles of Economies; 8th Edition, MacMillan London (1964)

4.   Saxenian A.L.; Regional Advantage; Culture and Competition in Silicon Valley and Route 128; Harvard University Press, 1994.

5.   See The Economist: A Survey of Silicon Valley, March 29, 1997

6.   The Economist; December 2-8, 1995, p.21.

7.   DEC was founded by Ken Olsen, a student of Jay Forrester who came to MIT in 1939 and was instrumental in shaping its computer science program.

8.   Raytheon was established in 1920 by Vannevar Bush, an associate professor of electrical engineering at MIT. Bush, in his war time role with government brought the British radar team to MIT, which gave further impetus to Raytheon.

9.   See Saxenian op cit.

10.   See Saxenian op cit.

11.   KPMG, DRI/McGraw-Hill/IMPAX Policy Services International; Building Technology Bridges; Cluster-based Economic Development for Western Canada (June 1996)

12.   Castells M. and P. Hall: Technopoles of the World; Routledge, 1994, p 158

13.   Castells M. and Hall P.: op cit. 97.

14.   Smilor et al: Creating the Technopolis: High-Technology Development in Austin, Texas; in Creating the Technopolis Ballinger Publishing Co. 1989.

15.   Castells & Hall, op cit p 139

16.   Nordicity Group Ltd; Prospects for Knowledge-based Cluster Development in Atlantic Canada; Prepared for the Atlantic Canada Opportunities Agency, August 1997.

17.   See for example (a) Nordicity Group Ltd.; Regional/Local Industrial Clustering: Theory and Lessons from Abroad prepared for National Research Council; February 27, 1996 (b) Roger Voyer and Jeffrey Roy; "European High-Technology Clusters" in Evolutionary Economics and the New International Political Economy (ed. J. de la Mothe, G. Paquet) Pinter 1996

18.   Roger Voyer and Patti Ryan; The New Innovators; Lorimer, 1994, p219

19.   Bank of Boston: MIT, the Impact of Innovation, March 1997

20.   Acs Z., Varga A and Anselin L.; Entrepreneurship, Geographic Spillovers and University Research: A Spatial Econometric Approach; March 1997, (mimeo)

21.   See Endnote #17

6 CONTRASTING U.S. METROPOLITAN SYSTEMS OF INNOVATION

Zoltan J. Acs
University of Baltimore
Felix R. FitzRoy and Ian Smith
University of St. Andrews

"Whenever and wherever societies have flourished and prospered rather than stagnated and decayed, creative and workable cities have been at the core of the phenomenon ... Decaying cities, declining economies, and mounting social troubles travel together."

Jane Jacobs

6.1 INTRODUCTION

As we close the final chapter on the 20th century, it has become clear that many cities, communities, and even whole regions have declined in the U.S. as industry after industry has succumbed to fierce international competition and rapid technological obsolescence. However, the same forces that swept away the past are transforming the landscape as new high technology clusters are emerging all over America (Acs 1996).

These new high-technology clusters –what some sociologists call "global nodes"– are connected to the international economy by global markets and multinational corporations, and anchored into their regional economies by production networks. This high technology revolution became identified, first and foremost, with Silicon Valley, which is in and around San Jose, California. Other high-technology clusters are dispersed throughout the 'sunshine belts' of the South, the South-West and the West.

It is well known that both Silicon Valley and Route 128 near Boston owe much of their success to their proximity to large research universities, especially Stanford University and MIT. Research laboratories of universities provide a source of innovation-generating knowledge that is available to private enterprise for commercial exploitation. These R&D spillovers are facilitated by the geographic proximity of university research and industrial activity within regions (Anselin, Varga, and Acs, 1999 and Varga, 1997).

Localized clusters of high technology firms, such as California's famous Silicon Valley or Boston's Route 128, are of significant interest to policy makers and economists alike (Saxenian, 1995). For in addition to favorable effects on international competitiveness, such clusters generate considerable regional benefits in terms of jobs, growth and economic development. Understanding the determinants of the spatial distribution of high technology activities is therefore important for both regional and industrial policy (Braunerhjelm and Carlsson, 1996).

However, spillovers from university research to commercial innovation are not the only effects of relevance to theory and policy. The ultimate economic interest lies chiefly in the impact of spillovers on the product and labor markets. Knowledge based economic activity is largely subject to increasing returns. Products like computers, pharmaceutical, missiles, aircraft, software and fiber optics are complicated to design and manufacture. They require large investments in research and development. As production increases, unit costs continue to fall and profits increase. Experience gained with one product or technology can make it easier to make new products incorporating similar or related designs. Increasing returns from R&D spillovers has led to divergent development both within and between cities, and regions (Krugman, 1991).

This is a question of considerable policy importance (Business Week, 1997). While high technology wages have been climbing, average wages have been stagnant for a decade, and inequality continues to climb. The aim of this paper is to test for the existence of such spillovers from university R&D to local high technology employment, and examine the effect of this on high-technology wages. The discussion is organized into six sections. The first outlines discursively the theoretical background. The second section provides a preliminary analysis of the data. The model is specified in section three and, wages are examines in section four. In the fifth section, the econometric results are reported and discussed. A final section concludes the paper.

6.2   THEORETICAL BACKGROUND

There are two related hypotheses explaining the development of high technology clusters in the vicinity of major university R&D activity. The first explanation of clustering argues that university research is a source of significant innovation-generating knowledge which diffuses initially through personal contacts to adjacent firms, especially those based in a science park. Since both basic and applied university research may benefit private enterprise in various ways it induces firms to located nearby. Lund (1986) in a survey of industrial R&D managers confirms the proximity of university R&D as a factor in the location decision due to the initial spillover from neighboring university research to commercial innovation. Of course, as research results are used and disseminated, the learning advantage created by close geographic proximity between local high technology activity and the university would fade but these learning lags may be long. Information flows locally therefore, through a variety of channels discussed below, more easily and efficiently than over greater distances.

There is a growing body of evidence which supports this hypothesis, especially in the United States[1]. Spillovers form university R&D to patent activity in the same state have been identified econometrically by Jaffe (1989). In addition, papers by Jaffe, Trajtenberg, and Henderson (1993), and Almeida and Kogut (1994, 1997) demonstrate the significant degree of localization of these knowledge externalities with respect to patent citations. Acs, Audretsch and Feldman (1992, 1994) and Feldman 1996) reinforce this result with, instead of patents, a more direct measure of economically useful knowledge production, namely the number of innovations recorded in 1982 by the US Small Business Administration from the leading technology, engineering and trade journals. More recently the analysis have also been carried out at the appropriate MSA level by Varga, 1997, and Anselin, Varga, and Acs (forthcoming). Likewise, Nelson (1986), using surveys of research managers finds university research to be a key source of innovation in some industries, especially those related to the biological science where he finds some degree of corporate funding of university projects. University research spillovers may be a factor which explains how small, and often new, firms are able to generate innovations while undertaking generally negligible amounts of R&D themselves[2].

Despite the presumed advantages of geographical proximity for receiving spillovers, the mechanisms by which knowledge is transferred are not well understood. Information flow are usually attributed to the use of faculty as technical consultants and post-graduate students as research assistants, the use of university facilities, informal communication between individuals at trade shows, industry conferences, seminars, talks and social activities, or joint participation in commercial ventures by university and corporate scientists through contracted research projects. The latter has grown in importance since the late 1970s as the universities established formal Offices of Technology Transfer (or Licensing) to foster interaction with industry and the commercialization of research results.

The second university-based explanation of clustering highlights the provision of a pool of trained and highly qualified science and engineering graduates. The high level of human capital embodied in their general and specific skills is another mechanism by which knowledge is transmitted (Beeson and Montgomery, 1993). To the extent that they do not migrate, such graduates may provide a supply of labor to local firms or else a supply of entrepreneurs for new start-ups in the high technology sector (Link and Rees, 1990). Some evidence for this latter link is provided by Bania, Eberts and Fogarty (1987, 1993) who, using cross-section data, find a significant effect of university research expenditure on new firm start-ups. University scientists themselves, of course, may provide the entrepreneurial input, working part-time as directors of their own start-up companies, or even leaving academia to take a position in a high technology firm. Parker and Zilberman (1993, p.97) report, for example, that MIT has incubated about 40 biotechnology firms since the late 1980s. Lumme et al (1993) in their study of academic entrepreneurship in Cambridge (England) identified 62 high technology companies whose business idea was based on the exploitation of knowledge developed or acquired in either a university of a research institute.

However, even if university research is either negligible or irrelevant to industry, university training of new industrial scientists alone may be sufficient to generate local labor market spillovers. Nelson (1986, p.187), for example, motes that industrial interest in academic departments of physics is confined mainly to their output of potential industrial scientists rather than to their research results.

A university and its associated science park may also play an important signaling role in locational choice (Shachar and Felsenstein, 1992) in the sense that they signal the presence of local technological capacity. Thus firms may be attracted even if the university spillovers are not in fact that great.

6.3   THE DATA

The Bureau of labor Statistics (BLS) supplied employment, or ES-202 series data, for the years 1988 through 1991. Employment data re reported to the BLS by State Employment Security Agencies (SESAs) and cover 98% of civilian employees. While an initial year prior to 1988 may b desirable for terms of increasing the length of the period of analysis, we chose 1988 to avoid complications caused by the reclassification of SIC codes in 1987. Labor force data are also supplied by the BLS and reflect data gathered as part of the Local Area Unemployment Statistics (LAUS) program.

Thirty-two 3-digit SIC industries were chosen and grouped in 6 high technology industries as identified by Acs (1996). These industries are: Biotechnology and Biomedical, Defense and Aerospace, Information Technology, Energy and Chemicals, High Technology Machinery and Instruments, and High Technology Research.

A sample of 37 metropolitan statistical areas (MSAs) was selected using the criteria of high technology development, venture capital, innovation, and university R&D. For instance, urban centers like San Jose, Boston, and San Francisco have large venture capital communities while New York, Baltimore, and Los Angeles have a high concentration of university research. The 37 MSAs represent 52.9% of U. S. high technology employment. On an industry basis, the sample ranges from 30.5 % for energy and chemicals to 77.6% for information technology and services.

Table 6-1 ranks clusters by high technology importance and growth. Between 1988 and 1991, 17 of the MSAs in the sample have more tan 1% of the U. S. high technology employment. Los Angeles leads the sample with 6.17% of U. S. high technology employment. The cities with the highest employment growth rates are Austin (36.8), Charlotte (21.5), San Francisco (21.6), and Raleigh/Durham (17.8). For an analysis of high technology growth by regional competitiveness see Acs and Ndikumwami (1998).

The relationship between university R&D and high technology employment can be analyzed in a preliminary fashion using scatter diagrams. Figure 1 plots aggregate high technology employment in 1989 against university research expenditure in 1985 for all 37 MSAs. Both variables display great variation across metropolitan areas though there is a clear positive association between them. The simple correlation coefficient is 0.60. University research expenditure and high technology employment are both high in the major cities of Los Angeles, Boston, New York and Baltimore.

Figure 2 plots a scatter diagram of high technology employment against the number of scientists and engineers per 100 workers by MSA. The motivation is that the stock of university science graduates with good general and specific skills influences the location of a high technology cluster. Empirically the association with high technology employment is not that strong. The simple correlation coefficient is 0.26. The plot illustrates that Austin, San Jose, Seattle and Raleigh have a high share of engineers and scientists relative to the level of high technology employment. In other words, the labor quality is relatively high in these MSAs. In contrast, the large number of employees in Los Angeles appear to be concentrated in low skilled occupations.[3]

6.4 HIGH-TECHNOLOGY WAGES

It is well known that hourly wages have not increased in real terms for the US economy since the late-1970s. By the early 1990s male real wages were falling for all ages, industries, occupations, and every educational group including those with post graduate degrees. The decline in earnings capacity have been particularly sharp for the young. Despite increases in average educational attainment, those twenty-five to thirty-four years of age took a 25 percent reduction in their real earnings. For year-round full-time make workers eighteen to twenty-four years of age, the percentage earning less than $12,195 (1990 dollars) rose from 18 percent in 1979 to 40 percent in 1989. Real starting wages were lower and the young were simply not getting the advancements that they could have expected in the past. At no other time since data have been collected have American median real male wages consistently fallen for a two-decade period of time. Never before have a majority of American workers suffered real-wage reductions while the real per capita GDP was advancing. (Thurow, 1996, p. 24).

However, this has not been the case in the high-technology sector. For the United States, the average wage in the high-technology sector in 1989 was $34,436. Between 1984 and 1991 wages in the high-tech sector increased by about 6 percent a year in real terms. When high technology wages are compared to non-high-technology wages in the US economy, wages are 55 percent higher in the high-technology sector. For our sample, high-technology wages are a remarkable 79 percent higher than for those in non-high-technology employment. In San Jose, high technology wages are 168 percent greater than all wages, while in New York City that are only 33 percent greater than high-technology wages (Acs 1996, p. 193).[4]

6.5 EMPIRICAL SPECIFICATION

In spite of data limitations, we estimate a parsimonious structural labor market model. The main missing variable is a proxy for product demand faced by high technology firms, such as sales, which is not available for individual MSAs. Our initial specification for the employment equation is written down in natural logarithms as:

$$(1) \quad EMP_{mit} = a_0 + a_1 W_{mit} + a_2 RD_{mi} + a_3 POP_{mt} + a_4 HK_m + a_5 INNOV_m + aX + u_{1mit}$$

where 'm' indexes MSA, 'i' indexes industry, and 't' indexes time: m=1,...,37; i=1,...,6; and t=1988,...,1991. EMP_{mit} refers to high technology employment and W_{mit} is the corresponding annual real wage per employee, defined as nominal wages deflated by the appropriate industry producer price index. Since the panel includes only four years of annual data, cross-sectional variability dominates[5]. For this reason, attempts to estimate equations specified in terms of employment growth rates proved fruitless.

For reasons of data availability, RD_{mi} university R&D, is specified for only a single year, 1985. Given the time span of our data set, it seems reasonable the use of R&D inputs dated in 1985 provides and appropriate lag for the knowledge externality to be transmitted into commercial products and employment. Edwards and Gordon (1984), for example, find that innovations made in 1982 resulted from inventions made on average 4.2 years earlier. The R&D data include industry funded university research, a component which rarely exceeds 10% of the total and is usually considerably less. Notice that RD_{mi} varies by both MSA and industry. Total university R&D spending in each city is desegregated by broad science department and allocated to each of the six industries. This is appropriate given substantial differences in the commercial applicability of university research across academic departments. Thus employment data by industry sector are linked to the relevant component of university research expenditure. The assignment of university department to industrial sector is close to Jaffe (1989) but it is doubtless not the only plausible allocation.

POP_{mt} refers to city population and controls for local market size. Of course, the market extends beyond MSA boundaries but we do not have a more appropriate measure of demand. The number of scientists and engineers as a proportion of the labor force of each MSA represents the potential human capital, or quality of the labor force, available for employers, HK_m. Data are only available for a single year, 1989. $INNOV_m$ is a simple count of the number of innovations by MSA in 1982, the year for which this variable has been collected. It attempts to control for the effect of pre-existing commercial innovation, that leads to product development and marketing with substantial time lags, on subsequent employment levels. Finally X represents a vector of industry, state and annual time dummies. These control for effects specific to each which may not have been captured by the continuous variables.

Since employment and real wages are jointly determined in the labor market, equation (1) should be estimated by a simultaneous method. The corresponding real wage level equation is given by:

$$(2)\quad W_{mit} = b_0 + b_1 W_{mi,t-1} + b_2 EMP_{mit} + b_3 HK_m + b_4 CW_{mit} + bX + u_{2mit}$$

This includes the average MSA hourly wage CW_{mit} and a lagged dependent variable in addition to human capital. Rank and order conditions indicate that both the wage and employment equations are overidentified. Two stage least squares (2SLS) is therefore adopted as the method of estimation.

Table 6-1 presents the main summary statistics by variable for the 37 MSAs in aggregate between 1988 and 1991. There are several trends evident. The mean high technology wage is $30,920. while the mean MSA wage is around $22,000 a year. However, this includes the high-technology wages. Without the high-technology wage, the all city wage is below this level.

Table 6.1 - Summary Statistics by Variable

Variable	Maximum	Minimum	Mean	Standard Deviation	Coefficient of Variation
EMP_{mit}	219500	67	15537	24713	1.59
W_{mit}	60114	7822	30920	7564	0.24
POP_{mt}	8978000	636000	2341200	1931900	0.83
RD_{mi}	479314	0	46655	73905	1.58
HK_m	6.39	0.76	2.72	1.14	0.42
$INNOV_m$	384	4	54.76	76.58	1.40
CW_{mit}	14.22	7.07	11.35	1.29	0.11

6.6   RESULT

Aggregate Equations

Table 6-2 reports the 2SLS estimates of equations (1) and (2). OLS estimates are listed for comparative purposes, though it will be noted that the coefficients do not differ much from their 2SLS counterparts.   Student t-ratios are in parentheses.   The coefficients on the fixed effects are not tabulated but for the employment equation, their joint significance cannot be rejected by an F-test. Taking the fixed effect groups separately, none fail a variable deletion test[6].

Each 2SLS equation was estimated in natural logarithms over the three year period 1989 to 1991 using 666 observations (3 years x 6 sectors x 37 cities). The coefficients should therefore be interpreted as elasticities. Statistically the wage and employment equations are satisfactory, with one notable exception, have the expected signs though not all are significant at conventional levels.

The central result is a positive and statistically significant coefficient on the R&D variable in the employment equation.   Although the magnitude of the employment elasticity is small (0.08), this is evidence of a direct spillover of university research on the high technology employment. In unreported regressions, we found the same result when the dependent variable is specified in terms of high technology employment share, in striking contrast to the statistically insignificant coefficient reported by Beeson and Montgomery (1993)[7].

A further result is that real wages and employment are positively related ceteris paribus. This is counter to our theoretical priors based on the perfectly competitive model. Dropping the wage variable did not markedly affect the signs and significance of the remaining regressors so this outcome does not vitiate our spillover story. Neither reestimating as single equations using OLS nor as random effects models produced major differences in the results. So the estimates are robust with respect to estimation technique.

Table 6.2 - Aggregate High Technology Employment and Wage Function Estimates with Industry, States and Time Fixed Effects

	(1)	(2)	(3)	(4)
Dependent Variable	EMP_{mit} **OLS**	EMP_{mit} **2SLS**	W_{mit} **OLS**	W_{mit} **2SLS**
constant	-20.40 (-10.34)	-23.19 (-9.29)	1.26 (5.85)	1.29 (5.45)
W_{mit}	2.52 (14.21)	2.78 (12.28)		
RD_{mi}	0.08 (7.89)	0.08 (6.73)		
POP_{mt}	0.37 (3.80)	0.39 (3.46)		
HK_m	0.004 (0.03)	-0.03 (-0.21)	0.01 (1.00)	0.01 (1.00)
$INNOV_m$	0.25 (3.11)	0.22 (2.39)		
EMP_{mit}			0.01 (3.89)	0.01 (1.82)
$W_{mi,t-1}$			0.86 (47.03)	0.85 (32.54)
CW_{mit}			0.07 (1.11)	0.07 (1.15)
R^2	0.62	0.62	0.93	0.93
σ	0.851	0.849	0.069	0.069
n	888	666	666	666

Notes: (i) t-statistics are in parentheses; (ii) all variables are in natural logarithms; (iii) R^2 is the adjusted multiple correlation coefficient, s is the estimated standard error of the regression, and n is the number of observations; (iv) unreported dummy variables for industry, time and state are also included in each of these regressions.

At first blush, this result is quite surprising. However it is quite consistent with two important features of high technology industries. First, output markets with continual product innovation and imperfect information are far from the traditional model of perfect competition. It follows that some proxy for product demand should be included in the employment equation but such a variable was not available. Thus we are estimating a reduced form rather than a true structural demand model. Second, specialized skills are often required in high technology sectors. Locational advantages that attract high technology firms may also generate shortages of skilled workers that lead to higher wages. Other wages typically follow to maintain differentials. The positive correlation between high technology employment and wages thus probably reflects the crucial shortages and imperfect mobility of skilled labor that has been the subject of much policy discussion and concern.

Equally plausible, and without relying on market imperfections, it may simply be the demand for products produced by the most skilled and highly paid workers that has grown most rapidly.

The University based labor market spillover story has weaker support. The proportion of engineers and scientists in a city, the human capital variable, is statistically insignificant in both the employment and wage equations. Our population and technical innovation variables, however, are both well determined.

Disaggregated Equations

We have also estimated employment functions separately for each of the six high technology industries. The mean MSA employment, wage and R&D in each sector is listed in Table 6-3 and the equation estimates are provided in Table 6-4. The R&D coefficients are positive and statistically significant (or near significant) in all but the Energy and Chemicals industry. A plausible explanation is the Energy and Chemicals represents a rather traditional industry dependent on both raw materials and products that are much more costly to transport than the inputs of other sectors. Access to port facilities, and other transport infrastructure, is thus likely to be much more important in the location decision, weakening considerably the role of the R&D variable.

Table 6.3 - Mean MSA Employment, R&D and Wages by Industry

Industry	Employment	R&D ($1000)	Wages($)
Biology and Biomedical	5934	68937	25412
Defence and Aerospace	22101	33948	33374
Energy and Chemicals	8215	15459	38237
High Technology Research	7886	127444	26616
Information and Technology Services	33082	7109	34396
High Tech Machinery & Instruments	16004	27032	27483

Table 6.4 - Disaggregated 2SLS High Technology Employment Function Estimates with State and Time Fixed Effects

	EC	DA	ITS	HTR	BB	HTM
constant	-13.84	-90.48	1.89	-9.68	-3.84	-39.47
	(-1.27)	(-6.83)	(0.55)	(-4.62)	(-1.09)	(-5.81)
W_{mit}	1.75	9.75	0.30	0.89	0.85	4.13
	(1.60)	(7.49)	(0.88)	(4.49)	(2.52)	(6.75)
RD_{mi}	0.005	0.25	0.01	0.14	0.08	0.09
	(0.23)	(8.47)	(1.55)	(15.49)	(4.28)	(6.02)
HK_m	-0.43	-1.57	0.54	0.68	-0.33	-0.54
	(-1.66)	(-3.94)	(5.02)	(7.78)	(-1.97)	(-2.94)
POP_{mt}	0.71	-0.12	0.17	0.85	0.24	0.71
	(3.12)	(-0.49)	(2.19)	(13.0)	(1.89)	(4.53)
$INNOV_m$	-0.08	-0.42	0.78	-0.04	0.32	0.15
	(-0.51)	(-1.75)	(11.97)	(-0.72)	(3.18)	(1.37)
R^2	0.86	0.79	0.94	0.97	0.85	0.86
σ	0.530	0.769	0.231	0.196	0.383	0.356
n	111	111	111	111	111	111

Notes: (i) t-statistics are in parentheses; (ii) all variables are in natural logarithms; (iii) R^2 is the adjusted multiple correlation coefficient, s is the estimated standard error of the regression, and n is the number of observations; (iv) unreported dummy variables for time and state are also included in each of these regressions.

Key: EC: Energy and Chemicals, DA: Defence and Aerospace, ITS: Information and Technology Services, HTR: High Technology Research, BB: Biology and Biomedical, HTM: High Technology Machinery and Instruments

Although the results for labor quality and prior innovation are mixed in terms of both sign and statistical significance, it is the university R&D effect, our key variable of interest, which is most consistent. The coefficients are largest in the Defence and Aerospace and High Technology Research sectors.

6.7   CONCLUSIONS

Previous empirical work on R&D spillovers has focused on their relationship with innovation and patent counts at the level of individual U.S. states. With new data for 37 American Standard Metropolitan Statistical Areas including the main university R&D centers we have found a statistically significant and robust spillover to employment in five high technology sector, after controlling for state fixed effects. This confirms the popular view of high technology clusters and provides the first quantitative evidence that academic research has a positive local high technology employment spillover at the city level. A further result is that innovation was also strongly related to high technology industry employment after a long time lag, again a plausible but hitherto untested proposition.

These results are clearly of relevance for regional policy. They provide support for the importance of high technology clusters in the U.S. and possible lessons for Europe and Japan where such clusters are much less well-developed and where there is no evidence of the localization of knowledge spillover, at least in the semi-conductor industry (Almeida and Kogut, 1994). In spite of dramatic declines in the costs of information transmission, local spillovers underline the importance of personal contacts and face-to-face communication in transferring scientific progress into jobs and products. Clearly more research is required on the nature of the transmission process as well as on the skill composition of high technology employment and the relationship of training and skills to wages and employment in local labor markets.

This partly reflects pressure applied by US government agencies to universities, for economic growth reasons, to hasten technology transfer from their laboratories to the private sector (Parker and Zilberman, 1993). Federal Acts passed in the early 1980s also promote knowledge spillovers. The Stevenson-Wydler Technology Innovation Act of 1980, for example, encourages cooperative research and technology transfer and the 1981 Economic Recovery Tax Act gives tax discounts to firms that provide research equipment to universities. Some universities have created industry consortia to help fund research. Firms pay membership fees to join these consortia and in return benefit from access to the research output and have some voice in the research agenda. Such channels would be expected to flourish given that universities as public institutions do not face the same incentives as private corporation to keep research results secret. In both the San Francisco Bay and Boston areas, for example, the introduction and growth of the biotechnology industry is a direct result of university R&D spillovers. Presumably, the chief benefits of geographical proximity to the spillovers source consist in a reduction in both the transactions costs of knowledge transfer and in the costs of commercial research and product development. As a caveat, it ought to noted that we do not argue that proximity is a necessary condition for spillovers to occur, only that it offers advantages in capturing them.

The second and initially somewhat surprising result is the strong positive correlation between wages and employment in high technology industries. This association is apparent in all equations and appears to be quite robust. Our view is that the positive partial correlation in the employment equation could arise from two sources which are not mutually exclusive, omitted variables and skill aggregation.

The most important omitted variable is the absence of a demand or sales variable in the employment function. This equation therefore essentially captures a labor supply relationship. For example, if the demand for products produced by the most highly skilled, and paid, labor grows fastest due to innovation, government procurement or whatever, employment and real wages would be positively related in a regression which did not control for demand effects. Likewise, if there are shortages or bottlenecks of key skilled personnel, their wages would rise with demand for their products or services to include a scarcity rent.

A complementary explanation relates to the heterogeneity of skill categories in the high technology sector. It is obviously rather crude to estimate employment demand implicitly assuming a single, homogenous category of labor. There exists both skilled and relatively unskilled employment in the high technology sector and the demand for these categories may move in different directions.

Naturally, university knowledge spillovers are not the only reason for high technology clusters. Other forces for localization are quite strong. They would include the development of specialized intermediate goods industries, economies of scale and scope, and network externalities. With respect to the latter, innovations by different producers may be complementary, yielding related new products or processes when combined. On these questions too, further research is called for.

ENDNOTES

1.   Shachar and Felsenstein (1992) report evidence from studies conducted in Europe and Japan which show very few benefits arising from the close physical proximity of high technology firms to a local university.

2.   It should be noted that R&D is not a good measure of small firm inputs into knowledge production since such inputs often arise informally without the support of an R&D laboratory.

3.   Because we have sampled on the dependent variable the data may be subject to sample selection bias. Acs, FitzRoy and Smith (1998) test for this and find no evidence of selectivity bias.

4.   These trends appear to have continued into the 1990s. While we do not have comparable data, a special issue of Business Week on Silicon Valley (1997) reported that in 1996 real wages in Silicon Valley grew by 5.1 percent while the U. S. average was less than 1 percent. In 1966 the average wage in the Valley was $43,510., 55 percent above the national average.

5.   Inclusion of a lagged dependent variable in (1) yielded a coefficient of almost unity and impaired the explanatory power of most other variables, suggesting a relative lack of movement in employment over time.

6.   All equations were also estimated with coarser regional instead of state fixed effects, yielding very similar results, though the R&D employment elasticity was smaller in this model. White standard errors to control for heteroscedasticity differed little from the reported results.

7.   Note that Montgomery and Beeson omit real wages and prior innovations from their employment equation. They do, however, include several variables to control for the effects of other area attributes on local labor market conditions, which we capture using state dummies.

REFERENCES

Acs, Z., and Ndikumwami, A. (1998). High-Technology Employment Growth in Major U. S. Metropolitan Areas, *Small Business Economics*, 10(1).

Acs, Z. (1996). US High Technology Clusters, in J. de la Mothe and G. Paquet (eds.), *Evolutionary Economics and the New International Political Economy*, London: Pinter, 183-219.

Acs, Z., Audretsch, D.B. and Feldman, F. (1994). "R&D Spillovers and Recipient Firm Size", *Review of Economics and Statistics*,100 (1).

Acs, Z., Audretsch, D.B. and Feldman, F. (1992). "Real Effect of Academic Research: Comment", *American Economic Review*, 82(1), 363-67.

Almeida, P., Kogut, B. (1997). "The Exploration of Technological Diversity and the Geographic Localization of Innovation", *Small Business Economics*, 9(1), 21-30.

Almeida, P., Kogut, B. (1994). "Technology and Geography: The Localization of Knowledge and the Mobility of Patent Holders", Department of Management,Wharton School, University of Pennsylvania.

Anselin, L., Varga, A. and Acs, Z.J. (1999). Local Geographic Spillovers Between University Research and High Technology Innovations, *Journal of Urban Economics*.

Bania, N., Eberts, R. and Fogarty, M. (1993). "Universities and the Start-Up of New Companies: Can We Generalize from Route 128 and Silicon Valley?" *Review of Economics and Statistics*, 75,761-66.

Bania, N., Eberts, R. and Fogarty, M. (1987). "The Role of Technical Capital in Regional Growth", Presented at the Western Economic Association Meetings.

Beeson, P., Montgomery, E. (1993). The Effects of Colleges and Universities on Local Labor Markets, *Review of Economics and Statistics*,75, 753-61.

Braunerhjelm, P., and Carlsson, B. (1996). "Industry Clusters in Ohio and Sweden", The Research Institute for Industrial Economics, Sweden.

Business Week (1997). Sharing Prosperity, September 1, 1997, 64-70.

Business Week (1997). Silicon Valley, Special Issue, August 18-25. 1997.

Edwards, K.L., Gordon, T.J. (1984). "Characterization of Innovations Introduced in the US Market in 1982", The Futures Group.

Jaffe, A.B., Trajtenberg, M. and Henderson, R. (1993). "Geographic Localization of Knowledge Spillovers as Evidenced by Patent Citations", *Quarterly Journal of Economics*, 577-98.

Jaffe, A.B. (1989). "Real Effects of Academic Research", *American Economic Review*, 79(5), 957-70.

Krugman, P. (1991). *Geography and Trade,* MIT Press.

Link, A. N., Rees, J. (1990). "Firm Size, University Based Research, and the Returns to R&D", *Small Business Economics*, 2, 25-31.

Lumme, A., Kauranen, L., Autio, E. and Kaila, M. M. (1993). *New Technology Based Companies in Cambridge in an International Perspective*, Working Paper No. 35, Small Business Research Centre, University of Cambridge.

Lund, L. (1986). *Locating Corporate R&D Facilities*, Conference Board Report No. 892, Conference Board, New York.

Nelson, R.R. (1986). "Institutions Supporting Technical Advance in Industry, American Economic Review", *American Economic Review*, 76(2), 186-89.

Parker, D.D., Zilberman, D. (1993). "University Technology Transfers: Impacts on Local and U.S. Economies", *Contemporary Policy Issues*, XI(2), 87-99.

Shachar, A., Felsenstein, D. (1992)."Urban Economic Development and High Technology Industry", *Urban Studies*, 29 (6), 839-55.

Thurow, L.C. (1996). *The Future of Capitalism*, New York: Norton.

Varga, A. (1997). *Regional Economic Effects of University Research: A Spatial Econometric Perspective*, West Virginia University, Unpublished Doctoral Dissertation.

7 CONTRASTING REGIONAL INNOVATION SYSTEMS IN OXFORD AND CAMBRIDGE

Helen Lawton Smith
Coventry University
David Keeble, Clive Lawson,
Barry Moore and Frank Wilkinson
University of Cambridge

"... contemporary regional economic development is inseparable from cultural, social and institutional accomplishment"
(Cooke and Morgan 1994, 91)

7.1 INTRODUCTION

This opening epigraph by Cooke and Morgan encapsulates recent thinking on regional innovation systems. While national regulatory frameworks in their broadest sense provide the overall operating context, the regional or local environment is where firms live and learn. This is also the geographical scale at which the nexus of actions by individuals, business intermediaries, universities, and society can make a difference to the evolution of economic development. Thus the world is composed of a 'hierarchical mosaic of densely-developed regional economies with specific resource endowments, assets, institutions, co-ordination mechanism, know-how, rules of conduct and cognitive frameworks' (Asheim and Dunford 1997, 451). The specific characteristics of regions arise from the interaction of geo-historical events, and increasingly by 'the elaboration of new forms of globalization in the organisation of industrial activity' (Amin 1993, 447).

Thus, on the one hand, there is enthusiasm for the position that economic growth is becoming increasing localised (Ettlinger 1996), and on the other there are reservations about the self-determining and self-sustaining capacity of regional innovation systems. First, we are told that globalization limits the power of national states to deal with global capital (Hudson 1997). Second, it has not been proved that the regional milieu is more important than the national environment (Asheim and Dunford 1997, 449). Moreover there are different types of innovators in regions–market orientated, production, software–which do not have the same propensity and need for localised linkages (Wilkinson et al 1997). It should not be forgotten that sectors are organised through space–nationally and internationally. The electronics industry is the same industry whether it is in Colorado or Timbuktu. Regional sectoral clustering of firms is only one manifestation of the organisation of the production filieres of particular industries.

This chapter takes into account both local and external processes contributing to the development of two elite 'knowledge-complexes' in the UK, the Oxford and Cambridge regions. They are both important centres of public and private investment in innovation. Both have premier universities, concentrations of high-tech industry, industrial and national laboratories and high quality labour forces. Yet both display quite different evolutionary dynamics.

The chapter begins by considering the context within which spatially organised learning takes place. Some of the more interesting developments in analyses of 'learning regions' have moved away from the stereotypical high-profile areas such as Baden-Wuttemburg to looking at a range of different types of regional innovation systems. The second part focuses on the characteristics of the two regions, highlighting the particularities of geography, resource endowment, high-tech activity and institutional initiatives and responses. This is followed by a discussion of findings from a recent survey of fifty high-tech firms in each of the Oxford and Cambridge regions. The intention is to examine the differences in the two regional innovation systems by comparing the mechanisms by which social cohesion is generated.

7.2   CONTEXT DEPENDENT, SPATIALLY ORGANISED LEARNING

It is a general tendency for firms which are most innovative to cluster in certain places, and for these clusters to contain a pre-dominance of particular sectors and/or activities. In the European Union, ten "islands" constitute the prime concentrations of innovation activity. These clusters, which are South East England; Ille de France; Frankfurt; Munich; Turin; Rotterdam/Amsterdam; Rhein-Ruhr; Lyon/Grenoble and Milano, represent 80% of the research laboratories and enterprises which participate in transnational R&D co-operation (EC 1994, 203). This observation describes a particular surface pattern. The evolution of these clusters has resulted from the interaction of a raft of general and place specific developments, which are 'fundamental processes of mutual interaction and moulding' and 'the combination of layers of investment and through human agency' (Massey 1994, 321).

Evolution of High-Tech Regions

Garnsey (1996) has developed a framework for analysing the evolution of new local industrial ensembles. Drawing on systems theory, she argues that the influence of initial conditions, chance events and re-inforcing, cumulative, processes are at work in the dynamics of complex systems through feedback effects whereby consequences feed into further outcomes. Inter-dependence is the emergent property of the system. Processes of self-organisation help to create a critical mass of interdependent activity and form a centre of attraction and resource generation. While similar initial conditions and resource availability pre-dispose developments in certain directions, a series of local events which result from chance occurrences provide the impetus for further developments. This approach allows for the interplay of political, cultural factors and the actions of individuals to influence local developments.

While this analysis attempts to conceptualise the underlying processes that are at work in the evolution of regional complexes in which shared learning is a central element, other studies have focussed on classifying different types of districts in order to examine their internal dynamics. Two major generic types are 'innovative milieux' (GREMI group), and Marshallian Industrial Districts (see Amin and Thrift 1992). Lawson (1997) points to the distinction between these two generic types. In the latter interdependence takes the form of transactions between firms in sequential stages in supply chains and associated other linkages particularly in the supply and recruitment of labour in sectorally specialised areas. There is a special focus upon the particular forms of co-operation which takes place in these districts including sharing technical information. In innovative milieux the local environment or milieu is the unit of analysis. The emphasis in particular is upon the importance of a set, or complex network, of mainly informal relationships (Camagni, 1991). The focus is on the ability of the milieu to foster or facilitate innovation but not sectorally specialised innovation, where links are often primarily horizontal rather than vertical (Lawson 1997). Camagni distinguishes between the static transaction costs and Marshallian external economies approach (for example in the work by Scott 1988) and the dynamic concept of the milieu, which facilitates collective learning and reduces dynamic uncertainty. Successful milieux are seen as facilitating learning mechanisms in the innovation process such as learning by doing and learning by using and interacting (Todtling 1994, 72). The caveat to this line of argument is that inter-organisation innovation networks are not necessarily spatially bounded and do not necessarily result in localised linkages (see also Malmberg 1996,395; Curran and Blackburn 1994). Moreover, not all firms depend on localised learning or knowledge embedded in the region of location (Maskell and Malmberg 1995, 13).

Markusen (1996) extends the industrial district model and argues that in addition to the new industrial district (Marshallian Industrial Districts) which based is on small, innovative firms, there are at least three other types of industrial region: hub-and-spoke; satellite industrial platform, and the state-centred district. These have quite different embedded patterns of social interaction. In the core type of Marshallian industrial districts there is an evolution of unique local cultural identity and bonds whereas in the satellite industrial platforms this is absent. An earlier typology of districts: 'technology-oriented complexes' (Steed and De Genova 1983) also identified archetypal regions such as those resulting from large scale government expenditure. Both approaches identify the kinds of initial conditions or static moments which represent a base-line from which evolutionary developments are considered, the moment of current comparison and the possibility of evolution from one type of district to another.

Storper's (1995) approach straddles the two generic types. He focuses on the non-traded aspects of regionally focussed technological change and learning in the form of interactions as 'untraded inter-dependencies'. Storper's model identifies the social glue of familiarity, shared experiences, and reciprocity generated in local networks. These relationships do not fit the industrial district transaction costs models, but involve social networks in line with the milieu approach. Earlier Storper (1993) argued that localised rules, institutions, practices explain both the geographical concentration and technological performance of technologically dynamic, export-orientated networks.

Information Resources

Learning is a resource issue which has political, economic, cultural and technical components. Learning is also spatially constructed. Information resources, including technical, professional and financial, such as the scientific and engineering competence of other firms and universities, like all resources, have a specific temporal economic value. They are associated with demand created by other developments in the production system (markets, technological advances etc.). Although information available at different spatial scales, for leading-edge science based firms in particular, it is localised through various co-operative mechanisms involving inter-organisational linkages such as subcontracting and research collaboration. The pattern of learning, like the evolution of districts, is path dependent, influenced by prior developments and shaped by further local developments (Garnsey and Lawton Smith 1997). The level and quality of demand for *local* resources may change over time. Keeble *et al* (1997, 1) argue that linkages within an innovative milieu may well be particularly important during the early stages of development of small technology-intensive enterprises, many of which tend to be established by spin-off of individuals, ideas or technologies from existing enterprises, universities or other institutions.

At the regional level, sets of common knowledge may be developed through accumulated technical competence, through sectoral specialisations, and through more general local knowledge and understanding of how the local system works. The latter becomes a regional resource and includes knowledge of the nature of local resources–which banks offer best service, who are the lawyers who specialise in intellectual property, who to approach to get information about what technical expertise is available in the universities etc. These sets of knowledge are developed by localised collective learning and 'untraded interdependencies' and result in efficient local production networks (Storper 1993, 1995; Morgan 1997; Amin and Thrift 1995; Asheim 1996).

Institutional Endowment

Maskell and Malmberg (1995,11/12) suggest that the institutional endowment of a region or country includes all institutions related to factors of production, efficiency of the market, quality of demand, governmental forms, the public sector, the political decision-making and implementation process, entrepreneurship, rules, practices, routines, conventions, culture, religion and other basic values characterising the region or the country. These require socialisation and valorisation of local economic communities. The institutional endowment, local resources and physical structures together constitute the region's capabilities, which are argued to influence the competitiveness of firms in the region (see Amin and Thrift 1995). However, as was mentioned at the outset there are limits imposed by national and international conditions to the extent to which regions can influence their own development path.

7.3   OXFORD AND CAMBRIDGE CHARACTERISTICS

> *"A Bishop of Ely in the 1970s is reported to have said that there would be no industrial chaplain appointed while he was Bishop because there is no industry in Cambridge"*
>
> (Pearson 1994)[1]

Therein lies the key difference between Oxford and Cambridge. Oxford indeed has an industrial past, while Cambridge was considered as something of a rural backwater as late as 1960 (Pearson 1994). While both have in common ancient universities dating back to the thirteenth century (Cambridge was founded by scholars from Oxford), are environmentally attractive places to live and are surrounded by agricultural land, are of similar size, similar distance from London and both are now leading foci of high tech industry, their evolutionary paths have been different. Differences between them arise from geography, the composition of the technological resource base, the extent and sectoral profile of high-tech activity, industrial history, the response of the universities to local development issues, and the institutional framework. As a result there are major differences in the kinds of industrial coherence and interaction.

Geography

The county of Cambridgeshire is on the face of it both larger in area and population than Oxfordshire. The county of Cambridgeshire covers an area of some 1300 square miles with a population of 684,000, while Oxfordshire has an area of 1000 square miles with a population of 598,000. However, this is primarily because the former contains a second major town, Peterborough (120,000), which is shortly to be granted independent, unitary, status. The population in Oxfordshire is more concentrated in the city of Oxford than Cambridgeshire's is in the city of Cambridge. Oxford has a population of 134,000 compared to only 106,000 in Cambridge.

Oxford occupies a more favourable centrally placed geographical location than Cambridge, which is in the more isolated East Anglian region. The construction of the M25 and the M40 extension has put Oxfordshire on major transport routes, north and south. Oxford has easy access to Heathrow airport, while the international airport local to Cambridge, Stansted, is much smaller and offers much less international access. While these geographical features may be incidental to the genesis of the high-tech communities, they may be more important in the longer term either as factors which influence whether firms remain in the area or for firms considering locating in either.

Resource Base

Both Cambridge and Oxfordshire are at the centre of national and international research networks. Oxford University, like Cambridge, is extremely successful in attracting research income. In 1995/6 for the first time, Oxford's overall research income broke the £100 million threshold, a rise of 10.4 per cent over the previous year. Research council income accounted for £42 million, UK industry £5.7m and overseas industry and contract research £4 million. A high proportion of industry income is for medical research. In 1993/4 Cambridge University had less research income from industry, £6.23m which amounted to 8.5% of total research income. The substantial difference between the two universities is that Oxford has an arts: science ratio of 7:5, as against 1:1 in Cambridge. Oxford also has a stronger emphasis on pure science, and its major scientific strength is in chemistry, whereas Engineering is the largest department in Cambridge, which also possesses internationally renowned physics, mathematics and computing departments (Segal Quince 1985,63).

Oxfordshire possesses a greater concentration of national laboratories, hospitals and medical research units than Cambridge, with four universities, nine hospitals of which seven have university research departments, and seven national/privatising laboratories, including two belonging to the United Kingdom Atomic Energy Authority (UKAEA) (Harwell and Culham laboratories) and the EPSRC Rutherford Appleton Laboratory. Cambridge has historic strengths in computing–it was chosen as the location of the government-funded Computer Aided Design centre in the 1960s–and more recently in molecular biology, with the Nobel prize-winning MRC Laboratory located alongside Addenbrooke's Hospital. Other major research institutes include Wellcome and CRC Cancer Research, the Plant Breeding Institute, the National Institute of agricultural Botany, the Babraham Institute of Animal Physiology, and the Abington Welding Institute. Multinational research laboratories are operated by Schlumberger, Smith Kline Beecham, Toshiba, Sony and prospectively Microsoft, whose first-ever R&D laboratory outside the USA is being established on the University's West Cambridge science campus in 1997. Oxford's industrial laboratories include the European Research laboratory of Sharp, and the research centres of Dow Elanco, Yamanouchi, and Esso,

As a result of the cumulative effects of investment and location decisions both regions have highly-skilled workforces. Oxfordshire's workforce includes a quarter of all R&D workers in the South East (8,200), which represents the largest number of people employed in research and development outside London (Jordan 1995, 12).

Cambridge has over 6000 employed in industrial R&D, with a locational quotient of 7.2.

High-Tech Activity

Both Oxford and Cambridge are located within the 'Western Crescent' (Keeble 1989), a band of high-tech activity which arcs around the west of London from Cambridge in the East to Southampton on the South coast. Cambridgeshire was the leading growth area in the UK in the 1980s for high-technology based employment (Keeble, 1994), while the Cambridge sub-region recorded an increase in the number of technology-based enterprises of 380 or 115% between 1984 and 1996. Oxford, however, is only a third tier location of high tech firm formation and employment growth (Keeble 1994).

The number of firms and employees in each area depends on the definition of high-tech used. Here the definition is based on the 1987 Butchart sectoral classification but in each region modified to exclude service activities such as those of British Telecom and to include only those which were known to be innovative. To be consistent, national laboratories were included in each sample. With few exceptions, their activities are commercially orientated. The data was provided by both Cambridgeshire and Oxfordshire County Councils, and in Oxfordshire by The Oxford Trust. The Cambridge Region is defined as the three local authority areas of the City of Cambridge, South Cambridgeshire and East Cambridgeshire together with the Royston Fringe. The Oxford Region excludes Banbury travel-to-work area to the north of the county and a rural area to the south of the City. On this basis, the population of high tech firms in the Cambridge Region in 1996 was 715 firms employing 24,024 while the much smaller population in the Oxford Region consisted of 179 firms employing 14, 619 employees.

Both regions are characterised by a diversity of technology-based sectors rather than by specialisation upon one particular sector. Cambridge is more service-orientated while Oxford has a greater orientation to manufacturing. In Cambridge, nearly two thirds of local technology-based firms are in services (63%) rather than manufacturing (Keeble and Moore, 1997). The biggest sector in terms of number of firms, computer software and services, accounts for only 29% of enterprises. Cambridge was a leading centre for the personal computer and software sectors in the 1970s and early 1980s when those related sectors were in the van of the micro-electronics revolution (Keeble and Kelly, 1986). However, the largest in terms of employment is research and development services, with 25% of total employment. Employment is marginally higher in manufacturing than services (51:49). However, while many Cambridge firms are in the forefront of technology they have been criticised in the past for their failure to acquire competence over a sufficient range of production activities to maximise value added and grow substantially (Saxenian 1988; Keeble 1989). There is certainly a higher proportion of small firms in the Cambridge region, with about 74% employing less than 20 as compared to 52% in the Oxford case. Conversely, only about 5% of Cambridge firms employ over 100 employees as compared to 22% in the smaller Oxford case. The former includes Domino (inkjet printing machinery), Uunet (internet services) and Ionica (telecommunications). Oxfordshire's large firms include those of the Oxford Instruments Group which makes a range of medical and industrial instruments. This has been the major success story and has been an important contributor to the development of the UK/Oxfordshire cryogenics industry in which the county specialises. The largest educational computer manufacturer in the UK, Research Machines, employing nearly 600 people, is also here. Both have sectors which the other does not. For example, Oxford also has an important motor racing industry which employs some 2000 people (Willis *et al* 1996), while Cambridge has an aerospace industry employing 1600 people.

One sectoral characteristic which they share is that both are leading growth centres for bioscience. In Cambridgeshire 76 firms and biotech research organisations were identified in 1996 (GOER, 1996). Oxfordshire now has some 40 organisations and firms employing 2,189 (Mihell 1996). Both cases reflect a university specialisation. Oxford University has 53 units undertaking medical research (Oxford University 1995), while Cambridge set up its Institute of Biotechnology in the 1980s.

An important development in Cambridge has been an increasing number of take-overs of formerly independent local firms. By 1992, 34% of surviving companies originally studied by Segal Quince (1985) had become subsidiaries, with 71% of the (small) biomedical sample within corporate groups (Garnsey and Cannon-Brookes, 1993, 180). This has, however, to be balanced against the considerable parallel increase in new independent start-ups in the region since 1985, which were not included in this study. External ownership has an important bearing on the character of the area and use of resources, and where technology is commercialised. In Oxfordshire, in the 1980s foreign ownership was already found to account for 10% of firms, but 30% of employment (Lawton Smith 1990). Most firms were from North America. Since then other foreign firms have set up research facilities but do not generally manufacture in the county. These include two Japanese firms: Yamanouchi (pharmaceuticals) and Sharp (electronics).

Industrial History

Oxfordshire has an industrial past in which motor car and component manufacturer, food industries and earlier blankets were leading sectors. More recently it has developed as the largest centre for educational publishing outside London. However, the origins of high-tech activity in Cambridge can be traced back a hundred years. Firms such as Pye and Cambridge Instruments founded in the 19th century, were important in the history of high technology industry (Garnsey and Lawton Smith 1997). The latter was set up specifically to design and manufacture scientific instruments for the University. It later diversified into other apparatus and was a key founder of the British instruments industry. In Oxford, some seventy years later, Oxford Instruments was founded in 1959 to supply cryogenic instruments for the university's physics laboratory (Clarendon Laboratory) and it too has had a profound impact on both the development of the region, and on instrumentation for both industrial and medical use. In 1994 the Group, plus their diagnostic imaging joint venture with Siemens Oxford Magnet Technology which makes whole body scanners, employed a total of 1,800 (1350 in the UK and 450 overseas).

The Response of the Universities

Universities are part of a national technological infrastructure. The relative autonomy of UK universities means that if they are involved in local economic development then it is usually because they have chosen to do so. Some of the older universities such as Warwick and Salford (Segal Quince Wicksteed 1988) have adopted explicit policies to engage in their local communities. The new universities, most ex-polytechnics, were funded until 1989 by local LEAs and always had a more explicit local focus.

The responses of the universities to developments occurring around them and to the exploitation of academic research have been the two most profoundly important differences between Oxford and Cambridge. First, Cambridge University has operated a 'benign supportive and non-interventionist policy' (Segal Quince Wicksteed, 1988, 17/22). On the other hand Oxford University's stance has been that it plays no part in local economic development.

The second major difference is in the internal technology transfer arrangements. It was not until the late 1980s that Oxford University adopted an active position in exploiting its technology. A full time industrial liaison officer was appointed for the first time in 1989. A year earlier, the University established ISIS Innovation as a wholly owned company of the University of Oxford. ISIS Innovation's main activities are handling downstream IP and dealing with patents and licenses, and managing the Oxford Innovation Society. This was formed so that members, (mainly multinational companies) can have a window on Oxford technology. By 1996, ISIS had only 3 managed spin-outs, while other firms had been formed by academics with or without formal university support. By the early 1990s Oxford University recognised that there have been problems with its internal procedures. In 1994, a committee was set up to conduct a 'Review of Technology Transfer Arrangements'. This was critical of the lack of co-ordination of industrial liaison activities, particularly between the research support office and ISIS Innovation.

In the Cambridge case, technology transfer from the University to local firms is encouraged by Trinity College's Cambridge Science Park, St John's Innovation Centre, and the University's own Industrial Liaison and technology Transfer Office, originally set up in 1970. The last exists to help academics commercialise their research, and operates Lynxvale Ltd, the University's own technology exploitation company. Cambridge Research and Innovation Ltd and Quantum Fund, in both of which the University is involved, are local investment funds for university scientists seeking to commercialise their technology.

A further aspect of this difference is the position adopted on intellectual property. Unlike in Cambridge, Oxford University has enacted a policy of *claiming ownership* of IPR generated by its staff and students in the course of or incidental to their studies (July 1995). This replaced the previous Statute under which the University "asserts its ownership of them". The new statute became effective in July 1995 (University of Oxford 1995). The University now has the obligation and responsibility to ensure that arrangements for exploitation of the IPR are efficient and easy to use. In contrast, in Cambridge University the inventor has the rights to intellectual property unless research is funded under some agreement, such as an industrial contract. The

University is positive about entrepreneurial academics (Cambridge University Reporter July 1990).

Institutional Arrangements

The planning regimes are also an explanatory factor in the trajectory of regional development in the two places In the 1950s and early 1960s, the traditional character of Cambridge which centred on the University and Colleges and the architectural splendour of their buildings in the city centre seemed threatened by population pressure and recent industrialisation, particularly during the war. In Cambridge it was felt that Oxford had suffered as a university city from the development of the motor industry at Cowley on its outskirts. The Holford Report in 1950 recommended that industrial expansion in and near Cambridge should be limited, and that large scale production activities should be discouraged anywhere in the county. In Cambridge, IBM was refused permission to establish its European R&D laboratories in the city in the 1960s. Other firms were either banned from expanding in Cambridge and relocated to small towns outside the city, or forced out of the area altogether (Segal Quince 1985,18).

However, developments in the 1960s became seen as advantageous to research with the potential for long-term funding by industry of academic research of industrial applications. A sub-committee, under the chairmanship of Sir Nevill Mott, then Head of the Cavendish, physics, Laboratory was set up by the University, supported by the City Council, and reported in 1969. The Mott Report's recommendations led to an easing of the Holford planning restrictions, and to a policy of actively strengthening the interaction between teaching and scientific research on the one hand and industrial applications on the other. It recommended that a 'science park' would provide a suitable environment for science-based industry and be accessible to university departments. The Mott Committee estimated that 20-25% of the research and technical staff of the University were already involved in applied industrial research supported by outside funding, in addition to the use of University equipment and support funds for this purpose. In its review of the county development plan, the County Council accepted the Report's recommendations (Pearson 1994,4). Segal Quince (1985,21) describe the Mott Report as being widely regarded as the 'watershed' in the evolution of the University's official attitude to industrial development and to collaboration with local authorities. There has been no equivalent 'town and gown' approach in Oxford.

Similar planning restrictions operated in Oxford. The Structure Plan for the South-East in the 1980s recommended that Oxford City and areas to the south of the city should be areas of restraint in order to protect the environment, character and agricultural resources of the County by restraining the overall level of development and to promote the 'country towns of Banbury, Bicester, Didcot and Witney' as preferred locations for new development, while limiting development in the rest of the county' (Structure Plan for Oxfordshire, Written Statement, May 1987). A key element in this the protection of the green-belt around the city. It was not until 1987 that restrictions were relaxed which enabled a science park to be built on the edge of city. The Oxford Science Park was established by Magdalen College, two decades after Trinity College founded the Cambridge Science Park in 1971. This delay in Oxford hindered the development of the same kind of high profile image which developed in Cambridge in the 1980s.

A major difference between Oxford and Cambridge is in terms of the nature of innovation support activities. On the one hand, Oxford has a more advanced and coherent network of providers than Cambridge. In Oxford a major driver of initiatives is a local charitable organisation which interacts with local government and national government agencies. The Oxford Trust was formed in 1985 by the founder of Oxford Instruments, Dr (now Sir) Martin Wood and his wife Audrey. This charitable Trust was established to encourage science and technology enterprise in Oxfordshire. It now has an innovation centre with incubator units, runs seminars series on different aspects of firm development, and the Oxford Innovation network which puts firms and business 'Angels' in touch with each other. In Cambridge, a significant share of the more limited support for industry which exists involves university organisations, probably the most important of which is the St John's Innovation Centre (Reid and Garnsey, 1996).

Other forms of innovation support which involve social contact are the business clubs which operate in each area. Cambridge has a far more extensive set of clubs than Oxford. They include the Cambridge Europe and Technology Club (mainly high-tech SMEs) and the Cambridge High tech Association of Small Enterprises (CHASE) for high tech firms up to 5 employees. In Oxfordshire, the Thames Valley Electromagnetic Compatibility (EMC) Club is based at the Oxford Science Park. This was created to help firms meet EU EMC regulation but has now adopted a wider role of helping firms comply with standards. Another important difference, which indicates the different levels of maturity and extent of social/business cohesion in the Cambridge high-tech complex, is the greater number and range of technological consultancies and specialist services. Major technology consultancies, from which numerous spin-offs have occurred, include, Cambridge Consultants, PA Technology, Scientific Generics, and the Technology Partnership. The similarities and differences in the regions are summarised in Table 7-1.

7.4 THE STUDY

This paper analyses the results of an original interview survey of 100 firms randomly sampled in the Oxford and Cambridge regions from the total population of high technology firms by the Cambridge University ESRC Centre for Business Research. The survey was conducted between October 1995 and May 1996. The sample was stratified to reflect the composition of the population, reflecting the balance between manufacturing and services in each area, and to include larger and small firms. The two largest groups interviewed were, not surprisingly, service and manufacturing firms employing less than 100 people. Virtually all the larger firms surveyed employ fewer than 500 employees. In Cambridge only one exceeded this figure. Table 7-2 shows the achieved sample and response rates. The lower Cambridge response rate reflects acute 'survey fatigue' by high-technology firms in this high-profile area.

Table 7.1 - Summary of Similarities and Differences between Cambridge and Oxford Regions

similarities	differences
historic universities	powerful local identity in Cambridge, less so in Oxford
Similar size populations	science parks - two decades apart
Western Crescent	airport access - Heathrow v Stansted
planning restraints	industrial history and manufacturing tradition
equal distance to London	more high tech firms in Cambridge but higher share of larger firms in Oxford
diversity of high-tech activity	universities science v arts ratios favours Cambridge
	stronger formal innovation support in Oxford
	sectoral specialisations different.
	Cambridge greater range of service support

Table 7.2 - Achieved Sample and Response Rates

	Cambridge		Oxford	
Total contacted	130		75	
Achieved sample:		response rate		response rate
Services <100 employees	22	42	17	71
Services > 100 employees	7	65	6	54
Manufacturing < 100 employees	16	33	22	65
Manufacturing > 100 employees	5	29	6	55
Total interviewed	50		50	
response rate	38		66	

7.5   MECHANISMS OF INTERACTION

This section examines evidence for a coherent regional innovation system as indicated by interactions between firms and universities, and between local firms themselves, in terms of various types of links, staff recruitment, social interaction and spin-outs. The comparison of both local and non-local linkages provides an indication of the embeddedness of firms within their regions, compared to their participation in non-local national and international networks (Archibugi and Michie 1996, Keeble et al 1997).

Inter-Firm Activity

The survey showed a much higher incidence of inter-firm linkages in the Cambridge than in the Oxford region. Just over three-quarters of the 50 firms in Cambridge claimed to have close links with other local firms, compared with slightly less than half of the firms in Oxfordshire. The most frequent inter-firm linkages are to be found in the manufacturing sector in Cambridgeshire, where 80% (17) of firms have links with other firms, and the least frequent is for firms in the service sector in Oxfordshire were 39% (9) have close inter-firm links.

On the other hand, firms' perceptions of the importance of such local links are greater in Oxford than in Cambridge (Table 7-4) in terms of the proportions rating particular links as especially important to them (5 or 5 on a scale of 1 to 5). On this basis, a higher proportion of firms with links regard them as important in Oxford (87%) than in Cambridge (66%). Most highly-rate local links are with suppliers and contractors and with service providers. Research collaboration with other firms rated more highly in Oxford (39%) than Cambridge (11%). There were very few horizontal links, that is with firms in the same line of business, in either location.

Table 7.3 - Close Inter-Firm Links between Small High-Technology Firms by Sector (Number and %)

Inter-firm links in	High-Technology Manufacturing	High-Technology Services
Cambridge	17 (34)[1]	21 (42)
Oxford	14 (28)	9 (18)
Number of firms in Survey of which		
- Cambridge	21	29
- Oxford	27	23

A related question asked about the importance of proximity in the establishment and maintenance of links. A greater proportion of Oxford high-tech firms (83%) than Cambridge firms (61%) regard geographical proximity as important to such links.

Proximity was particularly important for suppliers/subcontractors and providers of services rather than customers in both locations.

Table 7.4 - The Nature of Local Inter-Firm Links (Number and Percentage of Firms with Links)

Type of Link	Cambridge	Oxford
customers	8 (21)[1]	9 (39)
suppliers/subcontractors	17 (45)	13 (57)
Firms providing services	12 (32)	9 (39)
research collaborators	4 (11)	9 (39)
Firms in your line of business	4 (11)	3 (13)
Others	1 (3)	0 (0)
Total	25 (66)	20 (87)
[1] percentage ranking 4 or 5 of all firms with links in each local area		

University Linkages

The survey showed that there is a strong local clustering of university-firm linkages in both Oxford and Cambridge (Table 7-5). Cambridge firms had slightly more research links with universities in general since formation (84%) than Oxfordshire (75%). In both places exactly half had links with local universities or government laboratories. The level of local interaction is the same as that found by Foley and Watts (1996) in the mature industrial region of Sheffield. The major difference between the two samples was that Oxford firms are more likely to have formal links, whereas Cambridge firms are more likely to have informal links. This may reflect the difference in institutional formality of Oxford compared to Cambridge. It may also be a temporal phenomenon in that social interaction alongside formal linkages has existed for longer in Cambridge than Oxford and the University there is more embedded in the local social milieu.

The most frequent form of university or government research laboratory interaction in both places is through collaborative research projects (Table 7-6). Oxford firms have much more contact with local government laboratories, of which there are of course more in Oxfordshire than in Cambridge. Other ways in which the level of interaction differs are in the kinds of interaction and the use of formal procedures. For example, nearly twice as many Oxford firms have academics on boards of companies, and four times as many have licensed or patented university inventions. Indeed with the exception of research consortia and clubs, Oxford firms outscore Cambridge firms on every measure.

Table 7.5 - Forms of Interaction between Oxfordshire and Cambridgeshire Firms and Universities

Kinds of linkages	Oxford	Cambridge
Formal	37 (97)[1]	28 (66)
Informal	23 (60)	39 (93)
Both	22 (58)	25 (83)
Total	38 (100)	42 (100)
[1] percentage of firms with close university links		

Table 7.6 - Forms of Interaction between Oxfordshire and Cambridgeshire Firms and their Respective Universities, and with other Universities

	Oxford University Number (% of sample)	Other Universities Number (% of sample)	Cambridge University Number (% of sample)	Other Universities Number (% of sample)
Academics on board	11 (22)	5 (10)	6 (12)	1 (2)
collaborative projects with universities	16 (32)	24 (48)	14 (28)	18 (36)
collaborative projects with government research labs.	12 (24)	13 (26)	3 (6)	7 (14)
part-time secondment by academics	9 (18)	11 (22)	7 (14)	8 (16)
research consortia or clubs	4 (8)	12 (24)	5 (10)	8 (16)
university staff acting as consultants	15 (30)	15 (30)	12 (24)	13 (26)
licensing or patenting university inventions	10 (20)	9 (18)	2 (4)	5 (10)
training programmes run by the university	6 (12)	6 (12)	2 (4)	3 (6)
Other	6 (12)	7 (14)	4 (8)	2 (4)
Total	29 (58)	29 (58)	19 (38)	24 (48)

Source: CBR Survey

While in Cambridge there is now no consistent sectoral pattern of linkages, in Oxford there are distinct differences. Some sectors such as electronics and motor racing have no links, but in others such as instrumentation, computer software and biosciences links are strong. Previous research showed that the factors most closely associated with linkages were either that the founder(s) had graduated from Oxford University, or had been or were employed there (Lawton Smith 1990). In Cambridge, historically it is clear that there are important links between local firms and the University, particularly as regards the growth of the computer hardware sector in the early 1970s and 1980s. This reflected the expertise of the University's computer laboratory together with the CAD Centre. In addition, scientific instruments firms have historically been closely linked to the University. In the 1990s, however, links are much less clear cut with little or no direct connection between the University and the growth of the telecommunications micro-cluster nor with the development of the computer software and services sector, much of which is linked to the growth of the internet and related applications. Overall links are more general and diffused rather than specific to particular sectors. However, the biotechnology cluster which has developed in the 1990s, as in Oxford, is in part linked to the activities of the University and the Cambridge Medical Research Council scientists and Cambridge's international reputation in such fields as monoclonal antibody research.

Table 7.7 - Opportunities to Mix Informally with Managers or Professionals from Local Universities, National Laboratories and other Local Firms (Number and %)[1]

	Never	Occasionally	Frequently
Cambridge University	23 (46)[1]	21 (42)	6 (12)
Oxford University	18 (36)	20 (40)	12 (24)
National Labs Cambridge	37 (74)	12 (24)	1 (2)
National Labs Oxford	23 (46)	17 (34)	10 (20)
Other local firms Cambridge	11 (22)	26 (52)	13 (26)
Other Local Firms Oxford	13 (26)	29 (58)	8 (16)

Social Interaction

A key indicator of the cultural and social cohesion which underpins and sustains regional innovation systems is the degree to which personnel from the different constituencies (firms, universities, national laboratories) interact outside the work environment.

In both regions, over half the sample have either occasional or frequent interaction with the local university, but there is a much higher level of informal contact with national laboratories in Oxford, and there is a similar level of either occasional or frequent interaction with other local firms. The most surprising statistic here is that twice as many of the sample mix frequently with Oxford University than Cambridge University, although the level of occasional contact is about the same. The frequency of frequent inter-firm contact is appreciably higher in Cambridge than in Oxford. The low level of contact with the national laboratories operating in Cambridge is consistent with the smaller population and that many of them have become commercial organisations. Therefore formal and informal contact would be counted in different categories.

These findings suggest that social networks are expanding in Oxford. Part of the success of the 'Cambridge Phenomenon' has been explained by the level of local social interaction (Segal Quince, 1985). Elsewhere Garnsey and Lawton Smith (1997) have argued that the nature of interaction in the university and city of Cambridge encouraged innovation. Social networks were interlocking and threw together people with specialist and diverse expertise, who together could spot new opportunities. The city was small enough for face-to face interaction and free of the inconveniences of large city living. The small size of the city and the collegiate structure of the university brought together people of diverse backgrounds. If technology transfer is about the fusion and diffusion of knowledge, this was greatly facilitated by the nature of interaction in Cambridge.

Labour Markets

A key mechanism for the development of social glue and for technology transfer is the movement of people between firms and between firms and research centres such as universities and national laboratories (Lawton Smith 1997). The level of local interaction is indicated here by local recruitment patterns.

A slightly higher proportion of firms in Oxford (18%) than in Cambridge (14%) recruited research staff from their respective universities (Table 7-8). The most important difference was that Cambridge firms more frequently recruited both research and managerial staff from other local firms and organisations. In both places recruitment of research staff was twice as high from non-local – and often overseas – universities and firms/organisations. Management staff were overwhelmingly recruited from outside the region.

The survey also reveals (Table 7-9) that far more Oxford firms (36) have an explicit policy to recruit locally than is the case in Cambridge (24). In Cambridge, local recruitment is more often justified in terms of region-specific factors such as high quality or having appropriate skills. Relocation costs were a slightly higher factor in Oxford than in Cambridge.

Table 7.8 - Percentage of Firms Reporting at Least One of their Last Three Research/Management Staff from Local and Non-local Sources

Sources of recruitment	Cambridge		Oxford	
	research staff	management staff	research staff	management staff
Local university	14	4	18	4
Other local region firms/organisations	26	22	14	18
Other universities	28	8	44	4
Other firms/organisations	34	48	36	50

Table 7.9 - Local Recruitment Patterns

	Cambridge	Oxford
Firms having a conscious policy to recruit locally	24 (48)	36 (72)

main reasons for pursuing local recruitment:

to avoid relocation costs or related problems	12 (50)	21 (58)
specific reasons e.g. high quality/relevance, or prior knowledge of, skills, initiation of beneficial linkages etc.	22 (91)	22 (60)

The CBR survey also reveals that intra-regional inter-firm mobility of key research staff is high, with frequent resultant formal and informal inter-firm links. However, while Oxford firms more frequently report links because of the movement of personnel between local firms, Cambridge firms clearly value such links more strongly (Table 7-10).

Table 7.10 - The Percentage of Firms that Maintain Links with other Firms because of Personnel that Have Moved between these Firms

	Cambridge	Oxford
Percentage of firms with links to other firms which exist because of personnel that have moved between firms	46	58
Percentage of these firms reporting that such links have been important or crucial to their firm's development	77	56

Spin-Out

Incubation is a well-known mechanism for generating local multiplier effects (Cooper 1971, 2). Spin-outs from other firms and from universities and national laboratories affect recruitment patterns, the creation of a collective image, thereby contributing to the shaping and re-shaping of the local industrial milieu through time (Lawton Smith 1991, 176). In Cambridge 16% of founders of those CBR firms (44) which had begun life as new start-ups came from Cambridge University, a similar proportion to that in the Oxford sample (18%). More generally, in the 1980s, some 55 firms had their origins in Oxford University, formed by academics, technicians and graduates (Lawton Smith 1990). In Oxford five of the CBR sample had their origins in an Oxfordshire national laboratory. However, the government laboratories in Oxfordshire have not been a major source of new firms. In the 1980s only 8 firms had been established by current and former employees (Lawton Smith 1990). The recent data gathering exercise revealed a few more, but less than ten in total. Oxford University itself has engaged in the spin-out process through ISIS Innovation. The most important is the Oxford Molecular Group formed in 1989, which was floated on the stock market in 1994. It has also traded intellectual property for a share in the business as in the case of Oxford GlycoSciences and directly invested in companies, such as Oxford Asymmetry, founded in 1991, which makes chemical compounds that pharmaceuticals companies need to develop new drugs (*Observer* 27 April 1997). In Cambridge, estimates suggest that perhaps 4-6 Cambridge University spin-outs a year since 1980, including biotechnology firms such as Cambridge Pharmaceuticals.

Other evidence suggests that important vehicles for the dissemination of technology and innovation in the Cambridge area are the technical consultancies that are closely linked with the University. Thus Cambridge Consultants (now a subsidiary of A.D. Little) was established in 1966 as a spin out from the University and in turn has spun out PA Technology which spawned Scientific Generics, Technology Partnership and Symbionics. Between them these consultancies employ 1200 highly qualified research and technical staff in the local area and themselves are the source of a number of new technology based firms. It is the view of Walter Herriot (St John's Innovation Centre) that these and other incubators such as Acorn and the CAD Centre are now the most important engines of technology dissemination and transfer in the local area through the formation of and support for new businesses (Moore 1996,4)

The CBR survey supports the view that spin-outs from local firms are now a major source of new enterprise creation and high-technology growth in both regions, but perhaps especially in Cambridge. Just under half of firms in Oxford and Cambridge (Table 7-11) reported that personnel had left to set up another firm. All of the Cambridge start-ups had remained in the Cambridge area. A higher proportion of firms in Oxford had diversified into markets different to that of their parent company. Some have chosen to concentrate on complementary activities. The most interesting difference is that many more of the Cambridge firms report continuing both formal and informal linkages, suggesting active processes of synergy and information exchange.

Table 7.11 - Firms Reporting that at Least One Person Had Left to Set Up their Own Firm, the Number (%) if these Based Inside that Local Region and the Nature of Any Continuing Links Involved

	Cambridge	Oxford
firms reporting that at least one person had left to set up their own firm	24 (48)	22 (44)
New start-ups based in the local region	24 (100)	18 (82)
Nature of local start-up		
the same or similar line of business	16 (67)	6 (33)
complementary (subcontracting, freelancing, servicing etc.,)	4 (16)	4 (22)
unrelated and other	4 (16)	8 (44)
Nature of continuing links with people involved		
formal links	15 (63)	8 (44)
informal links	18 (75)	8 (44)
both formal and informal	15 (63)	3 (16)
neither formal links or informal links	6 (25)	5 (16)

7.6   CONCLUSIONS

The two regions represent examples of how technology-focussed regional economies have developed and changed of Oxford and Cambridge over the last thirty years. The description of the historical background to developments indicated that despite similarities, there are many differences in the nature and operation of the two regional systems of technological innovation and high-technology firm growth. It showed how the inter-dependence of local and non-local geo-historical factors have produced different types of innovators, demands for resources, institutional responses, mechanisms for creating social glue and consequently regional outcomes. The presence of historic, nationally and internationally important universities is only a small part of the story of how the two differ in their evolutionary paths. The actions of entrepreneurs, key firms and philanthropic individuals have helped to shape local cultural identities. It also shows how non-local factors such as government decisions to establish particular institutions and then change their status has had important local consequences. Likewise the development of sectors such as bioscience, and in Oxford the cryogenics sector, which grew strongly in the 1980s, are responses to general processes of technical change and sector development world-wide, and the learning/technology transfer processes associated with internationally important centres of expertise, underpinned by national government funding through research council support and other national funding mechanisms.

Against those conditions, the Oxford and Cambridge region clusters differ in interesting ways in the nature and extent of local inter-firm, and firm-university and national laboratories linkages. What we perhaps see here is a particular stage in the evolution of the regions, with Cambridge being more advanced in its evolutionary trajectory than Oxford. It is likely that their progress will remain different. There is evidence of networked economies in both regions but in the case of Cambridge, with its longer history of development and greater volume of firms, the model is much more like an 'innovative milieu' with the University originally playing an instrumental role in development. The division of labour is 'flatter'. Oxford perhaps accords more to the Marshallian Industrial District type with the division of labour more vertically constructed. The university has generally played a role by default rather than design.

This proposition is supported by the evidence from the recent study. High-technology SMEs in Cambridge show a greater propensity to form close inter-firm links than do high-technology SMEs in Oxford. This may well reflect the much greater number of SMEs in Cambridge than in Oxford and therefore a greater opportunity for SMEs in Cambridge to identify suitable local links. However, a greater proportion of the fewer SMEs in Oxford which do possess links regarded their local links as especially important. This shows up in several indicators including the greater importance attached by Oxford firms to geographical proximity in their local links.

The level of interaction with the local universities is similar but differs distinctly in character, being much more likely to be formal in Oxford. This indicates a much more structured set of relationships in Oxford than in Cambridge. The greater density of government laboratories in Oxford is indicated by the level of both formal collaborations and informal contact.

The labour markets also appear to have differing characteristics. The evidence suggests that there is much more of a local firm-focussed scientific labour market in Cambridge based on recruitment within the high-tech sector than in Oxford, where slightly more have recruited from the local university. While more of the Oxford firms have a conscious policy to recruit, more of the Cambridge firms gave local-specific reasons for local recruitment.

Spin-out of new high-technology enterprises from both the university and from firms is an important mechanism generating new firms and sets of relationships. In Cambridge the extent of continuing linkages suggests a much more socially well-adjusted system than in Oxford. However, a basic difference which will have long term effects on the evolution of the two areas is that of the service versus manufacturing orientation. Oxford's bigger manufacturing firms are more embedded in the local economy and over time, assuming that they continue to adapt to changing realities, are likely to have greater long-term benefits for the economy as a whole. In Cambridge, while the interaction of specialist service providers to the mainly service firms will sustain a particular set of occupations and activities, this may not in the long term provide for a more integrated economy which can avoid wider pressures or cope with them in a superior manner.

The policy implications are that bringing together firms and universities and national laboratories is easier in some local environments than others. This is because of the potential to find matches of interests between firms and other local organisations. In Oxfordshire, The Oxford Trust has actively pursued a policy of sectoral initiatives such as motor racing and biotech rather than attempting across the board help for high-tech firms. This appears to have had some success, and in the case of motor racing, has led to an engineering degree course established at Oxford Brookes University. In Cambridge, similar initiatives could be developed for the biotechnology sector, but whether they would be appropriate for the software sector depends on the commonality of interests in that group of firms. In general then, policy needs to be based on a clear understanding of what kind of system is operating locally and its evolutionary stage, and where the local system fits into the broader context of national and international conditions. As has been shown here, demand conditions are quite different in Oxford and Cambridge.

To conclude, various mechanisms identified have been shown to produce somewhat different modes of learning in specific spatial contexts. The know-how gained informally from non-traded information exchanges with other local actors in the economy is itself a local subsidy. While analyses based on systems thinking help to identify where to look for information inter-dependencies in order to understand the trajectories of regions, it is also necessary to take account of how local, national and international processes work upwards to affect the way that the 'global is constructed' (Lovering 1995, 125).

ENDNOTE

1.   Pearson became the third industrial chaplain in Cambridge in November 1993. The first was
        appointed in 1980.

REFERENCES

Amin, A. (1993). "The Globalization of the Economy: An Erosion of Regional Networks?" In G.Grabher (ed) *The Embedded Firm,* Routledge, London/New York.

Amin, A. and Thrift, N. (1995). 'Globalisation, Institutional "Thickness" and the Local Economy', in Healey, S. Camerson, S. Davoudi, S.Graham and A. Madani-Pour (eds) *Managing Cities: The New Urban Context* Wiley, Chichester.

Amin, A. and Thrift, N. (1992). 'Neo-Marshallian Nodes in Global Networks' *International Journal of Urban and Regional Research* 16, pp 571-587.

Asheim, B. and Dunford, M. (1997). "Regional Futures", *Regional Studies 31* (5) pp 445-456.

Asheim, B.T (1996). "Industrial Districts as 'Learning Regions': A Condition for Prosperity?" *European Planning Studies 4* (4), pp 379-400.

European Commission (1994a). *The Community Innovation Survey: Status and Perspectives.*

European Commission (1994b). *Evidence from Europe and North America on "Intangible" Factors Behind Growth, Competitiveness and Jobs,* Industry Panorama, August.

European Commission (1994c) *The European Report on Science and Technology Indicators* EUR 15897, Luxembourg

Foley, P. and Watts, D. (1996). "Production Site R&D in a Mature Region", *Tijdschift voor Economische en Sociale Geografie* 87 No. 2 136-145

Garnsey, E. and Lawton Smith, H. (1997). 'The Science-Based Industrial Complex: Diverse Paths; Common Processes', *Research Papers in Management Studies*, WP 12/97 Judge Institute of Management Studies, University of Cambridge

Garnsey, E., Cannon-Brookes, A, (1993). "The Cambridge Phenomenon Revisited; Aggregate Change Among Cambridge High Technology Companies since 1985", in *Entrepreneurship and Regional Development*, vol. 5 no 1.

Hudson, R. (1997). "Spatial Futures: Industrial Restructuring New High Volume Production Concepts and Spatial Strategies in the New Europe", *Regional Studies 31* (5) pp 467-478.

Keeble, D., Lawson, C., Lawton Smith, H., Moore, B. and Wilkinson, F. (1997). *Internationalisation Processes, Networking and Local Embeddedness in Technology-Intensive Small Firms*, ESRC Centre for Business Research, University of Cambridge, Working Paper 53.

Keeble, D. (1994). "Regional Influences and Policy in New Technology-Based Firm Creation and Growth", in Oakey, R. (ed), *High Tech Industry in the UK,* London: Paul Chapman.

Keeble, D. and Kelly, T. (1986). "New Firms and High Technology Industry in the United Kingdom: The Case of Computer Electronics", in D. Keeble and K. Wever (eds), *New Firms and Regional Development in Europe,* London: Croom Helm.

Lawson, C. (1997). *Territorial Clustering and High-Technology Innovation: From Industrial Districts to Innovative Milieux*, ESRC Centre for Business Research, University of Cambridge, Working Paper 54.

Lawton-Smith, H. (1990). *The Location and Development of Advanced Technology Industry in Oxfordshire in the Context of the Research Environment,* Unpublished DPhil thesis, University of Oxford.

Lawton-Smith, H. (1990). *The Location of Innovative Industry: The Case of Advanced Technology Industry in Oxfordshire.* School of Geography, Oxford University. Oxford.

Lovering, J. (1995). "Creating Discourses Rather than Jobs: the Crisis in the Cities and the Transition Fantasies of Intellectuals and Policy Makers" in P. Healey, S. Camerson, S. Davoudi, S.Graham and A. Madani-Pour (eds) *Managing Cities: The New Urban Context,* Wiley, Chichester.

Markusen, A. (1996). "Sticky Places in Slippery Spaces: A Typology of Industrial Districts", *Economic Geography.*

Mihell, D. (1996). "Biotechnology Sector in Oxfordshire", Oxford Innovation Ltd, September, Oxford.

Moore, J.F. (1996). *The Death of Competition - Leadership and Strategy in the Age of Business Ecosystems,* New York: HarperCollins.

Morgan, K. (1997). "The Learning Region: Institutions, Innovation and Regional Renewal", *Regional Studies 31* (5).

Reid, S. and Garnsey, E. (1996). "Incubator Centres and Success in High-Technology Firms: the Work of St John's Innovation Centre", Judge Institute of Management Studies, University of Cambridge.

Saxenian, A. (1988). 'The Cheshire Cat's Grin: Innovation and Regional Development in England', *Technology Review* February/March, 67-75

Segal, Q. (1985). *The Cambridge Phenomenon: The Growth of High Technology Industry in a University Town,* Segal Quince and Partners, Cambridge.

Steed, G. and De Genova, D. (1983). 'Ottawa's Technology-Oriented Complex' *Canadian Geographer* XXVII,3,1993 pp 263-277.

Storper, M. (1995a). "The Resurgence of Regional Economies, Ten Years Later," *European Urban and Regional Studies* 2(3).

Storper, M. (1995b). "Territories, Flows and Hierarchies in the Global Economy," *The Swiss Review of International Economic Relations (Aussenwirtschaft)*. (June).

Storper, M. (1993). "Regional Worlds of Production: Learning and Innovation in the Technology Districts of France, Italy and the USA" *Regional Studies*, 27, 5.

Todtling, F. (1994). 'The Uneven Landscape of Innovation Poles: Local Embeddedness and Global Networks' in A. Amin and N. Thrift (eds), *Globalization, Institutions and Regional Development in Europe*, Oxford, Oxford University Press.

Willis, T., Kingham, D., Stafford, J. (1996). 'Oxfordshire's Motor sport Industry: Building on Local strengths' A Report commissioned by Heart of England Training and Enterprise Council Oxford Innovation February 1996.

8 TELECOMS IN NEW JERSEY: SPATIAL DETERMINANTS OF SECTORAL INVESTMENTS

Cliff Wymbs
Rutgers University

8.1   INTRODUCTION

Telecommunications is a critical sector which dramatically shapes New Jersey's role in a globalizing economy from both a service and a technology perspective. New Jersey has universal telephone service, more than 500 service providers and the lowest telephone rate in the nation and in virtually all developed countries (Perone, 1996). On the technology side, (State, NJ, 1996) reported that New Jersey has more engineers and scientists per capita than any other state in the United States, more software engineers than any state except California, and more privately funded basic and applied research in telecommunication related areas, e.g., software, microprocessors, voice, video and data compression, etc., than any other state and all but a few nations in the world. New Jersey also has the highest density of cable TV users and is a global center for the research in High Definition TeleVision (HDTV) and the establishment of its standards (Sherman, 1995). In addition, New Jersey is the home of the second largest provider of paging and personnel communication services in the United States and a lead player in the Caribbean and Canadian markets (MobileMedia, PR, 1996).

Telecommunications as an enabling technology reduces the space and time constraints associated with modern business. Multinational Enterprises (MNEs) increasingly are considering the locational parameters of telecommunication features, price and quality in making new investment decisions. Firms in New Jersey are designing and engineering these global offerings through cross-border alliances which, in effect, place the power of information with the MNEs who, in turn, are using it to extract better terms from local governments. Not surprisingly, the areas of the world which foster competition also have the most advanced telecommunication services. New Jersey is one of those markets.

The first section of the paper discusses the Macro and Micro Determinants of Telecommunications Policy to highlight the complexity of firm, government and industry dynamics. Section (ii) entitled Industry Evolution provides the analytical framework necessary for a State specific analysis and this is followed by a Key Enablers section which addresses the State's locational characteristics that aid the telecommunications sector. Section (iv) - Information Corridors - argues that all areas of New Jersey do not benefit equally from telecommunications activity, but rather, corridors are created around major transportation arteries and major areas of investment in Central New Jersey. The intersection of spatial variables, pro-competitive policies, firm initiatives and demanding customers creates an environment conducive to the creation of a Knowledge Hub along these information corridors; this issue is discussed in Section (v). As with most centers of innovative activity, New Jersey both attracts foreign direct investment (FDI) and is a major provider of telecommunication investment around the world; these issues are analyzed in Section (vi). Today, telecommunications as a global industry is rapidly changing from a government-owned/ private monopoly to a competitive industry. These political and institutional changes coupled with rapid technology development are causing (vii) Global Fusion of many different industries, i.e., voice, data and media, thereby fundamentally changing the competitive parameters of the industry. Section (viii) of the paper - Implications for New Jersey in The 21st Century, speculates how New Jersey's telecommunications industry and its users will both affect and be affected by an increasingly global economy.

(i)   Macro and Micro Determinants of Telecommunications Policy

Before analyzing the current role telecommunications plays in New Jersey's globalizing economy, a discussion of the historical importance of government/industry relations in this sector is instructive. Ettlinger (1992) stated that economic development increasingly is a local, bottom up phenomenon, but he highlights the importance of national context to define the parameters of the production system in which firms and workers operate. Both play important roles in the study of telecommunications in New Jersey. For more than one hundred years, "telecommunications [has] occupie[d] a very special place in the spectrum of government-industry relations. . .it has enjoyed privileged treatment from the state in practically every capitalist country, with the result that the Telecom 'market' has been highly insulated from competitive pressures" (Cawson, 1990, p77). Service was provided by monopolies and their requisite equipment was provided by closely allied manufacturers (AT&T in the US, NTT/NEC in Japan, Deutsche Telekom/Siemens in Germany, France Telecom/Alcatel in France, and Bell Canada/Northern Telecom in Canada). Equipment side entry was restricted by large economies of scale of production and the need for close R&D collaboration between the service providers and the equipment suppliers due to nationally specific standards and network features (Cawson, 1990). Justifications for the resulting "Telecom Club" are based on economic theory of natural monopoly and socio-political mandates of uniform service irrespective of cost requiring cross subsidization between high and low density routes.

Increases in globalization and the spatial distribution of economic activity are placing increased pressure on inefficient, cross-subsidizing, nationally regulated, government controlled monopoly providers of telecommunication service. The following factors all positively influence privatization, liberalization and general introduction of competition to the sector. Technology changes have increased scale economies of manufacturers and have both increased and decreased scale economies of service providers. Business customers are demanding advanced telecommunications services and have the ability to spatially adjust their operations and investment accordingly. Supra-governmental organizations desire to promote free trade. This competitive renaissance affords a few large equipment manufacturers and service providers a vast global economic opportunity to be pursued via foreign direct investment (FDI) and/or with alliance partners.

At the same time, the very competitive forces that are creating the foregoing opportunities have caused the second largest telecommunications company in the world to break apart. After years of denial, AT&T's management came to the realization that AT&T could not compete directly with Local Exchange Carriers (LECs) - Regional Bell Operating Companies (RBOCs) and Independent Telecommunication Companies such as GTE - in the domestic market and PT&Ts (post, telephone, telegraph administrations) in foreign markets and hope to have another division sell equipment to the same PT&Ts and LECs. Previously AT&T had divested several New Jersey-based companies including New Jersey Bell (now called Bell Atlantic of New Jersey) in 1984, an employer of 13,000 residents; BELLCORE in 1984, part of Bell Labs which employs 6,000 scientists in support of the RBOCs (formerly part of AT&T); UNIX Systems Laboratories in 1993, the provider of the popular UNIX operating system; and AT&T Capital Corp. in 1996, the largest lessor of AT&T products. In contrast to most theories of regional districts and agglomeration which have firms concentrating to operate more efficiently, this breakup is driven by both core competencies and government initiatives.

8.2   INDUSTRY EVOLUTION

In 1994 Rugman and D'Cruz presented a theory of business networks where suppliers, customers, competitors and members of the non-business infrastructure are all linked together through the common global strategy of the flagship firm (a lead multinational enterprise). In their paper, they observed that flagship firms were finding it necessary to divest some of their business system activities, either for strategic reasons or because the internalization of the activity was more costly than the sum of external and transaction costs. An earlier analysis by Porter (1980), though somewhat more general, highlighted the need for a firm to focus on one of three generic strategies: Cost leadership, differentiation, or cost/differentiation focus. In the mid-1990s, with pending telecommunication re-regulation in the United States and international liberalization, AT&T could not be the global flagship firm nor execute a successful differentiation strategy because of a severe internal identity conflict. One of AT&T's business units, Communications Services, was increasingly coming in direct competition with the LECs and PTTs around the world while the main customers of another of AT&T's business units, Network Systems (today Lucent Technologies), were those same RBOCs and PT&Ts.

AT&T has struggled with this problem ever since divestiture (the split-up of AT&T from the RBOCs mandated by the United States Department of Justice) in 1984. Jim Olson, AT&T's new Chairman in 1986, identified managing RBOCs' relations as a key strategic initiative. This became an increasing irritant during Mr. Allen's decade of leadership. In September 1995, AT&T chose to separate its operations into three lines: computers (NCR), services (AT&T) and equipment (Lucent Technologies).

In Enright (1992) the author concludes that regional clusters differ from the business networks in that business networks involve communication and cooperation among firms that need not be located in close physical proximity. However, what is unique for the New Jersey business network is that all four partners have significant spatial presence in New Jersey's Telecommunications hub. In fact, a large part of the regional cluster and business network are coterminous.

In a recently published paper Markusen and Gray (1996) have stated that one variant of the industrial district structure is the hub-and-spoke form, where an industry and its suppliers cluster around one or several core firms. They further argue that the strength of the regional economy will remain with the hub organization's position in the national and international markets. The split-up of AT&T and Lucent Technologies will replace vertical integration, the traditional form of a hub-and-spoke industrial structure, with embedded inter-firm relationships based on long term contracts, sharing of proprietary information and strategic alliances. Lucent Technologies' global manufacturing operations can be classified as spoke activities while its corporate planning and R&D activities are hub activities that will continue to reside in New Jersey. Similarly, AT&T's R&D and planning functions will remain at its central New Jersey hub. Historically AT&T's international spokes have been handled by foreign correspondents. The current trend, however, is to obtain greater control over these relationships. The creation of alliances with foreign PTTs is becoming the governance modality of choice.

Using the Rugman and D'Cruz (1994) business network approach, AT&T could be viewed as a flagship firm; Lucent and Bell Atlantic will remain key suppliers of equipment and access respectively. Large business and residence users will be key customers; MCI, Sprint and the RBOCs will be key competitors; and the government and the regulators will be the major non-business players.

Because telecommunications is a network service, the flagship firm may, in fact, be a constellation of firms that develop interactive strategies (Gomes-Casseres, 1996). The brightest of the associated star firms is determined by the power it exerts over others in the constellation. In the AT&T alliance case, World Partners, AT&T is clearly the largest and most powerful firm. It appears that asymmetric power in a constellation will result in the creation of a new shared equity alliance like World Partners. In other flagship firm constellations where power is spread more evenly, a joint venture with relatively equal cross equity ownership of the parent companies is more likely. This has occurred in the MCI/British Telecom arrangement called Concert.

The merger of Bell Atlantic headquartered in Philadelphia and NYNEX headquartered in New York City will solidify New Jersey as the center of gravity for a regional business network. In this arrangement, Lucent Technologies will be in a better position to share proprietary information with the newly merged RBOCs. Of course, Northern Telecom will also remain a key RBOCs equipment supplier. AT&T and other long distance providers will compete with Bell Atlantic in both the local and long distance markets. Similar to AT&T's business network, large businesses and residence users will be key customers and government and regulators will be players in the non-business infrastructure.

In addition to the service provision networks, a separate manufacturing network is likely to emerge with Lucent Technologies as the flagship company. Lucent's key customers will be AT&T, the RBOCs and local telephone companies around the world. Lucent's competitors will be other global switch manufacturers, i.e., Nippon Electric Company (NEC), Siemens, Alcatel, Ericsson, and Northern Telecom. Many of Lucent's suppliers are wholly owned subsidiaries and/or recently divested companies, e.g., power systems. Lucent Technologies' manufacturing facilities in Research Park Triangle, North Carolina and Massachusetts will likely remain Satellite Platforms and become increasingly vulnerable to global pressures to relocate to areas of rapidly expanding demand. (Markusen and Gray, 1996).

As a result of the RBOCs' merger and AT&T's separation, two separate, but related hubs, are likely to be formed in New Jersey: A "Service Hub," which will be dominated by AT&T and possibly Bell Atlantic/NYNEX, and an "Equipment Hub" which will be controlled by Lucent Technologies. Competition for voice services within New Jersey will become fierce between the two indigenous service firms. Because of its historical starting point, Bell Atlantic/NYNEX's employment is expected to be initially more closely tied to regional performance than AT&T's employment. With regard to the "Equipment Hub," Lucent Technologies' performance is directly related to extra-regional (national and international) demands. A key objective of Lucent Technologies is to penetrate the rapidly expanding infrastructure markets, particularly China and India. As a market entry condition, these governments require some indigenous research and production. Over the next few years, this will likely result in reduced United States domestic employment by Lucent.

It is possible that some European equipment manufacturers may choose to locate a listening center in New Jersey to take advantage of this skilled workforce and knowledge spillover, much like the Japanese have already done. However, the largest concentration of these global manufacturers' United States presence will remain in Research Triangle, North Carolina and Richardson, Texas. (More on these locations in Section V.) Markusen and Gray (1996) stated that the hub-and-spoke variant may render a region vulnerable to cyclical and sector decline, the potential for the hub organization to externalize and abandon commitment to the local economy and/or crowding out of noncompeting, newer sectors. It would be reasonable to argue that labor turnover is low and workers are committed to the dominant firm, which offers better pay and more stable employment. This makes New Jersey employment and economy potentially vulnerable to firm-specific actions such as downsizing and global downturns in the economy and politically motivated procurement decisions.

8.3   KEY ENABLERS

Demographic

New Jersey's population and commercial density facilitates telecommunications engineering. New Jersey's public policy encourages competition and New Jersey's economy is significantly expanded by the telecommunications industry. New Jersey has several demographic characteristics that have aided the efficient provision of information; as cited by Deloitte & Touche (1991). They included:
*   The high density of the State is clearly an advantage in the deployment of a telecommunications infrastructure. The population density of New Jersey is 1,034 persons per square mile, as compared to an average of 70 persons per square mile for the total US.
*   There are 531 individuals employed per square mile in New Jersey, as compared to average of 33 individuals per square mile for the nation.
*   There is an average of 28 business firms per square mile in New Jersey, as compared with an average of less than two business firms per square mile in the total US.

Pro-Competitive Policy

Since approximately 96% of New Jersey residences have basic telephone service, the traditional goal of universal service has effectively been achieved. As New Jersey moves closer to an information/service-based economy, the non-fiber local exchange carrier network still constrains users' ability to participate in the "Information Age." A key telecommunication policy issue is how regulators should encourage LECs to accelerate the deployment of advanced telecommunications technology to support broad band availability of higher bandwidth services.

In 1990, the State Board of Public Utilities commissioned a study entitled "The New Jersey Telecommunications Infrastructure Study" aimed at identifying the relationship between telecommunications and the New Jersey economy. The main conclusions of the study were that "there is a direct causal link between an advanced telecommunications infrastructure and economic development" (Salmon, 1994, p9) and that investing in new diverse communication services would benefit all sectors of the economy, producing jobs and thrusting New Jersey into the forefront of the "Information Age." The study also found that the focus of future economic development in the State will be on the service-producing sectors of the economy, such as the finance, real estate and insurance industries. Many other states are also targeting these telecommunication intensive segments. For example, Massachusetts is targeting the information intensive industries of financial services, medical services, technology and education (Weld, 1996). Therefore, it is essential for New Jersey's telecommunication network to be able to support state-of-art applications for these sectors. The study finally suggested that the deployment of advanced telecommunications technology can be significantly accelerated at minimal cost relative to the base of local exchange carrier intrastate revenues.

In response to this study, the State Legislature passed the New Jersey Telecommunications Act of 1992 that eliminated unnecessary regulatory barriers, defined a framework for competitive standards, freed interexchange carriers from pricing regulation, and permitted Local Exchange Carriers to seek alternatives more efficiently focused of regulation, i.e., price caps. The Act was finally approved on May 6, 1993 and was to run through December 31, 1999, with a commitment by New Jersey Bell to complete statewide fiber optic deployment by 2010.

In February 1996, the United States Congress passed legislation to further redefine state and federal regulation of telecommunications, the Telecommunication Act of 1996 (the "Act). The Act permits RBOCs to provide interexchange services outside their home region and, after meeting FCC guidelines, to provide such service within their home region. In exchange for granting the RBOCs an opportunity to participate in the long distance marketplace, the Act preempts state and local government law that prohibits firms' entry into the local telecommunications service market. Negotiations on access and interconnection between local and interexchange carriers still need to be worked out. In this process consultants, lawyers and economists will present the facts to the advantage of their client's economic position. (Virtually every large consulting firm already has a telecommunications practice group in New Jersey.) This pluralist process is well entrenched in American culture and raises many stakeholder perspectives.

New Jersey is well positioned to implement the Federal pro-competitive mandate. On June 19, 1996, the State Board of Public Utilities approved five local access provider companies to offer basic services in competition with Bell Atlantic of New Jersey. By January 1997, New Jersey's 4.7 million telephone customers will be able to choose between competitors, and by May 1997 callers will have to dial a five-digit access code to connect to discount carriers to make local toll calls. Opening the local market to full competition will likely result in significant price reductions.

Economic Impact

Telecommunications as defined by Standard Industrial Classifications (SIC) 4820-Telegraph and other Communications, 4810-Telephone Communications, 4890-Communication Services, and 3660-Communication Equipment, plays a dominant role in New Jersey's economy. Britt (1996) stated that in New Jersey, AT&T is responsible for $6B of economic activity, has 326 facilities, 20 million square feet of office space leased and locations in 19 of the 21 New Jersey counties. In 1995, AT&T employed more than 48,000 employees in New Jersey and paid $3.8 billion in wages ($70,000+/employee). The average tenure of employees is fifteen years. AT&T purchases $2 billion of goods and services each year from 11,300 businesses in New Jersey. In addition, AT&T has more than 17,500 retired employees in New Jersey. Also in 1995, Bell Atlantic was New Jersey's second largest employer with 13,617 people and its wages are estimated to be in excess of $1B.

Because of AT&T's dominant presence in New Jersey, forty-eight telecommunications consulting companies are headquartered there (Duns, 1996). AT&T high tech and software competitors, firms such as Anadigics (which makes integrated circuits for cell phones for Nokia and Ericsson and satellite TV systems for General Instrument and Scientific Atlanta), are growing rapidly and chose New Jersey because of the talented pool of engineers, designers and support personnel located around AT&T's knowledge hub (Perone, 1996). However, the spatial decision by MobileMedia Corporation, the second largest provider of paging and personal communications services in the United States (MobileMedia Corporation, PR, 1996) to locate its headquarters in New Jersey is probably due to customer demographics rather than AT&T's presence.

8.4   DEVELOPMENT OF INFORMATION CORRIDORS

The Communications Act of 1936 clearly defined AT&T's domain as a dominant provider of universal telephone service for the United States market. In the Fordist tradition of the time, AT&T became a fully integrated manufacturer and distributor of telecommunications service. Regulators both at the State and Federal level replaced the market as the creator and enforcer of contracts between AT&T and its consumers. On the provision of services, AT&T embarked on a classic multi-domestic strategy (if one can for the moment assume that each state in the United States represents a foreign country). For most large states, AT&T created separate subsidiaries to deal with the state public utility commissions and created a separate company for the provision of long distance service.

In support of its rapidly growing manufacturing presence in New Jersey, AT&T moved more and more of its research facilities there. In the 1930's, AT&T had more than 13,000 people employed at its Western Electric's Kearny, New Jersey plant making switch boards and copper cable (Brooks, 1976). The Edison Labs which had already invented the carbon button transmitter (the device which eliminated the need to shout into the phone), electricity and a variety of audio/video products were well established and located not far away in Menlo Park. AT&T began radio broadcasting in 1922 and was immediately challenged by GE and RCA. In 1926, AT&T agreed to withdraw from broadcasting and sell its existing stations to RCA (Stone, 1991). Though separated in a commercial sense, AT&T and RCA relied on the same basic technology, the radio spectrum. The David Sarnoff Laboratory was formed in Princeton by RCA in 1942. Its initial focus was on components design until the discovery of the transistor by Bell Labs. RCA subsequently lost its manufacturing competitive advantage to the Japanese and was sold to General Electric, one of its original owners in the 1920s (Brooks, 1976). Also, Princeton University's physics department has contributed to the understanding of radio research as far back as the early 1800's (World Book, 1994).

After World War II, AT&T's regulated monopoly was confronted with a series of attacks on both its vertical and integrated network structure. The Federal Communications Commission (FCC) began handing down decisions that eroded AT&T's boundary through the customer premise (Carterfone decision in 1968 which permitted interconnection of non-AT&T equipment to the network), computers (Computer Inquiry I and II which attempted to separate computer and communication technology), and specialized common carriers (EXECUNET service which permitted

MCI to gain switched access). MCI was challenging AT&T in the courts and in 1974 the United States Department of Justice (DOJ) sued AT&T and charged it with engaging in monopolistic, anticompetitive practices (Cohen, 1992). Unmistakably, a pro-competitive thrust was sweeping the United States telecommunications industry.

Competition meant that AT&T could no longer function solely as an engineering company. This led to the formation of marketing, product management, brand management and advertising departments. To improve internal communications and coordination, in the early 1970's, AT&T relocated the headquarters of its long distance subsidiary, Long Lines, and its associated worldwide network control facility from New York City to Central New Jersey. Other Corporate functions, e.g., target setting, finances, etc., were also relocated from New York to New Jersey; however, day to day operating management remained in the location of each operating company.

Because of locational advantages associated with New York City, advertising, investor relations and public relations were the last major corporate functions to move. With improved information technology, AT&T has demonstrated that the important support industries of advertising and financial services can now be out of New Jersey. In 1995, AT&T spent more than $2B on advertising and over the last few years financed takeovers totaling more than $17B (AT&T Annual Report, 1995). Unlike Enright's (1994) definition of regional clusters as a group of firms in the same industry, or in closely related industries that are in close geographic proximity to each other, government regulation provided AT&T the opportunity to create hierarchical regional clusters for most areas of telecommunications industry, and it located them in Central New Jersey similar to the hub and spoke clustering of Markusen and Gray. New Jersey served as both the product and service hub for most central AT&T activities. The spokes on the service side were seven regional organizations (approximately equal in size and collocated) with the seven RBOCs, while the spokes on the product side corresponded to manufacturing facilities located throughout the United States and the world. Consistent with AT&T's manufacturing approach, Audretsch and Feldman (1994) found that there is some tendency for innovative activity to take place outside the location where the bulk of the production is located. This appears true for AT&T whose main manufacturing operations take place in North Carolina and Massachusetts. However, Lucent Technologies announced at the end of 1996 that it plans to design and build more than three million digital wireless phones in New Jersey (Rosenbush and Marsico, 1996). About 300 jobs will be new to New Jersey, and 150 will be transferred from other Lucent sites. New Jersey has agreed to provide $500,000 to help train the new workers.

The growing information intensity of United States markets, the desire by many businesses and carriers to replace their analog switches with state-of-art digital switches and the desire by developing economies to update their telecommunications, all contributed to the rapid expansion of AT&T. To keep pace, AT&T expanded these clusters and organized by market: one for end users (PBX, telephones, services, computers) and one to provide carrier equipment (switches, transmission equipment, cables). The product management for these areas was clustered along information corridors in central New Jersey. Krugman (1991, p. 57) makes the important point that "States aren't really the geographic units, because of the disparities in population size and lack of concordance between economic markets and political units." This analysis builds upon this notion and focuses on a fifty-mile elliptical shaped area in central New Jersey crisscrossed by five important information corridors: Route 287 - Somerset/Morris County/Middlesex, Route 10 - Morris, Route 78 - Union County, the Garden State Parkway - Holmdel/Monmoth County, and Route 1 - Princeton. The first corridor contains most of the remaining AT&T. The second through fourth corridors are associated with Lucent Technologies, while the fifth corridor is mostly made up of companies with no prior Bell System affiliation.

Along the first corridor, Route 287, are located the de facto corporate headquarters of AT&T and its global network communication headquarters. Between these two facilities is located the corporate headquarters of Bell Atlantic cellular operations. Also, located on Route 287 are the worldwide product management organizations for business equipment and consumer products. Before the creation of Lucent Technologies, AT&T had jointly sold business and consumer products with its long distance service and located facilities near each other. Further down Route 287 in Piscataway is the main facility of BELLCORE, the RBOC cooperative research facility. Along the second corridor, Route 10, is located AT&T Capital Corporation's headquarters, close to the business equipment unit whose products it originally leased. However, today the vast majority of AT&T Capital Corporation leases are not telecommunications related. It too is being divested. Also, located on Route 10 in Whippany is the main Bell Laboratory research facility for wireless services.

Just off the third corridor, Route 78, is located the corporate headquarters of Lucent Technologies. Co-located here is the Murray Hill branch of Bell Laboratories. Worldwide product planning facilities for microelectronics, part of Lucent Technologies, are also nearby along with the worldwide headquarters for UNIX Systems Laboratory Inc., a spinoff company from a Bell Laboratory software product. Along the fourth corridor, the Garden State Parkway, are Bell Laboratories' main transmission research and software facilities, and many Lucent Technologies Corporate functions are located in Holmdel, the largest telecommunications complex in New Jersey. Many related Lucent R&D facilities are located within a few miles of Holmdel.

Along the last corridor, Route 1, is located the RCA David Sarnoff research laboratories. Also located here are research facilities for many Japanese companies, e.g., Toshiba, Matsushita, Sony, NEC, and Hitachi.

A more quantitative way of measuring telecommunication employment density of New Jersey is the location quotient. It is a measure of the total employment for a sector in a particular region, e.g., county or state, divided by the ratio of national employment for that industry to total national employment (Markusen and Gray, 1996). For New Jersey the LQ is approximately 1.6; however, for Somerset County the LQ is almost 8.0. An alternative way of interpreting the LQ is that an employee in Somerset County is eight times as likely to work in the telecommunications sector than an employee in the United States and an employee in Somerset County is almost five times as likely to work in the telecommunications sector than an employee in the State of New Jersey.

Over the last ten years, two other major information corridors have emerged in the United States, Richardson, Texas and Research Triangle, North Carolina. A brief description of each follows. More than 200 companies, mostly high tech computer and telecommunication related, have located in a five square mile area of Richardson,

Texas. In the 1970's, Texas Instruments (TI) and Rockwell had employed more than 20,000 in Richardson and it became known as "The Electronic City" (Robinson, 1996). AT&T divestiture opened the U.S. telecommunications market and international players began searching for entry points. Rockwell and TI had already had experience on the defense side of telecommunications. Northern Telecom bought Danray, a start-up founded by engineers who left TI. Alcatel came when it acquired a major portion of Rockwell's non-defense telecommunications business. MCI came to Richardson because Rockwell was a major supplier of its products. Fujitsu came because it was one of the few international manufacturers to sell to MCI (Robinson, 1996). Key factors positively influencing the Ericsson location choice were availability of land along major transportation arteries, the availability of highly skilled labor, close cooperation between technology industries and the City, outstanding quality of life features and deregulation of the telephone industry (Telecommunications Infrastructure, 1993). Major telecommunication employers are: Alcatel-2,184 people, mostly manufacturing; Ericsson-2,254 people; Fujitsu-900 people mostly manufacturing; MCI-3900 people; Northern Telecom 4,500 people; and Southwestern Bell-450 people(Richardson, 1996). Similar to New Jersey's telecommunication corridor, this corridor is also located along highways, US 75, known as Central Expressway, is the north-south freeway and SH-190, the east-west freeway. The term "telecom corridor" was first used to characterize this region in 1988 by the *Dallas Times Herald*. Today, there is a Telecom Corridor Technology Business Council which actively markets this region.

The other information corridor is Research Triangle, North Carolina. More than one-half of jobs in this area are associated with the manufacturing activities of digital switching, fiber optics and PCS. Major employers include IBM - 18,000, Northern Telecom - 8,500 and AT&T - 6,400. Most MNE telecommunications manufacturers, i.e., Alcatel, Siemens, AT&T, Northern Telecom, Ericsson and Fujitsu, have some presence in North Carolina. Massachusetts has the highest concentration of telecommunications manufacturing employment in the country. The telecommunications industry employs more than 75,000 people, accounting for approximately 17% of the high tech manufacturing jobs in the state (Governor's Council, 1992).

8.5   KNOWLEDGE HUB

The intersection of technology, planning and customers creates an environment conducive to creation of a localized center of innovation. Markusen (1994) has demonstrated that as high tech industries expand, investments are drawn to nationally dominant cities where skilled labor, information and business services are most available. Zucker, Darby, and Brewer (1994) have shown that organizational boundaries serve as informational envelopes within which valuable information characterized by natural excludability is much more likely to be used within rather than outside the organization. Audretsch and Feldman (1994) found that high tech, innovative activity tends to cluster around the three sectors: electronic components, switching apparatus and telephones.

The above literature provides the theoretical foundation for the concentration of telecommunications research in New Jersey. Almost since Bell Labs' inception in 1925, AT&T has been relocating its research out of New York City to New Jersey (Mahon, 1975). Bell Labs opened the following New Jersey facilities: Whippany-1926, Chester-1928, Holmdel-1930, Murray Hill-1941 and in 1967 terminated its research operations in New York. In the 1970s, AT&T made a similar decision to begin moving its management to New Jersey. New Jersey had a tax advantage over New York and a much better commute for a majority of its workers. (See Attachment 1 which details the information corridors which developed in New Jersey.) Today, central New Jersey is the operating headquarters of AT&T (even though it maintains a New York address); is the network operations center of AT&T's worldwide network which processes 175 million calls each business day; and is the home location of Bell Labs, the innovative arm of AT&T and Lucent Technologies.

In 1996, AT&T was split into three companies: AT&T which will provide telecommunication services, Lucent which will sell equipment, and NCR which will market computers. The R&D arm of Lucent Technologies currently has 21,000 R&D employees in eight states (with approximately sixty percent located in New Jersey) and 21 other countries. The 5,000 employee research arm of AT&T, AT&T Laboratories, will be located in Florham Park, New Jersey and consist of mostly mathematicians and computer scientists (MacPherson, 1996). Nearly 4,000 of laboratory employees possess doctoral degrees. About 1,500 of Bell Labs' employees are involved in basic research and $220M has been budgeted for this activity for the next three to five years. Bell Labs' annual budget is $3.4B. Bell Labs is known for worldwide leadership in the three key sciences of the information age: microelectronics, photonics and software, and for the engineering of these sciences into the basic technologies of network computing, wireless, messaging, visual communications, and audio processing. In fact today, Bell Labs focuses 70% of its resources on software development, compared to less than 20% only eight years ago (MacPherson, 1996). Bell Labs is part of the grand alliance of electronic firms now developing the standard for HDTV (Bell Labs, 1996).

A Bell Labs spin off at divestiture, BELLCORE, has provided technical support for the RBOCs since divestiture. It currently employs 6,100 telecommunication professionals of which 5,700 are in New Jersey. BELLCORE has approximately $950M of revenues with a work force concentrating on communication software (60%) and telecommunications consulting (40%).

Other large multinationals, e.g., RCA, ITT, Lockheed, etc., as well as world class universities contribute to New Jerseys' talent pool of engineers, designers and software professionals. Princeton University has world renowned physic and chemistry departments while universities, such as Stevens Institute of Technology, New Jersey Institute of Technology and Rutgers University, have telecommunications related research centers. However, because of Bell Labs international reputation, it has recruited successfully recruited from institutions around the world and has brought those people to New Jersey to live and work.

8.6 FOREIGN DIRECT INVESTMENT

The growth of the information economy, the desire of developing countries to modernize their telecommunications infrastructure and the liberalization of the procurement policies of developed governments have all contributed to the increase in the sector's FDI. As a telecommunication knowledge hub for the world, New Jersey has been attracting foreign direct investment (FDI) mainly from Japan and has become a major exporter of telecommunications FDI throughout the world.

Inward FDI

Inward FDI in New Jersey in information technology related activities is primarily strategic asset seeking, i.e., MNEs engaging in acquiring assets "to promote their long run strategic objectives - especially that of sustaining or advancing their international competitiveness" (Dunning, 1993, p60.). The number of United States-located Japanese R&D facilities in the exceptionally diverse electronics field increased fivefold over the past five years, from 22 facilities in 1987 to 116 facilities in 1993 (Florida & Kenny, 1994). The largest concentration of Japanese electronics R&D facilities is in Silicon Valley, but the Princeton, New Jersey area maintains a significant concentration of such facilities. Unlike the Japanese who seek out new sources of high information technology (Florida and Kenny, 1994), large scale European Community firms developing telecommunication related projects such as high definition TV have not chosen to seek out U.S. knowledge but, rather, conducting that development in Europe.

In 1989, NEC established the NEC Research Institute in Princeton, focusing on advanced software development, artificial intelligence, and machine learning. The facility employs 100 basic researchers, 40 of whom hold Ph.D.s, many of whom have been recruited from leading industrial labs (Bell Labs and Sarnoff Labs) and university research facilities (Princeton University, NJIT, Rutgers University and Stevens Institute of Technology) in the U.S. In 1990, Matsushita created an Information Technology Laboratory in Princeton to carry out basic research in computer graphics, document processing and system software. One year later, Hitachi established a high definition television laboratory in Princeton. Sony and Toshiba have electronic research centers in New Jersey. Toshiba America Consumer Products, Inc. conducts research in advanced TV technology in Princeton, New Jersey. Unlike the large NEC research lab in Princeton with more than 95% of the employees of Japanese background, all research scientists of the Hitachi small multimedia research lab in Princeton are non-Japanese (JETRO, 1996). At this point, one can only speculate if this dramatic difference in percentage of Japanese employees influences the Labs basic mission. In addition to Japanese investment, Siemens, the large German telecommunications equipment manufacturer, has a fiber optic components group in Northern New Jersey.

Japanese companies, the largest foreign telecommunications investors in New Jersey, focus on R&D related activities and prefer greenfield to acquisition modes of entry. New Jersey is a hub of information technology research and a key place to learn about emerging standards. Only California and Michigan have more Japanese R&D facilities than New Jersey. The Japanese are particularly interested in standards associated with HDTV because they have failed in trying to establish their own universal standard in the area and because HDTV is likely to become an important consumer electronic product. The Dalton and Serapio (1993) study in Kenney and Florida (1994) identified ten major location factors in the selection of R&D sites by Japanese electronic firms. The most important was availability of scientists and engineers, followed by nearness of customers, proximity to universities, and proximity to other private R&D facilities. Clearly, New Jersey is well positioned in each of these attributes.

The Japanese view U.S. R&D as a supplement and not a replacement and they desire to employ U.S. scientists and engineers to acquire technology, keep abreast with technology and localize R&D (Kenney and Florida, 1994). Howells (1990) has pointed out how this type of inward FDI may not lead to agglomerative economies. He stated that there is a growing concern that foreign located R&D establishments of Asian firms in the UK and Europe are part of a wider technology leakage in which the economic benefits of research undertaken in these laboratories do not reside within the local or national economy but are transferred and exploited abroad.

In addition, many of the consumer electronic firms which directly benefit from the above research are also located in New Jersey. Sharp Electronics with $200M of capital employs 2,200 people in New Jersey (of which 95% have Japanese background). Toshiba employs more than 1,300 people (of which all but 43 have Japanese background) in New Jersey and has invested $71M.

To date, European and Canadian telecommunication investment in the United States has tended to concentrate outside of New Jersey in three southern areas: Research Triangle, North Carolina, Boca Raton, Florida, and Richardson, Texas. History plays a critical role in explaining Siemens location in Boca Raton because that is where Rolm, a large PBX manufacturer was located before IBM sold it. High tech companies in the area include IBM and Motorola. Research Triangle, with its relatively low factor costs when compared to New Jersey and its high quality of life is a large manufacturing center for AT&T, Northern Telecom, Japanese and European telecommunication companies. Northern Telecom and others chose to co-locate their manufacturing and research and development facility there.

In the late 1980s, Richardson, Texas has become the location hub for manufacturing and R&D facilities, particularly wireless communications. In the late 1980's, the cellular market was growing rapidly and no manufacturer dominated; in fact AT&T lagged the market. Ericsson, the Swedish firm, Northern Telecom and Fujitsu all choose Richardson as a hub for part of their North American operations. Factors contributing to this hub include the location of a large regional MCI facility (a major purchaser of these companies' products), knowledge spillovers (associated with the closely aligned microprocessor industry), and agglomerative economies (associated with skilled workers), and high quality of life (Chamber of Commerce, 1996). Even though these regional hubs are relatively large, employing thousands of people, each of these companies have home country hubs which are larger and exert considerable more control over their worldwide operations. Therefore, these regional locations are not as secure as the home base but are better positioned with the company than pure satellite manufacturing sites.

Outward FDI

Outward FDI by New Jersey information technology companies, particularly AT&T, appears to be more diverse. The leading type of investment is market seeking where companies attempt to gain access to countries as they liberalize their telecommunication markets. The second is resource seeking where companies obtain plentiful suppliers of low cost and well-motivated unskilled and semiskilled labor. The third is efficiency seeking where companies consolidate worldwide production in unique, factor endowed areas. However, the fastest growing investment type appears to be strategic asset seeking where companies seek to form global network alliances (both equity and non-equity) to compete against other networks of alliances.

From the late 1800s onward, AT&T mainly capitalized on its ownership advantage derived from patented technologies to gain entry in foreign markets. With the exception of Canada where Alexander Graham Bell gave the patent rights to his father (Newman, 1996), AT&T assisted a foreign partner with the creation of a factory to manufacturer Bell proprietary telecommunications products in exchange for equity ownership in the venture. By 1914, AT&T's International Western Electric Company had locations in Antwerp, London, Berlin, Milan, Paris, Vienna, St. Petersburg, Budapest, Tokyo, Montreal, Buenos Aires, and Sydney. Due to mounting political and regulatory forces, in 1925, Walter Gifford, Chairman, decided that AT&T and the Bell System should concentrate on its stated goal of country-wide telephone service in the United States (AT&T, 1996).

With the exception of its interest in Canada, AT&T sold its international operations to ITT in 1925 for $33m. AT&T retained 44% ownership of the allied Canadian equipment manufacturer for almost 50 years. In 1962, after the FCC barred AT&T from giving third party exclusive access to patented technology, AT&T sold its share (of what is known today as Northern Telecom) back to Canada's monopolist service provider, thereby reestablishing national order in North America (Ammesse, Seguin-Dulude & Stanley, 1994). One could make the case that these forty years of common heritage and tacit knowledge accumulation between Northern Telecom (NT) and Western Electric facilitated NT's entry into RBOCs equipment markets.

In 1982, AT&T settled its eight-year challenge from the Department of Justice and agreed to divest its twenty-two operating companies and in return was freed from the 1956 consent decree which prohibited it from entering businesses outside of telecommunications. In December 1983, AT&T and Olivetti reached an agreement where AT&T purchased equity in the Italian office automation company. AT&T attempted to use the venture to gain entry to the European PBX market and to make its domestic computer operation profitable. Neither goal was achieved and AT&T sold back its shares to Olivetti (Pisano, Russo, and Teece, 1988). In 1983, AT&T and Philips NV, an electronics conglomerate headquartered in the Netherlands, formed a venture to produce and sell AT&T central office switching in Europe and other global markets. AT&T and Philips had shared a long history of cross-licensing, since AT&T

was prevented by law from selling telecommunications equipment abroad until the antitrust settlement (Pisano, Russo, and Teece, 1988). Philips had distribution channels but lacked products. AT&T, although having a competitive product, found out the realities of protective country-based procurement policies. The joint venture did make sales in the Netherlands, where AT&T located an R&D facility, and in Saudi Arabia where AT&T has a strong military alia. AT&T finally purchased the entire venture and Lucent Technologies now uses it as the corner stone for its European platform. Also in the 1980's, AT&T used its technology to create joint ventures in Asia, e.g., AT&T-Taiwan and AT&T-Korea. More recently, AT&T had entered in an increasing number of smaller joint ventures and alliances in developing parts of the world, particularly China (AT&T Going Global, 1996).

In 1995, AT&T had 25,000 employees in Europe engaged in a wide variety of high tech work, including the development and manufacture of microelectronics, computer and telecommunications products, helping multinational businesses design, install and manage their networks. (The majority of these employees came from acquisitions of Phillips and NCR as opposed to greenfield investment.) AT&T has more than 5,000 employees in Japan; however, all but 800 were associated with NCR which has been divested.

Lucent Technologies, the equipment manufacturing spinoff of AT&T, is headquartered in New Jersey and in 1995 had revenues of $21B, assets of $20B, employed 131,000 workers, and derives approximately 20% of its revenues from outside the US. It has offices or distributors in more than 90 countries and territories around the world. It ranks second in the world in the public network infrastructure market, with its flagship product 5ESS 2000 Switch employed in 49 countries. It is a world leader in operating system software and power supply systems and is the largest producer in the largest market in the world in seven other categories: Large central office switching machines, transmission equipment, wireless networks, business communications equipment, corded and cordless phones, answering machines and digital signal processing chips. Lucent alone employs 7,000 people and has 20 joint ventures in 13 countries in the Asia Pacific region (AT&T Press Release, June 3, 1996).

The remaining AT&T entity has relationships and jointly provides international telecommunication services with the governments of 274 countries and territories all over the globe. AT&T World Partner alliance provides a seamless corporate communications network in 27 countries.

Two likely accelerants for both inward and outward FDI, not directly accounted for in the above analysis, are related to firm actions that trigger government responses. The first is the BT proposed purchase of MCI. To help the BT acquisition go through U.S. regulatory hurdles, the British government will likely make the British market more competitive for foreign entry, i.e., encourage foreign investment. The Transatlantic Business Dialogue meetings between U.S. and European business and political officials' main purposes are to reduce impediments to trade and investment. Any successes here in telecommunication will increase the U.S. flow of both inward and outward investment.

8.7 GLOBAL FUSION

Kodama (1991) has put forth the idea of combining existing technologies into one fusion technology. He asserted that high levels of R&D are required to stay at the leading-edge of changing technologies, develop new products and product technology. This is clearly happening in New Jersey where billions of dollars of research are being spent on fast packet switching, photonics, compression technology, HDTV, microprocessors and software. Advances in many of the above technologies will likely make most of the following activities possible in the next five years:

*   Today, cable TV can be used to download millions of bits per second of data per second. Research is being conducted to efficiently pass information upstream to the head end and route the information to the desired location. Once accomplished, INTERNET video downloading will be greatly improved by a factor of up to 1,000. New direct sales and other information intensive applications will most certainly follow.
*   Today, Integrated Switched Digital Networks (ISDN) are offered in many areas of New Jersey. Within the next ten years the RBOCs have plans to deliver fiber to all customers in New Jersey offering high speed access. The competitive pressures which increased AT&T nationwide fiber deployment by 20 years, will likely have a similar expediting effect on Bell Atlantic fiber deployment when it sees local competition beginning next year.
*   Today, Internet uses a packet switching technology which works well for a primary data application, but it is too slow for real time voice. Breakthroughs are expected in Asynchronous Transfer Mode (ATM) technology, fast packet, which will permit real time voice transmission over INTERNET.
*   Today, microprocessor technology is producing a in dramatic reduction in processing and storage costs. Factor reductions in price are expected each year for the foreseeable future, increasing the power of the home office.
*   Today, we have different devices for telephone, computers and television. Prototypes are available that do all three.
*   Today, regulation is being replaced with competition. Consumer desires, rather than a regulatory edict, are being introduced to the market for the first time.
*   Today, there is tremendous uncertainty concerning where technology fusion will occur, what different application users will want and which alliances will serve as the best hedges. In the near future, more information will be available to help answer the technology and alliance questions. Uncertainty associated with user applications will remain because technological changes influence what is possible and, ultimately, what is desired.

Technological, competitive and market forces make fusion inevitable. New Jersey with the greatest population density in the US, one of the areas of highest disposable income, proximity to the largest business hub in the world, the preeminent research concentration of all areas of information technology (HDTV, microprocessors, switching, transmission, software), and the defacto headquarters location of the largest international carrier in the world will lead the way in testing fusion applications and implementing policies and procedures to encourage their development.

The rapid growth, enormous size and global scope of the information industry attracts new entrants and encourages existing competitors to broaden their offerings. Technology change, which creates new markets, shortens product life cycles and causes the convergence of different areas, is fueling this industry turmoil. Alliances, joint ventures, mergers and acquisitions both affect and are affected by regulatory and legislative decisions, further changing the competitive landscape. Current and potential competitors in telecommunication services include local telephone companies, other long distance carriers, cable companies, INTERNET service providers, wireless service providers and existing large business customers (AT&T Annual Report, 1995).

One can think of telecommunications infrastructure providers as being in the middle of the information value chain. At the top end are the content suppliers and at the bottom end are the device suppliers. For most of the history of telecommunications, the content suppliers have been individual users talking to each other and their chosen device was the telephone. However, today the content providers are just as likely to be film studios, TV channels, software houses, electronic data bases and the device of choice could be a TV or a computer. The network operator and network equipment supplier are still in the middle (Tarjanne, 1995). Firms in each of the value chain parts are attempting to influence standards and seek alliances to gain access to both upstream and downstream capabilities. Worldwide alliances are also encouraging global standards. The only thing that is certain is that no one today can accurately predict where the fusion of technology will occur (Kodama, 1991).

8.8   IMPLICATIONS FOR NEW JERSEY IN THE 21ST CENTURY

The New Jersey Telecommunications Infrastructure Study found, and recent business activity confirms, that the focus of future economic development in the State will be on services-producing sectors of the economy, such as finance, real estate and insurance industries. Many states are targeting these telecommunications-intensive segments. The locational winner will have to remain at the technological cutting edge; therefore, it is essential for New Jersey's telecommunication network to be able to support state-of-art applications for these sectors.

Demographic characteristics and competition in the provision of telecommunication services will likely keep New Jersey the lowest cost service area in the world. These low rates coupled with digitally switched, fiber-based services and the most heavily cabled area in the nation will enhance the development of the communications-intensive services sector. Because of software technology that offers all customers in a particular service area the benefits of the advanced infrastructure, small businesses will be permitted to enjoy the same telecommunication service benefits as larger users, e.g., high speed digital access and transport and advanced software network features. Global expansion and an increasing number of alliances by AT&T, Lucent and Bell Atlantic will result in less absolute and proportional employment in New Jersey and more overseas. A brief description of a few representative alliances shows the general direction in which each of the companies is heading.

In 1993, AT&T entered into an alliance, Uniworld, with four European PTTs: PTT Telecom (Netherlands), Telia (Sweden), Swiss PTT(Switzerland), Telefonica(Spain) and another alliance, WORLD PARTNERS, with eight Asian PTTs: TCNZ, Telstra, Hong Kong Telecom, Korea Telecom, Unitel, PLDT, Singapore Telecom, KDD-Japan (Financial Times, 1995). AT&T's alliances compete with the MCI/British Telecom alliance Concert and the Sprint, Deutsche Telekom, and France Telecom alliance called Phoenix.

Lucent Technologies intends to "utilize a combination of joint ventures and direct investments" (Prospectus, 1996, pp40) to achieve its international goals. The Business Communications Division of Lucent has entered into alliances with Lotus Development Corp., to enable multimedia messaging with Lotus Notes, and with Novell, Inc. to extend computer/telephone integration. It was also one of the founders with IBM, Apple Computer, Inc. and Siemens AG of VERSIT, an industry consortium organized to ensure the interoperability of multi- vendor multimedia applications. Network Systems International is a holding company for many other Lucent alliances throughout the world, e.g., AT&T Taiwan Telecommunications, Goldstar Information and Com (Korea), AT&T- ISTEL(Italy), etc. As developing countries, particularly, China and India, enter the information world, an increasing number of joint ventures with local partners will occur. Many of these countries will require equipment suppliers, such as Lucent technologies to locate both manufacturing and R&D facilities within their countries in order to obtain a greater percentage of knowledge spillovers.

In 1995, Bell Atlantic invested more than $1B to acquire a 41.9% economic interest in Grupo Iusacell, S.A. de C.V., a leading telecommunications company in Mexico whose primary business is the provision of cellular telephones (Bell Atlantic 10-k, 1996). Bell Atlantic located the headquarters of its cellular division in central New Jersey. This New Jersey headquartered Division also has telecommunication investments in Italy, Slovakia and the Czeck Republic that consist of joint ventures to build and operate cellular networks.

Even with these alliances and a greater globalization of the telecommunications business, New Jersey, because of its embedded innovative capacity and corporate center of gravity, will remain a significant global telecommunication, television, microprocessors, and software research center. It is likely that global equipment manufacturers located in Europe and Canada may set up research facilities in New Jersey (much like many Japanese companies have already done) to monitor activities.

Lucent Technologies will likely require Bell Labs to focus more on applied rather than long term basic research. If the Federal Government desires to keep basic telecommunications related research, it will likely have to provide a subsidy to Lucent Technologies or create a government laboratory. In an era of reducing the size of government, the most likely alternative is no government action. This would likely disadvantage America's future productivity if foreign governments, particularly the Japanese, retain both government and private sector funding. Alternatively, Lucent Technologies' competitive position and free cash flow necessary to fund research could be significantly enhanced if European government subsidies in the form of preferential procurement and government financing are phased-out and eliminated. Lucent Technologies will not likely achieve penetration rates in the fast growing Southeast Asian markets equal to its world share due to the aggressive targeting of these markets by Japanese companies which have a better understanding of the culture. The net result to New Jersey is that there will be considerably fewer people employed, maybe up to ten thousand, but the main area of applied research will stay in New Jersey, as Lucent business becomes much more a global player with spatially dispersed operations.

AT&T will likely become an anchor of a strategic business alliance network which will have facility presence in all triad markets. AT&T, because of its historical relationship and accumulated tacit knowledge associated with using Lucent's equipment, will retain Lucent as its key network supplier. Large business customers will likely decide in increasing numbers that provision of telecommunication services is not a core competency and choose to thrust this activity with a telecommunications service provider such as AT&T. New Jersey being the center of global network management for AT&T network will benefit from this trend.

A merger of cable, telecommunication and data service and device providers would result in a rapidly changing marketplace. Through alliances global players likely will have to participate in the three areas, voice, video and data. They will also likely establish physical presence in key competitors' home R&D locations in order not to fall behind in technology and will successfully anticipate and/or copy follow-on products. The AT&T anchored alliance, possible with a cable franchise partner, will remain a significant worldwide player, located in New Jersey. The tough decisions by the New Jersey and United States regulators necessary to orchestrate a transition to full competition will be behind them by the 21st Century. New Jersey business and residence customers will likely have the ability to choose from a great number of low priced, advanced, information exchange services.

The spatial patterns for the information sector are rapidly evolving. A limited number of networks of alliances providing seamless global information access to business as well as residential consumers could emerge in the next year or so. The technology used to provide the backbone for these services will be fiber based, the switching will likely be fast packet, and the access medium will likely be through the air. The number of equipment suppliers will likely be reduced by two or three, but Lucent will remain viable. National governments will have reduced power, customers through markets will determine fused technologies and supra-governmental organizations may be required to prevent cartels of networked alliances.

A less optimistic view could have the major global communication providers using their global market power to gain monopolistic profits. Collusion among the oligopoly players (rather than competition) would reduce the rate of innovation, lower foreign investment and once again bring outcries from business for government or supra-government involvement.

8.9   CONCLUSION

We have shown that historically Central New Jersey has been a center for telecommunication innovation in the United States. We have reported that state owned and operated telecommunications monopolies are being replaced by private competitive environments. We assert that liberalization will encourage globalization by the largest equipment manufacturers and service providers and the creation of networks based on alliances and joint ventures that are able to quickly provide global communication to users. Evidence indicates that the innovative centers of gravity for flagship firms of the competing networks will remain where they have historically been. Therefore, New Jersey will likely remain an advanced global center for telecommunications innovation and implementation. As such, New Jersey will remain a critical access and destination point to most facets of the globalizing economy. The connectivity and ease of linkages to the number of spatially distant islands in the global business archipelago will be greatly influenced by action of executives located in Basking Ridge, New Jersey, AT&T's corporate nerve center and Murray Hill, Lucent Technologies' headquarters.

Because of New Jersey's demanding customers (a large number of Fortune 500 companies, high disposal income residence consumers and population and business density), telecommunication providers have built information superhighways in New Jersey based on fiber optic transport and ATM and electronic switching. Many of the vehicles (products and services) using the information highway are also created in New Jersey. Entrepreneurial manufacturing and relatively small service businesses receive the benefit associated with this advanced infrastructure. In effect, these small firms get a free ride due to a positive technology externality of larger firms in the area.

The design of major components of much of the world carrier and end user switching and transmission equipment occurs in New Jersey. However as the regulatory shackles are lifted from the telecommunications industry around the world and procurements are no longer based on the origin of the supplier's flag but rather technology excellence, parts of Lucent Technologies research, design and planning will likely follow markets around the world. This global scanning capability, particularly in an industry which could experience paradigmatic technology shifts, will likely be required by all global industry players in the next century and result in additional foreign research being conducted in New Jersey.

In today's global economy, alliances between major players in the oligopoly structure will continue to occur. These tendencies will be more pronounced in telecommunications because a century of regulation has created a greater path dependency of user requirements and supplier capabilities in each major market. Also, the desire by carriers to provide a seamless global service to customers quickly, necessitates alliances as the most efficiency organizational modality to meet large customers' demands.

In conclusion, the New Jersey telecommunications industry is a major designer, engineer, planner, and researcher of the information bridge to global economy and will remain so as we enter the 21st century.

REFERENCES

AT&T Going Global (1996). www.att.com/global/asia4.html.

Bell Labs Home Page (1996). *General Information About Bell Labs*, http://www.bell-labs.com:80/geninfo/history/.

Britt, R. (1996). "What AT&T Means to NJ", *The Star Ledger*. P37.

Dalton, D. and Serapio, M. (1993). *U.S. Research Facilities of Foreign Companies*. Washington, D.C.: U.S. Department of Commerce, Technology Administration, Japanese Technology Program.

Deloitte and Touche (1991). *New Jersey Telecommunications Infrastructure Study*, January.

Dunning, J.H. (1993). *Multinational Enterprises and the Global Economy*. Addison-Wesley, NY.

Florida, R. and Kenney, M. (1994). "The Organization and Geography of Japanese R&D: Results of a Survey of Japanese Electronics and Biotechnology", *Economic Geography*, pp. 344-369.

Gomes-Casseres, B. (1996). *The Alliance Revolution*. Harvard University Press, Cambridge, Mass.

Howells, J. (1990). "The Internationalization of R&D and the Development of Global Research Networks". *Regional Studies. 24* (6) 495-512.

Kodama, F. (1991). *Analyzing Japanese High Technologies*. London Pinter.

Markusen A. and Gray, M. (1996). "Industrial Cluster and Regional Development in New Jersey", *CIBER Working Paper Series*. No. 96.002.

Markusen, A. (1994). "Interaction Between Regional and Industrial Policies: Evidence from Four Countries". Proceedings from the *World Bank Annual Conference on Development Economics*.

Perone, J.R. (1996). "State Opens Field for Local Phone Service". *The Star Ledger*. P.1,22, 6/20/96.

Porter, M.E. (1980). *Competitive Strategy*. Free Press, New York.

Rosenbush, S.A. and Marsico, R. (1996). "Jersey to Gain 1,100 'High Salary Jobs". *The Star Ledger*. p.53.

Salmon, E.H. (1994). *The State of Telecommunications in New Jersey: Response to the Telecommunications Act of 1992*. January 1994.

Sherman, T. (1995). "AT&T Breakup Escalates Clash of the Titans". p.1,6,7 *Sunday Star Ledger*.

Stone, A. (1991). *Public Service Liberalization*. Princeton University Press.

Tarjanne, P. (1995). *Preparing the Telecom Infrastructure for the Information Society*, Technical University of Denmark. May.

Zucker, L.G., Darby, M.R., Brewer, M.B. (1994). *Intellectual Capital and the Birth of US Biotechnology Enterprises,* NBER, Working paper, No. 4653, February.

PART D

PERSPECTIVES ON CANADA'S LOCAL AND REGIONAL SYSTEMS OF INNOVATION

9 INNOVATION IN ENTERPRISES IN BRITISH COLUMBIA

J.A.D. Holbrook and L.P. Hughes
Simon Fraser University

9.1 INTRODUCTION

For most researchers and policy makers, the importance of science and technology to industrial development is unchallenged – the focus in the last half of the 20th century has been to understand the *process* of technological progress and industrial development, the better to manage it. Science and technology is an activity embedded in a complex system. In coming to grips with this, the discussion has shifted from science and technology to "innovation" and to "systems of innovation."

In 1986 the OECD sponsored its first formal meeting on the measurement of innovation and innovative activity, providing a forum for discussions that had been going on for at least three years previously (cf. Smith, 1992). The impetus for venturing out into this quantitative exercise was policy-makers' desires to be able to link the application of knowledge and skills to production with measurable increases in economic growth. While the success in quantifying this linkage may still be debatable, the utility of measuring innovation is well accepted, particularly as a tool that can forecast potentially competitive firms, industrial sectors and economies. Just as it is important to measure physical capital stocks in classical economic analyses, it is important to measure and follow the stocks of S&T knowledge or technological capital.

Current theories take a wide view of the innovative process, and recognize that R&D is only one of several inputs to wealth generation and social progress. The OECD, in a recent report examined this issue and concluded that investments in technology embedded in capital equipment, whether imported or produced domestically are equally important and should be included in assessments of the knowledge intensity of nations (OECD, 1997a, p. 127). In smaller industrialized nations and in developing economies, the actual level of R&D activities may be quite low, but the level of investment in related science activities may be substantial. Statistics on industrial investment on R&D miss those innovative industries which are not R&D intensive. For example, in British Columbia, Lipsett *et.al.* have found that while over half of some 13,000 firms had used government S&T programs and incentives, less than one third made use of tax credit program which is based on innovative activities within the firm. Firms profit from the larger pool of external knowledge by absorbing and adopting some of it to their own needs; the source can be a competitor, another industry, government, universities or another country.

Investment in S&T knowledge occurs primarily through three means:
- domestic development of new S&T knowledge through R&D programs
- investment in intellectual property through the purchase of patents and or by appropriate knowledge in the public domain
- acquisition of knowledge through technology embedded in imported capital goods (e.g. usually "high-technology" products)

These investments, while usually thought of in financial terms, also represent an investment in human capital. Thus of these investments must not only include the financial resources devoted to them but also the human resources assigned to these efforts. Analysts need to measure all three types of contributions to knowledge which contribute to economic growth, as economic structures vary widely.

It has been found convenient, when measuring stocks and flows of knowledge in an economy to describe the process of innovation in an economy as a system. The characteristics of a national system of innovation (NSI) model can be summarized as follows:
- A NSI consists of a network of public and private sector institutions whose activities and interactions initiate, import, modify and diffuse new technologies
- A NSI consists of linkages (both formal and informal) between institutions
- The linkages in a NSI define flows of knowledge and capital
- Learning is a key economic resource, and physical location is important.

The emphasis on institutions is the cornerstone of NSI analysis. Charles Edquist, in the introduction to his recent book on innovation (Edquist, 1997) analyzes the work of many authors, and notes that all of the NSI approaches emphasize the role of institutions. He notes: "Institutions are of crucial importance for the innovative process. . . .It is therefore a great strength of the systems of innovation approach that 'institutions' are central in all versions of it."

In analyzing NSIs, it is necessary to be able to measure the stocks and flows of knowledge among Innovation does not necessarily occur only in the private sector, but as yet there has been no procedure offered for the assessment and quantification of innovation in the public sector. The OECD has concluded (OECD, 1997a) that the study of NSI offers new rationales for government technology policies. Government policies in the past have focused on *market failures*; studies of NSI make it possible to study *systemic* failures.

9.2   WHY MEASURE INNOVATION ON A REGIONAL BASIS?

As defined by Schumpeter (1949), innovation refers to more than just technological innovation but also encompasses marketing, finance, management and other issues. Firms are elements of this system, as are labour markets, public institutions and government programs, and other entities that support innovative activity. Much effort has gone into designing policies to strengthen systems of innovation; cogent policy design requires first an understanding of the systems to be managed and, second, measurements of system processes to optimize policy design and monitor the impacts.

Earlier analyses focused on how national economies organize for innovation. However, much of the industrial dynamic takes place at a *regional* scale and, recently, the concept of *regional* systems of innovation has become important. As the National Research Council has noted: "Regions within nations are increasingly being recognized as the primary milieu where innovations occur and are spread and appropriate targets of government action."

The concept of regional systems of innovation has been discussed in detail by authors such as Acs *et. al.* (1996) in looking at particularly successful industrial clusters, such as the research triangle around Raleigh NC. The concept of "poles" of innovation has been used as part of the literature describing the geographic concentration of innovative industries in national systems of innovation (Voyer & Ryan, 1994). The question then arises: can one talk of a regional system of innovation in the Canadian context and measure innovative activities that may exist in a particular region?

Figure 9-1 shows a theoretical model of the institutions and linkages of the Canadian NSI. However, does it represent accurately the different regional systems of innovation in the major economic regions of Canada? Would a survey, based on the "Oslo Manual" of the OECD (1997b) give data on all of the linkages and institutions in a region such as British Columbia? Does it fully explain the role of small private-sector technology transfer agents such as retail computer stores and S&T service providers (of all sizes) such as consulting engineering firms?

While a survey based on the Oslo Manual might not describe all of the institutions and linkages (particularly public sector institutions - there is still no agreement on the measurement of innovation in the public sector), it should be possible to determine the major linkages and institutions from a short survey of enterprises. Thus, a survey of enterprises in BC should be able to determine:

- which institutions in the regional system of innovation are important to firms in BC, particularly small and medium sized enterprises (SMEs)?
- what are the relative strengths of these links?

From the information gained from such a survey, it should be possible to build a partial model of the BC system of innovation.

9.3 MEASURING INNOVATION IN A REGIONAL ECONOMY: BRITISH COLUMBIA

The Centre for Policy Research on Science and Technology (CPROST) at Simon Fraser University, in collaboration with the Centre for Policy Studies in Education at the University of British Columbia, has established a major multi-year project to study innovation in regional economies. The first element in this study was to carry out a preliminary survey of technological innovation in British Columbia. This survey was conducted with two goals in mind:

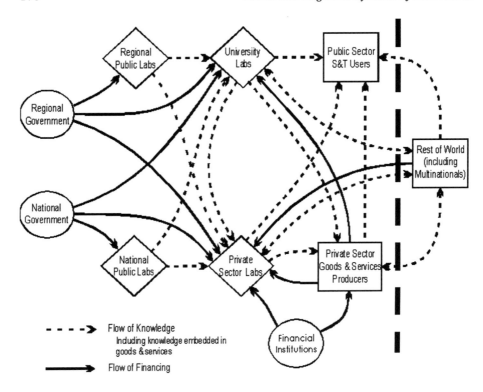

Figure 9.1 - A Regional System of Innovation

- To examine the structures and linkages of this regional system of innovation.
- To see if such an examination could be conducted economically, and in a manner that was relatively painless and non-intrusive to respondents.

Given the links between technology, innovation and economic growth, such a regional approach should work best if it is carried out in the context of a politically and economically distinct region. Such an economy usually has only one pole of innovation, and its GDP is usually small compared to national levels (e.g. less than $100 billion Cdn.). This does not necessarily differentiate between national and regional economies: the economies of New Zealand and BC are similar in size and to certain extent in structure, yet one is separate nation and the other a province in a federation.

British Columbia is an ideal laboratory for experiments in the measurement of innovation. The economy is simple: there is one large metropolitan area, where most of the innovative firms are located, supported by a hinterland whose primary outputs are in the natural resources sector. BC is a distinct geographic region so that external influences in the acquisition and adoption of technology are readily noticeable. Thus (in theory) economic measurements in BC should be relatively well-behaved and predictable. This preliminary survey concentrated on BC's Lower Mainland region (the city of Vancouver, its immediate suburbs, and the lower Fraser River valey). Other studies are underway in other regions of the province, including the Okanagan Valley and Vancouver Island.

Since BC is an expanding and vibrant regional economy in Canada, knowledge of how its innovation and technology transfer systems work is important, not just to regional policy makers, but also to national policy and program managers. Measurement of innovation on a regional basis provides an ideal test bed for larger national innovation surveys. Work done at a regional level can test approaches, concepts and outcomes, before a large-scale national project is started.

A questionnaire, based on the "Oslo Manual, 2nd Edition" (OECD, 1997b) was prepared for this survey. The Statistics Canada Survey of Innovation, 1996, also based on the Oslo Manual, was used as the initial source of questions. The survey was not intended to cover all aspects of technological innovation identified in the Oslo Manual, but it had to conform to the main points of the OECD standard.

Since the survey depended on private sector respondents taking their own time to complete the questionnaire, this task had to be made as painless as possible. Surveys based on the Oslo Manual tend to be quite long, and require considerable resources within a firm to complete. Most such surveys are conducted under the terms of national statistics legislation, which generally contain legal provisions requiring the respondent complete and return the questionnaire. The CPROST survey, on the other hand, was voluntary. In order to ensure an adequate response rate, questionnaire had to be short, no more than two sides of a single page, straightforward, and relatively easy for one person in a firm to complete.

9.4   TESTING THE QUESTIONNAIRE

In order to ensure that the requirements of brevity and validity were met, two focus groups were conducted to test the questionnaire. The discussion guide for these groups consisted of three main sections:

- A discussion of the concept of innovation, based on the models used in the Oslo Manual
- A brainstorming session in which participants developed a single sentence definition of "innovation"
- Completing and critiquing the draft survey questionnaire.

The focus group participants were all senior executives at their firms, either in R&D or in general management. A cross section of the BC economy was represented: telecommunications, electronics, software, multimedia, medical devices, biotechnology, forest products, ocean systems, and specialty agriculture. A third of the participants were from large corporations, while the remainder were from small- and medium-sized enterprises (SMEs).

Following the first section of the discussion guide, the concept of innovation was examined, with participants encouraged to draw on their own experiences. Some points that were made in this part of the discussion:

- innovation is more likely to happen in certain key industrial sectors

- innovation can range from "needs driven" to spontaneous innovation is essential to survival and growth firms innovate to provide solutions characteristics of innovative firms include: a nurturing environment; sensitivity to people's needs; an "identify a problem/find a solution" attitude; rewarding success but not punishing failure; taking advantage of creativity and serendipity
- innovation usually needs a champion within a firm and an receptive organizational culture
- there either is or is not a culture of innovation in a firm.
- innovation can be measured, but it is subjective and not easy

The focus groups noted that *the process of innovation in a firm is intellectual property in its own right. Often the process of innovation in a firm is more valuable than the innovations themselves.*

One of the more interesting findings from this section was the observation that large firms often transfer the risks associated with innovation to SMEs, for example, by sub-contracting research, and picking up the innovation when it is no longer risky.

For the brainstorming section of the discussion guide, each group was asked to prepare an agreed definition of innovation, using less than 25 words. The purpose given to the participants was to assist the researchers in the design of the final questionnaire, for which a readily-understandable definition of innovation was required. Participants were given Schumpeter's definition of innovation, and the Oslo Manual interpretation of it. The two definitions that emerged from their brainstorming efforts were:

- "Recognition and exploitation of a new, unique, idea, that can be applied to improve a product, service or process."
- "Creative and unique application of knowledge to effectively meet a need or (to) solve a problem"

Participants were then asked to complete the draft questionnaire. This task had been allocated ten minutes in the focus group discussion guide, and most participants were able to complete the survey in this time. Several participants noted that they did not have the information to answer all of the questions themselves - they would have to ask someone else in the firm - but agreed that this would not be a problem, for themselves or the other individuals. Participants agreed that the questionnaire was easy to complete, and asked most of the right questions in the right way. The participants confirmed that this survey would be relatively unobtrusive to those asked to participate, and only minor editorial changes were made to the questionnaire as a result of the focus groups.

The discussion then turned to how to conduct the survey. Upon being presented with a number of options, both groups agreed that the best method would be a combination of telephone and mail-out: phone the company to identify the appropriate individual and determine whether that person was willing to participate; mail the survey to the attention of that person; and follow-up the mailing with another telephone call.

The groups were asked who should be seen to be conducting this survey: a government agency or a university. There was almost universal consensus among the participants that this survey would be much better received if it came from a university. Several participants stated that government surveys of this nature are *perceived* as a tax on time - university research projects, particularly if students are seen to be involved, are viewed more favorably. The last issue discussed was who at a firm would be the appropriate individual to receive and complete the questionnaire. This produced a rather lively discussion in one group, where it was agreed that the chief financial officer (CFO) was probably the most appropriate person to receive the survey, for a number of reasons:

- The CFO is less likely than someone in R&D to be "married" to an innovation, and therefore better able to objectively judge its success or failure, and the process that went into it;
- The CFO is more likely to be objective about the corporate climate than either general or R&D management;
- a number of the questions have to do with finance, which would require input from the CFO in any event.

9.5 SURVEY METHODOLOGY

Based on findings from the focus groups and the recommendations of a workshop on measurement of regional systems of innovation[1] it was decided that two samples be drawn, one from the high technology sector and one from other areas of policy interest in the BC economy. The high technology sector was based on the BC Stats definition (1997), and included specific classifications under the standard industrial codes (SIC) for "Manufactured Products," "Computer Services," and "Technical Services." The other sample, referred to hereafter as the "policy sector," included food products, wood products, electrical products, transportation products and services, and construction. Firms in the sample were all larger than 10 employees. Figure 9-2 shows the distribution of the sample, and Figure 9-3 shows the distribution of the survey response.

9.5.1 Sector in Response

The final questionnaire consisted of questions grouped into four categories. The first category dealt with new products, and their effect on the firm. The second category dealt with sources of innovation, and their relative value to the process of innovation for the firm. The third category addressed factors affecting innovation, and whether these factors helped or hindered innovation in the firm. The fourth group was a general set of questions looking at such things as R&D spending, purchases of capital equipment, intellectual property issues, and personnel practices within the firm.

The third category, factors affecting innovation, was modified from the draft used in the focus groups to the final questionnaire, after a question was raised at the workshop concerning bias in these questions. As a result, the section is somewhat different than the practice outlined in the "Oslo Manual." Rather than asking separate questions about factors that assist and impede innovation, this questionnaire provided a generic list of factors and asked whether each of them helped, hindered, or had no effect on the innovative process within the firm.

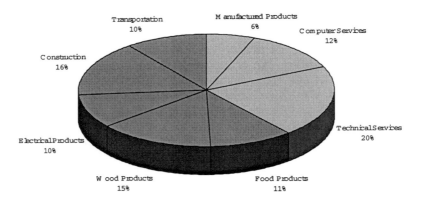

Figure 9.2 - Industry Sector in Sample

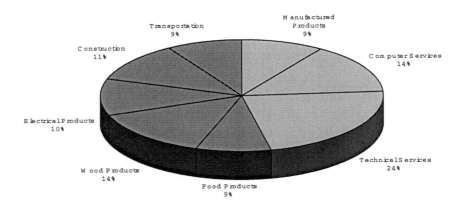

Figure 9.3 - Industry

9.6   SURVEY RESULTS

The survey was conducted in March 1997. Firms were contacted by telephone, to identify the appropriate person in the organization and asked if (s)he would participate in the survey. The questionnaire was then mailed or faxed to this person, along with a cover letter and a postage-paid return envelope. Firms that had not responded to the survey within ten days of the mailing were then re-contacted.

The response to the initial phone screening was most gratifying. Only 10% of all of the firms approached refused to participate. Reasons ranged from an outright "no!" to having a policy of not filling out any surveys. Many of the smaller firms cited the absence of the appropriate official from the firm during the mailback period as a reason for not participating. Most firms that gave a positive response indicated that the university affiliation was a factor in agreeing to participate. Many felt that they might not have much to contribute to such a survey in that they felt they were not innovative, but that they would respond to such questions as they could. On the second callback a more significant percentage of companies indicated that after having received the survey they felt it did not apply to them (approximately 20%). The response rate to the mail-out survey was 39%. Response rates varied by sector: the high technology sectors had higher response rates than the policy sectors (Table 9-1).

Innovativeness

Behaviour that is usually considered innovative (see for example, Johnson *et.al* 1997) appears to be common throughout BC industry. Almost four in five respondents to this survey (78%) reported having introduced a new product and or process in the past five years. However, there is more to *innovation* than product or process development. Inherent in Schumpeter's definition – as well as in the focus groups' understanding – of innovation is the idea of *uniqueness*. For a development to be innovative, it must not be simply new for the firm in question, it must be unique in that firm's competitive market. For the purposes of this study, a more stringent definition of innovation was therefore used – *unique* products or processes developed in the past five years. The questionnaire asked if the firm had introduced a new product or process, and if that new product or process was *unique in the industry*. Hopefully this definition captured the difference between firms that were staying ahead of their competition, and those struggling to keep up. A product or process unique in the industry *to the knowledge of the respondent* would likely be unique *in the firm's market* and thus innovative..

Of respondents reporting a new product or process, 60% claimed that it was unique in their industry. By this measure, only 47% of firms responding to the survey were considered innovative. These firms were divided almost equally between high technology and policy sectors. By this applying this definition, firms in the high technology sector could be, and were, considered non-innovative. The sample was divided into four segments based on this filter: high technology innovative; high technology non-innovative; policy sectors innovative, and policy sectors non-innovative. (Figure 9-4). Firms that introduced new products and/or processes, unique or otherwise, reported that these innovations affected a variety of areas, to varying degrees: profitability, cash flow, market share, competitiveness, productivity, and quality. The only areas that did not appear to be clearly affected by innovation were labour relations, and somewhat surprisingly, environmental impact.

Table 9.1

Response Rate By Sector

Sector		Response Rate
High Technology Sectors		**46%**
Manufactured Products	59%	
Computer Services	46%	
Technical Services	43%	
Policy Sectors		**34%**
Food Products	31%	
Wood Products	38%	
Electrical Products	41%	
Construction	28%	
Transportation	32%	
All Sectors		**39%**

Sources Of Innovation

At first glance, there were no real surprises in the data collected from this survey – the results parallel those of previous studies involving British Columbia firms, such as Baldwin *et. al.* (1994) and Lipsett *et. al.* (1995). In-house R&D, customers, and sales & marketing were the most important sources of new product or process ideas, while out-sourced R&D, competitors, and particularly trade shows were less important. This last finding should have some significance for the trade promotion activities of government agencies.

Results from the high technology sectors follow the overall results. However, analysis of data from the policy sectors yields some interesting findings. (Figure 9-5)

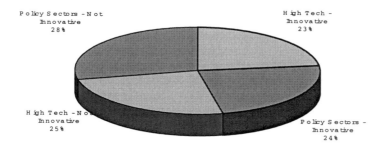

Figure 9.4 - Innovation by Sector

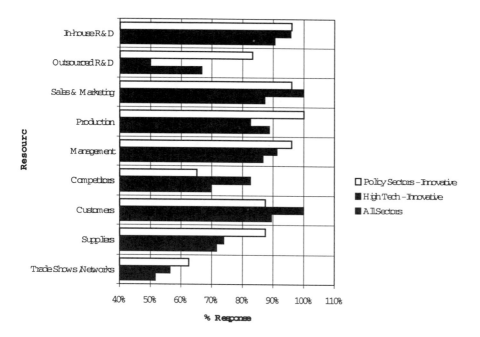

Figure 9.5 - Sources of Innovation

Innovative companies in the policy sectors ranked suppliers, management, production, and out-sourced R&D as significantly more important than the sample as a whole. This may indicate that these firms are more likely involved in *process* innovation – new ways of making traditional products rather than new products.

Factors Affecting Innovation

Customers, competition, and risk were identified by the firms participating in the survey as the main external factors "helping" innovation. The availability of raw materials, and environmental concerns were thought to have as having "no effect" on innovation. Government policies and programs were generally considered to have no effect, although more respondents felt that government "hinders" innovation than felt it "helps".

Factors that "hinder" innovation are much more difficult to isolate from the data. Roughly 30% of innovative firms identified development costs and the availability of personnel as impediments; for non-innovative firms, 40% identified these factors as hindering innovation.

There is a great deal of interest in the availability of financing as a factor affecting innovation. In this survey the results were mixed (Figure 9-6). For all segments, 28% replied that the availability of financing "hindered" innovation, while 38% replied that it "helped". For the innovative segments, 48% viewed financing as "helping" innovation. In the high tech non-innovative segment, on the other hand, only 17% replied that the availability of financing "helped" innovation. Those results are similar to those reported by Statistics Canada (Johnson *et. al.*, 1997).

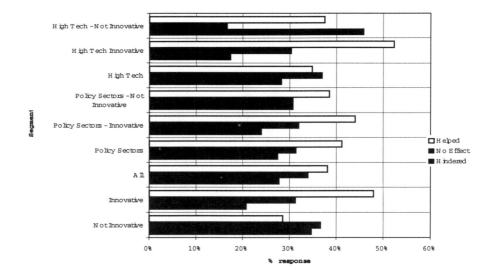

Figure 9.6 - Financing as a Factor Affecting Innovation

In general, there was more agreement in this section among innovative and non-innovative than there was among high technology and policy sector. Innovative firms, from either sector, tended to identify the same factors as helping or having no effect, in roughly the same proportions. The data on internal factors affecting innovation is suspect in this survey. Virtually all respondents claimed that corporate culture and management attitude helped innovation in their firms. However, the respondents to the survey were all upper management in their firms, who would be less likely to highlight internal problems to outside researchers.

Other Results

Several other observations can be drawn from the data collected in the survey. Research and development programs are an important component of innovative activity – 79% of firms surveyed reported having R&D programs now and/or in the past. 91% of firms that introduced new products or processes in the past five years have (or had) R&D programs, while 98% of innovative firms have R&D programs. However, while innovative firms have R&D programs, the presence of R&D does not necessarily indicate innovation. 64% of non-innovative firms also indicated having a R&D program.

The survey also sheds some light on the perception government's role in innovation. Only two in five companies reported having used a government incentive program (42%), however 65% of innovative firms reported having used these programs. Once again, there was no significant difference between innovative firms in the high technology and policy sectors. An interesting comparison can be made between the use of these programs and the perception of government as a factor in innovation: 65% of firms that government helped the innovative process in their firm had made use of a government incentive program, while 57% of those in that government hindered innovation, had not, only 30% of those who had used an incentive program indicated government programs as "helping" innovation.

Intellectual property protection does not appear to be particularly important to BC firms. Only 17% of respondents indicated having patented or licensed a product in the past three years: 56% of these were .

The last item of note is the relationship between innovation and exports, which confirms earlier studies (Lipsett *et. al*,1995; Baldwin *et. al.*, 1994) Almost three in five (57%) innovative firms reported more than half their sales outside BC, in contrast to only one in six (17%) non-innovative firms. 36% of innovative firms reported more than half their sales outside Canada, compared to 13% of non innovative firms. Once again, there was no significant difference between innovative firms in the high technology and policy sectors.

9.7   CONCLUSIONS

There are several important generalizations that can be drawn from the work carried out for this report. These conclusions are diverse and each (except perhaps the first one) suggest the need for further work in that particular area.

A key feature of the project was the collection data from a control group as well as from high-tech firms. The conventional image of innovation is that of ideas bubbling up in a high-tech environment, yet we know that non-high tech firms must also innovate to survive. Thus further analysis of the data may suggest important areas of policy differentiation between high-tech firms and more conventional enterprises.

Qualitative testing, such as the focus groups conducted in this survey, should be a prerequisite for any survey of innovation. The Oslo Manual language may not be the most appropriate for entrepreneurs; in any case it does not take into account regional differences in cultural, linguistic or economic circumstances. Additionally, interesting can be volunteered by focus group participants, that the researchers simply had not considered.

The focus groups emphasized the need to explore the issue of "culture of innovation" further. Earlier work has suggested that innovation occurs all of the time, yet there appear to be "sparks" of innovation when one firm exploits an innovation that some other enterprise may have looked at and rejected.

The focus groups also raised questions on the methodology of innovation surveys. Is a technology manager the best person to answer these questionnaires? How does one deal with firms who believe that they are not innovative, or who feel that innovation, as a concept, does not apply to them?

As a result of both the literature review and the focus groups, this survey uses a more stringent test of innovation than is standard practice. Innovation implies both newness and *uniqueness* - innovators are leaders, not followers. Within a regional system, innovations are those developments that are *new to that market*. Although this measure reduces the level of innovation obtained by other surveys, it identifies those firms that tend to introduce new technologies to their regions - this allows the flow of knowledge within the region to be more easily tracked.

The results of this survey indicate that the relationship between high technology and innovation may not be as strong as was previously thought. Innovative firms, whether high tech or not, appear to have more in common with each other than innovative and non-innovative firms in the same industrial segment. All firms, regardless of their industry, must innovate to distinguish themselves from their competition.

One of the important relationships not measured were the linkages between firms and other sources of. The analysis of innovation using the systems of innovation paradigm is on measuring stocks of knowledge and volumes of knowledge transferred of knowledge among the various of the innovation system.

Comparison of various responses may produce some serendipitous results which can only be guessed at this time. A key task would be to compare results from this survey, which was limited to the Lower Mainland of BC, with some other region of Canada. Factors affecting innovation, particularly those involving human capital, may well vary across the country; from the point of view of national policy makers this would be vital to understanding how Canada generates innovations and how they might be translated into increased productivity. In particular, as has been discussed by Holbrook (1997), BC has a unique relationship between the natural resources sector and the services sector. Many of BC's high-tech activities are in the services sector. These firms have built strong export markets based on their work for the natural resources industries that are the base of the BC economy. Thus innovative activities in technology-driven services may be linked to industrial clusters that can extend across several sectors of the economy.

This questionnaire is being used in other regions within BC, as part of an ongoing program, and the data will be amalgamated to form a larger whole. In particular, study of a sub-regional economy (such as that in the Okanagan Valley) may provide useful information, particularly about innovation in an area that does not have significant S&T infrastructure, yet, almost paradoxically has a measurable group of innovative industries, some of which fall into the high-tech category. This sub-region may provide one of the clearer examples of cluster-based innovation. The culture of innovation in these firms itself will be interesting, if only because of the additional constraints imposed upon the innovative process due to distance and lack of facilities in the Okanagan district.

Measurement of Systems of Innovation: New Approaches

Results of surveys, such as the one reported above suggest two lines of further inquiry. On the one hand the theoretical bases of a local/regional/national system of innovation need to be explored. It is a complex system and the interactions among the institutions within it behave in complex ways. In the mathematical and physical world complex systems can be described as systems operating on the edge of chaos. In this situation it is theoretically possible to determine a over-riding set of operating rules, even if the actual interactions are themselves chaotic. (See, for example Guastello (1995), Kelly (1994), Kiel & Elliott (1996), Merry (1995)). Thus, if the institutions, the members of the system, or "actors" are building blocks, as described by Edquist, they can form networks which have a higher order of operating rules. McKelvey (McKelvey, 1997) and Savoiotti (Saviotti, 1997) examine directly the applications of evolutionary theories to systems of innovation. This approach can lead into the study of the interactions among the actors, the "actor-network theories" attributed to Latour, and draws on the wide and ever-growing body of work in the area of the sociology of science (See, for example, Latour (1987, 196), Bijker (1995), and Law (1994)). Actor-network analyses look at the webs of relationships that grow up around the development of technologies, between individuals, groups and other enabling technologies. An offspring of this approach, socio-technical graphs, (Latour, Mauguin & Teil, 1992) represents an attempt to develop empirical models of actor-networks.

A second approach is from the bottom up. Again, starting with the premise that the building blocks of a system of innovation are the institutions that make up the system, and that most of these institutions are firms, there is a need to understand how innovation is managed within a firm, before moving to an understanding of the system. The human capital issues of corporate culture and management attitudes towards innovation are clearly important in promoting innovative activities, yet hard to measure. It is difficult to quantify the results into forms that would suggest policy options for governments, yet this is precisely what governments need, and hope to acquire through the process. Thus extensive review of the data, as it is affected by corporate culture and management attitudes is a must, and should form a separate study in its own right.

This survey suggests that corporate culture, management attitudes and attitudes towards risk support innovation. On the other hand, anecdotal evidence suggests that this may not always be the case - indeed, perhaps more often than is suggested by this survey, these factors can hinder innovation. Risk-averse management, entrenched personal interests and people in fear of losing their jobs do not generally create an environment conducive to innovative work. The work on innovation management - how to define, value and develop the knowledge assets of the firm - is a direct attempt to address this. (See Leonard-Barton (1995), Amidon (1997), Tid *et. al.* (1997), and Cheisa *et. al.* (196))[2] Indeed, in future surveys in BC, knowledge audits will be carried on at the same time as the basic survey work, to add depth to the basic statistics of innovation.

A synthesis of these approaches may yield new insights into innovation at the firm level. By analyzing a statistically significant number of case studies of actor-networks within and surrounding a firm, and evaluating the character of these networks as the basis for a complex emergence of innovation, new indicators of innovation that go beyond traditional econometrics may be found.

Acknowledgments

Elements of this project received financial support from the S&T Redesign Project of Statistics Canada, and the Social Sciences and Humanities Research Council of Canada. The authors would like to acknowledge the invaluable assistance of their research assistant, Judy Finch, throughout the entire project, and the significant contributions made by the students of Communication 438 (Ed Boroevich, Helen Boufeas, Heather Eynon, and Michelle O'Neil) in the focus group and survey activities. We would also like to acknowledge the assistance of Dr. Catherine Murray of the SFU School of Communication, who gave us valuable advice on methodology and focus group dynamics. Thanks is also due to the Business Register Division of Statistics Canada for their assistance.

ENDNOTES

1. The workshop, called "Innocom 97", jointly sponsored by CPROST and CPSE, was held at CPROST February 11-18, 1997.

2. Amidon provides a timeline of some of the major developments in the study of knowledge capital, see Amidon, D.M. (1997). *Innovation Strategy for the Knowledge Economy: The Ken Awakening.* Boston: Butterworth-Heiman.

REFERENCES

Acs, Z., de la Mothe, J., and Paquet, G. (1996). "Local Systems of Innovation: In Search of an Enabling Strategy". In P. Howett, *The Implications of Knowledge-Based Growth for Micro-Economic Policies*, (pp. 339-360). Calgary: University of Calgary Press.

Amidon, D.M. (1997). *Innovation Strategy for the Knowledge Economy*, Boston: Butterworth-Heiman.

Baldwin, J.R., Chandler, W., Le, C., and Papailiadis, T. (1994). "Strategies for Success" (61-523RE). Ottawa: Statistics Canada.

Bijker, W.E. (1995). *Of Bicycles, Bakelites, and Bulbs - Towards a Theory of Sociotechnical Change*. Cambridge: MIT Press.

Cheisa, V., Coughlan, P., and Voss, C.A. (1996). "Development of a Technical Innovation Audit". *Journal of Product Innovation Management*, 13, 105-136.

Edquist, C. (1997). "Introduction: Systems of Innovation Approaches - Their Emergence and Characteristics," In *Systems of Innovation: Technologies, Institutions and Organizations*. Pinter. C. Edquist (ed.), London. 1-35.

Guastello, S.J. (1995). *Chaos, Catastrophe, and Human Affairs: Applications of Nonlinear Dynamics to Work, Organizations, and Social Evolution*. Mahwah, NJ: Lawrence Erblbaum Associates.

Holbrook, J.A.D. (1997). "The Link Between the Natural Resources Sector and the Services Sector: The Case of Canada and the Pacific Rim" (CP 97-01). Vancouver: Centre for Policy Research on Science and Technology.

Johnson, J., Baldwin, J. and Hinchley, C. (1997). "Successful Entrants: Creating the Capacity for Survival and Growth" (61-524-XPE). Ottawa: Statistics Canada.

Kelly, K. (1994). *Out of Control: The New Biology of Machines, Social Systems, and the Economic World*. Reading, MA: Addison Wesley.

Kiel, L.D. and Elliott, E. (eds.) (1996). *Chaos Theory in the Social Sciences*. Ann Arbor: University of Michigan Press.

Latour, B., Mauguin, P. and Teil, G. (1992). "A Note on Socio-Technical Graphs". *Social Studies of Science*, 22 (1), 33-57.

Latour, B. (1987). *Science in Action: How to Follow Scientists and Engineers Through Society*. Cambridge: Harvard University Press.

Law, J. (1994). *Organizing Modernity*. Oxford: Blackwell.

Leonard-Barton, D. (1995). *Wellsprings of Knowledge*. Boston: Harvard Business School Press.

Lipsett, M.S., Holbrook, J.A., Lipsey, R.G. and deWit, R.W. (1995). "R&D and Innovation at the Firm Level: Improving the S&T Policy Information Base". *Research Evaluation*, 5(2), 123-129.

McKelvey, M. (1997). "Using Evolutionary Theory to Define Systems of Innovation". In C. Edquist (Ed.), *Systems of Innovation: Technologies, Institutions and Organizations*. London: Pinter.

Merry, U. (1995). *Coping with Uncertainty: Insights from the New Sciences of Chaos, Self-Organization, and Complexity*. Westport: Praeger.

Saviotti, P. (1997). "Innovation Systems and Evolutionary Theories". In C. Edquist (ed.), *Systems of Innovation: Technologies, Institutions and Organizations*, (pp. 180-199). London: Pinter.

Schumpeter, J.A. (1949). *The Theory of Economic Development (1911)*. Cambridge, MA: Harvard University Press.

Smith, K. (1992). "Technological Innovation Indicators: Experience and Prospects". *Science and Public Policy*, *19* (6), 383-392.

Tidd, J., Bessant, J. and Pavit, K. (1997). *Managing Innovation: Integrating Technological, Market, and Organizational Change*. Chichester: John Wiley and Sons.

Voyer, R. and Ryan, P. (1994). *The New Innovators - How Canadians are Shaping the Knowledge-Based Economy*. Toronto: James Lorimer and Co.

10 HOW DO SMALL FIRMS INNOVATE IN BRITISH COLUMBIA?

Hans G. Schuetze
University of British Columbia

10.1 INTRODUCTION

Innovative capability and behaviour of firms are crucial for their survival in the market and, from a macro-economic perspective, important factors of economic development and employment. As small firms tend to have certain size-related disadvantages, public policies of various kinds are designed to assist especially small firms to successfully innovate. In the past, R&D and the use of new technologies were seen as central to innovation, therefore most of these programs aimed at the strengthening of the R&D function (primarily through tax credits and research grants) and at enhancing the use of new technologies. The measurement of innovation consequently concentrated on R&D personnel and outlays, on patent data, and on the rate by which new technologies, especially information technologies, were adopted and used.

More recent research on innovation has shown that there are other factors besides in-house R&D that are driving knowledge creation and its application in the form of new processes and products - factors which might be equally or even more important to the innovation process, especially for small firms. In contrast to the traditional linear model which saw technical innovation as the result of R&D, innovation is now increasingly seen as a complex process which evolves in a non-linear mode, with feed-back loops between the various stages of the process (OECD 1997) and with linkages between the various actors within the firm (e.g. designers, production managers, and marketing people) but also with external ones. Innovation is therefore to a considerable extent dependent on the 'system of innovation' in which the firms operate, and on the linkages which firms have with the various institutions and networks of this system. Critical especially for smaller firms is the extent to which they are capable of using information and knowledge from external sources and turning them into new knowledge and applications (Lundvall 1995). The ability to recognize the value and relevance of external information and to assimilate it into the firm's knowledge, depends on related prior knowledge, the ability to learn, and an organization that supports and enhances the process of knowledge conversion and creation. According to this perspective, flexible forms of work organization, human resources management and continuous learning are main ingredients of innovation.

Such a perspective which sees innovation as a process which occurs within a system requires an approach to assessing and measuring innovation that is different from the standard innovation surveys as they are presently conducted (OECD 1997). It is the purpose of this paper to present elements of such a different approach to studying and measuring innovation and the innovation process typical for small firms. This will be done by first outlining the framework within which innovation is managed in small firms and, secondly, by presenting and discussing the methodology and results of a recent B.C. study that looked at some 125 firms in four industrial sectors combining case study and survey methodology.

10.2 SMALL FIRMS AND INNOVATION

Starting with Schumpeter (1942), economists have long believed that large firms have an inherent advantage over smaller ones with respect to their capacity to innovate. This belief is largely based on the assumed benefits from formal R&D departments and activities and the internalized specialized business services which provide resources to market product innovations (Feldman 1994). With the exception of very few, highly specialized firms, small businesses do not have such laboratories or business services of their own. Nevertheless, as research has shown (e.g. Acs and Audretsch 1990; Rothwell and Zegveld 1982), small firms do much innovative work and are in fact the source of innovation in certain industries. How do they do it?

One explanation is that small firm innovative activity relies more than larger firms on external sources of knowledge as inputs to the innovation process. Such external sources are of various kinds: They can be part of the 'technological infrastructure', i.e. public R&D organizations such as research universities or non-university research institutes, industrial R&D conducted by other, larger firms, or mediating institutions. However, there are also other types of knowledge besides those generated by R&D which are contributing to, and critical for the innovation process. Such knowledge comes from other sources such as venture capitalists, informed customers, and specialized business services that provide services that in larger firms are typically provided by specialized in-house units, for instance information, design, marketing and others. Using data from the US Small Business Administration, several American studies have found that the relative importance of inputs from these sources is higher in smaller than in larger firms, thus compensating for example for a less well-developed internal R&D capability through exploiting spill-overs from university and other industrial research (Acs and Audretsch 1990; Feldman 1994).

If much of the information and knowledge small firms require to innovate comes from external sources rather than from inside the company, access to the sources of such information is critical. Access can be sought both in a targeted mode and ad hoc, for example the search for a particular technical information or product that is needed in a particular stage of the development process for a new industrial product. Very often however information is required on a more permanent basis, for example for the purpose of monitoring the market and the competition, or of longer-term technological developments. In this case, networking is an appropriate mode of information and data collection. Through multi-directional flows information is gathered from, and shared with, business partners, competitors, public agencies, professional associations, and various other institutions. In addition, there are also personal networks operating at all levels of the organization, typically related to professional contacts, former colleagues, class mates and both professional and academic mentors.

However, information and data are just the raw material and in order to be useful have to be converted into knowledge or know-how. The ability of a firm to recognize, assimilate, and subsequently make use of new information from external sources is critical for innovation - indeed the firm's "absorptive capacity" (Cohen and Lowenthal, 1990) determines its innovative capability. As external knowledge and information are only retained when viewed as relevant and valuable, the absorptive capacity depends on a company's skills and knowledge base. A firm's the firm's "absorptive capacity" is therefore closely related to the experience, prior knowledge and diversity of backgrounds found within the organization's workforce. Human resources are thus the most important source of innovation.

Within the context of the innovation process, learning is therefore central. Learning occurs sometimes during the process of formal skill training but is often the result of a broader learning process which occurs as a part of everyday experience often linked to communication and interaction with others (Rubenson and Schuetze, 1993). Thus learning is only partly a planned and organized activity but to a large extent an incidental process that includes learning by doing, by using, and by interacting with others, both within and outside the firm, as well as by monitoring both the environment, including the competition, and one's own performance. Therefore, both the organization of the flow of information, and of the learning process are important and affect the innovative capability of the firm (Lundvall, 1995). Such learning occurs both at the individual and the organizational level. Individual learning takes place in the form of both formal and informal education and training, and organized or incidental on-the-job learning. Organizational learning builds on individual knowledge and skills. It is a process by which an organization, through its internal mechanisms "amplifies the knowledge created by individuals and crystallizes it as part of the knowledge network of the organization. This process takes place within an expanding 'community of interaction' which crosses intra- and inter-organizational levels and boundaries" (Nonaka & Takeuchi, 1995). Such learning at the level of the organization is the hallmark of 'learning organizations' which are characterized by a management style and work organization that enhance and support interactive learning, knowledge and innovation (Senge, 1990; Gjerding, 1991).

There is thus a close and symbiotic relationship of technical and organizational innovation, or the dependence of the innovation process on the way organizations manage information and knowledge. The internal organization of the firm which permits the 'absorption' of the information and knowledge coming from external sources is of critical importance. This capability of internalizing and using externally generated knowledge or information provided from outside the firm is dependent to a large extent on flexible organizational structures.

The trend towards knowledge-based economies has underpinned and increased the pivotal role of human resources. Numerous research studies and statistical surveys indicate that innovative sectors of the economy are associated with an increased level of skills (OECD, 1992; OECD, 1994). This does not only imply increased demand for higher levels of technical and/or job-specific skills, companies in these sectors are also looking in the personnel they hire for more broadly-based and more generic skills, for example in the area of communication, work ethics, problem solving, team work, and adaptability. From the viewpoint of innovative firms, education and training, and the institutions that provide it, namely schools, universities, technical institutes and colleges are therefore most important parts of the public infrastructure.

The 'upward skill bias' (OECD, 1996) has a number of consequences for enterprises and their investment in human resources development. Essentially firms are faced with two decisions, namely, what type of new skills they will hire, either through conventional hiring practices or making use of sub-contractors and consultants, and secondly, the extent of their 'human capital' investment regarding the continuing skill development of the workforce. The latter type of decision relates not only to formal training, whether undertaken in-house or by external agencies and individuals, but also continuous or recurrent learning opportunities of other kinds, ranging from participation in conferences and trade shows to quality circles, in-firm mentoring, systematic multi-skilling and job rotation.

Small firms are thought to be experiencing problems with regard to shortages of qualified personnel (OECD, 1996). The lack of clearly defined career paths, of social status associated with hierarchies, and of other incentives or rewards makes small firms often less attractive to highly qualified personnel than larger firms. But small firms appear also to have also problems with their own human resources management. With few exceptions surveys on enterprise-based or firm-sponsored training have found a negative correlation between size and (formal) training incidence, or, in other words, that smaller firms train less than larger ones (Betcherman 1997). On the surface, these findings indicate that not only are small firms at a disadvantage regarding the recruitment of suitably qualified staff, but they also under-invest in the development of their human capital. If human resources are indeed a critical factor of innovation, and if the findings are correct that they have problems finding qualified personnel and of providing their workforce with relevant continuing training, how do small firms manage to innovate and be successful in highly competitive markets?

There are several explanations. One is that studies on enterprise training and learning activities are not conclusive as they are still few in number, data especially on small firms are patchy, and their methodology is, on the whole, little sophisticated (Betcherman 1994). There is yet no appropriate statistical framework for the collection and analysis of training statistics (OECD 1992). Most surveys measure formal and structured rather than informal training activities, and training rather than learning activities. Almost all of such survey data are collected from employers rather than from employees. Moreover surveys tend to concentrate on data that are easily observable and measurable , for example the number of participants, the number of trainers or of available programs, and the outlays for training activities. To collect such data at the firm level requires some formal structure, such as units for human resources development, structures that small firms are often lacking. Also, human resources development in the context of a less structured and more informal environment might mean less 'training' and more directed, mentored and self-directed 'learning'. Therefore, what appears to be a different profile of skill development and learning activities within smaller companies may well be reflecting a contrasting style of organizing work and of innovation.

It follows from the above that a study on innovation in small firms needs not only to look into the linkages and interactions with the innovation system but also into work organization, skill development, and human resources management. Since it can be expected that there are differences between the strategy and the practice, and the blueprint and the actual pattern, the formal and the informal organization, a study must attempt to find evidence for the actual set-up, interactions and relationships between the various actors, both within the firm and with outside contacts, sources of information and networks. Rather than concentrating on in-house R&D and formal training activities, a study on innovation in small firms must focus on the larger framework in which innovation takes place.

Different from other studies on innovation in small firms (e.g. Baldwin 1994; Voyer & Ryan 1994), most of which concentrate on technological innovation, i.e. the design, implementation or production of new or substantially improved technological products and processes, the emphasis of the perspective applied here is on linkages with the environment, i.e. the system of innovation (or the 'technological infrastructure'), and on organizational innovation and learning. Such a focus on knowledge generation and management, on individual and collective learning, and on the utilization of skills and knowledge concentrates on the process of innovation rather than on the result of such activity.

This approach takes account of another insight of recent innovation research, namely that innovation is not a singular or one-off event, but rather an on-going process of 'learning, searching and exploring', which can be expected to eventually result in new products, new techniques, new forms of organization and new markets (Lundvall, 1995). It is neither primarily concerned with innovation inputs, nor with innovation outputs as they are conventionally measured (OECD 1997). Rather this perspective looks at the black box in between, i.e. the innovation 'throughput' by which tangible and intangible inputs are transformed into innovative outputs. While such an approach cannot yield data which are easily defined, measured and compared, it can be expected to provide insights into how small firms innovate by illustrating how external and internal factors contribute to the innovative capability of these companies.

If innovation means learning, and learning means the process by which relevant information and knowledge is absorbed and converted, the questions 'What do we want to measure?' and 'How can we measure it?' must be answered differently than in standard innovation studies where the role attributed to human resources, their management and development, and to learning is largely overlooked. Thus, to the extent that human resources management and development are mentioned at all in the standard innovation surveys, training activities are conceived as very narrow and instrumental, counting as 'innovation activity' only 'when it is required for the implementation of a technologically new or improved product or process' (OECD 1997 p. 61). Such an understanding is clearly inadequate when innovation is seen as a complex learning process that is dependent on prior knowledge and experience, on the context in which it occurs, and on social interaction with others.

Learning in this wider sense is not the result of formal education and training activities alone, but involves a variety of informal and incidental learning activities many of which resulting from interaction with others, as mentioned above. Not only innovation surveys concerned with measuring technical innovation fail to recognize the importance of these activities, but also most surveys on industry training tend to overlook or disregard them. While this may be partly due to the difficulties of measuring these activities, the main reason is probably that most training surveys still ask 'first generation questions' (Betcherman 1993), inappropriate for gauging the role of learning in the innovative process. That does not mean that the responses to some of these first generation questions (e.g. incidence of training; training budget; management responsibility for skill development) are entirely without value as they can serve as indicators of management attitudes and firm practice regarding formal training activities. But, as pointed out, these do not account for the bulk of learning activities required and therefore traditional studies of training at the enterprise level are not suitable for adequately measuring learning activities.

10.3 THE BRITISH COLUMBIA STUDY

Between 1995 and 1997, a study on innovation in small firms was conducted in B.C. which was based on the perspective developed above. Organized under the auspices of the Centre for Policy Studies in Education at the University of British Columbia and financed by grants from the Social Sciences and Humanities Research Councilas well as from the provincial and federal governments, the study focused on the sources of innovation in small firms and on the relationship between work organization, human resources development and management and innovative capability.

10.3.1 Methodology

The study of the innovative process in small firms is a fairly new field of research as most studies so far have focused on larger firms many of which with their own R&D activities and departments. Therefore research on small firm innovation lacks well developed concepts and coherent theoretical foundations. These circumstances militate against the use of surveys, or at least, the sole use of surveys, and suggest more intensive and qualitative research methods such as the case study approach.

The B.C. study applied therefore a combination of firm case studies using a fairly comprehensive interview guide, and a mail survey that was subsequently mailed to a random sample of firms in British Columbia. This combination of methods seemed to be appropriate for the study of small firms since it permitted not only the collection of relevant statistical data, but also qualitative data both in the form of interviews with owners or managers and employees, and useful anecdotal information from informal work site observations , which provided a broader context for the interpretation of interview and survey data. This type of contextual evidence was particularly valuable since the actual behaviour and practice of small firms is very much dependent on the personal attitudes and philosophies of founders, owners or managers. The survey provided a statistical picture of the industries under investigation, adding detail to the initial case studies, while positioning the case study firms within the larger spectrum of the industries to which they belong.

Strategies and practice of work organization, human resources management and learning are not only dependent on firm size, but are to a considerable extent, sector-specific. Since there is no 'typical small firm' which shares characteristics besides size across sectors, the study concentrated on a limited number of sectors. The choice of these sectors was determined by three general criteria: first, the role of small firms within the sector, secondly, the importance of these sectors for the Canadian, and in particular the British Columbia economy; and, thirdly, the level of innovation within these sectors.

The first two criteria were relatively straightforward to determine. In British Columbia, there is a dominance of small firms in almost all sectors, with the exception of the resource extracting sectors like forestry and mining. Economic importance of a sector was based on its export orientation, contribution to GDP, employment numbers. Given the structure of the British Columbia (and generally the Canadian economy) it was clear that the study would have to include the service sector which generates 70% of the GDP and 75% of the jobs. To look into service sector firms was also important for another reason. As most of the studies on work organization, HRD and innovation, have almost exclusively concentrated firms in the goods producing sector, the inclusion of the service sector was of considerable interest from a methodological point of view.

The third criterion, 'innovative capability', was more difficult to determine. Here, the level of modern technology made use of by a particular sector, along with the level of technical expertise, both in the form of product and process innovation, and also the degree to which a particular sector engaged in organizational innovation were the main elements. Unlike standard innovation surveys which measure whether or not firms be have been 'innovators' or 'non-innovators' over a fixed time period, typically the last three years before the survey (OECD 1997), a study on innovative capability, i.e. the role of organizational innovation and human resources in the innovation process, does not require to make such a distinction. In other words, for a study on the innovation process at the micro level the actual measurable outcomes of innovative activities, while an indicator for such activities, are less interesting than the process by which the various elements of the innovation process, both internal and external, are organized to interact.

There are two reasons why the result of this process is of less interest than the process itself, the time lag involved in innovative activities, and the dependence on the respective industrial sector. Since innovation is a complex process, concrete and measurable results take often time to materialize, or may materialize in indirect ways rather than in an innovative product or process. For example, according to the guidelines for innovation surveys (OECD 1997), aborted innovation activities are not to be included in the count. Yet such failed attempts provide the firm with valuable experiences and insights that may be crucial for other innovative activities. If the assumption is correct that innovation is a prerequisite for survival in the market (with the exception maybe of resource-based industries where supply and prices on the world market may be the determining factors of success), non-innovative firms will not survive for long. It therefore seemed sufficient to include firms in the sample that had been in existence for a number of years and had reached a certain size. While this excluded some new innovative firms , and may have included some others who are not innovative but have been able to hang on for one reason or other, actual market success and survival were considered as a primary test and indicator of innovativeness. Taking such a longer-term perspective has the advantage of avoiding some difficult conceptual problems, for example the assessment of how new and improved products and processes need to be in order to qualify as a innovation, or the problem, even more complex, of defining what constitutes an innovation in the service sector.

Also, information on actual innovations that companies have realized over a particular time period are a somewhat unreliable indicator of innovative capability because the speed of an innovative process depends not only on the individual firm, its strategy, management and organization, but also to a considerable extent, on the sector and its business environment. It is this sector-specific pattern of innovation that is often overlooked in studies that try to analyze the innovative capability of firms making the exercise of counting the population of innovative firms problematic.

Based on the three criteria mentioned, the following four sectors were selected, two from the larger category of manufacturing, and two from the service sector:

- secondary (value added) wood product manufacturing; (including cabinet, wooden doors and windows, and other wood industries);
- telecommunications equipment manufacturing;
- engineering consulting; and
- computer services.

The choice of sectors thus included two mature sectors, wood manufacturing and engineering, and two new sectors which have been in existence for only twenty years or less. This combination permitted not only a comparison of innovative organization and behaviour in young and older industries, but also of manufacturing and service firms, even if that distinction was somewhat blurred in the case of the high tech industries. (Figure 10-1).

Figure 10.1 - Sectors in the B.C. Study

	Mature industry	**New industry**
Manufacturing	secondary wood	telecommunications
Service	engineering consulting	computer consulting

The firms studied had 10 - 100 employees and had been in existence for five years or longer still thus eliminating those which are still in the start-up stage. Companies with fewer than ten staff were not included in the study on the grounds that very small firms have particular structures, modes of operation and problems that are not comparable to those of small firms of a larger size. Thus, companies of more than ten and under 100 employees were considered to represent the norm when looking at the organizational structure of small firms in most industrial sectors. Companies with more than 100 staff typically develop a more pronounced division of labour, in conjunction with a more layered, or hierarchical supervisory and management structure, likely to significantly impact on both work organization and human resources management. Size was also assumed to have an effect on the extent of participation in the kind of strategic partnerships or networks that link firms to customers, suppliers, 'spin-off' enterprises, competitors, government and industry-operated laboratories, universities and colleges.

While this concentration on small firms of 10 to 100 employees seemed appropriate for the purpose of the study, it had also a certain disadvantage, since it excluded the possibility of a systematic comparison between firms included in the study and very small companies, or larger corporations. But even within the size chosen there are significant differences between the smaller and the larger firms. In order to capture these differences, three sub-groups were distinguished within the sample, namely firms with 10 - 25, 26 - 60, and 61 - 100 employees, thus allowing for size-specific differences.

Sixteen case study firms were selected: four from each of the four sector mentioned. Interviews were conducted with firm owners and managers as well as with some of the employees. In the case of the wood manufacturing firms, an industry that is heavily unionized, interviews included also shop stewards or other union representatives. Together, these interviews lasted on average between six and ten hours per firm.

Following the completion of the case studies, a random sample of British Columbia firms within the four sectors were surveyed by means of a mail back questionnaire. The data base for the sample comprised lists obtained from B.C. Statistics, industry and professional associations and their publications. The survey was mailed out in November 1996 to a total of 234 firms, and concluded by the end of March 1997. Of the 118 responses, 109 or 46% were usable.

The survey questionnaire was based on a standard set of topics and questions and had been developed from the interview guide, used for the case studies. Besides information on the general profile of the firms and their business strategy, both instruments contained questions relating to three larger fields:

- the firm's environment, business relationships, external sources of innovation, and innovative activities;
- the organization of the workplace with respect to the management of information and knowledge
- employment, human resources management, skill demands and learning;

The case studies also mapped the trajectory of particular innovations in each firm in order to determine the origin of the idea, the sources of information and knowledge, the several stages of the innovation process (which more often than not was unplanned and non-linear), and the individual and organizational learning that was involved.

Together, case studies and survey yielded a relatively clear picture of the elements and processes dominant in small firms' innovation activities (Schuetze et al. 1998). In the following section, I will summarize this picture.

10.3.2   Summary of Findings

The profile of the 125 small firms who participated in the study confirmed those studies who have found small firms, especially those in advanced technology sectors, to be particularly dynamic. Overall, 75 % of small B.C. firms are active in foreign markets, whereas 92 % of the telecommunication manufacturing firms, and 80 % of the computer service firms, exported their products and services. As could be expected, activities in foreign markets was dependent on size: Within the group of companies studied, the larger ones (61 to 100 employees) were more active than the group of the smaller firms (10 - 25). The vibrant nature of these firms was also documented by employment growth. More than 50 % of firms had increased the number of their permanent employees over the last five years, over one third had grown by more than 50 %. The telecommunications sector proved to be the most dynamic: more than 75 % of the firms had hired additional employees over this time span.

10.4   EXTERNAL SOURCES OF INFORMATION AND KNOWLEDGE

The study confirmed the pattern, noted above, that small firms rely heavily on external information sources, using networks for sharing information on a regular basis with other firms, industry or related research organizations, and public infrastructure institutions.   Confirming the findings of other studies on the importance of producer-user relationships, the study shows that contacts with other firms are the most frequently used source of information. On the vertical axis of such relationships, these contacts involve suppliers and customers, while on the horizontal axis these contacts are with other firms in the same sector or the same geographical region. Relationships with supplier and customers are often individual, while contacts within horizontal networks are often mediated through industry or sectoral associations.

Hence, professional and sectoral associations are important sources of information for small firms. Approximately three quarter of firms are members of a professional or sectoral organization. Not surprisingly, given the strong influence of professional engineering associations reflecting a clearly defined professional field of theory and practice, the highest membership rate was found within the engineering firms (83%). When broken down by size, the data showed that larger firms are more active in these associations than smaller ones. Although this appears to contradict the assumption that especially smaller firms rely on networks, this result reflects the typically wider range of activities of larger firms and the greater availability of staff who can actively participate in such associations.

Associations are offering a number of services to member firms. Providing members with market- and technology-relevant information appears to be the most important of these. Especially offices of engineers are interested in influencing, through their professional associations, public norms in areas such as waste or water treatment, or environmental standards as these have a direct impact on the nature and volume of their business.

Many of the associations offer custom-tailored education and training courses of the kind that are unlikely to be offered by public institutions because they are either oriented to a very specific topic, e.g. training for a new type of software application, or a rather broad theme, for example an overview of the impact of NAFTA on exporting conditions. Such customer-tailored orientation, information and training activities are particularly in demand from the larger firm segment. They are also popular among secondary wood manufacturing firms of all sizes, a sector that in the past has not been served well by the public post-secondary education and training institutions in the province. The recent establishment of a technician and woodworker training school by the main industrial association (financed by the Province) is undoubtedly the main reason for this popularity.

When firms were asked specifically about the usefulness of a number of sources of information and knowledge for the development of innovative goods and services, trade shows and related activities were cited by more than three quarters of all firms. Trade shows, fairs or exhibitions typically include industry presentations, demonstrations or workshops, and provide firms with the opportunity to meet with suppliers of new equipment, systems, software applications and other related solutions. As these industry shows offer also an opportunity to gauge the competition, and to set up contacts with potential clients, they figure significantly in the marketing and development strategies of firms, especially in the wood manufacturing sector (84%).

Publications, ranging from scholarly journals to professional and technical periodicals or trade magazines of various kinds, were identified by the study as important sources of information for computer and engineering firms, where the monitoring of research and development figures prominently in their business strategies. Increasingly, the InterNet is used as an information gathering tool, particularly by firms in the computer sector. Developers and employees in the sales units routinely search for technical information, download free software, access the latest product prices and specifications, and share information with colleagues, within time frames and costs that can significantly bridge the resource gap between small and large organizations. In all sectors, the InterNet has gained quickly in importance as a search tool outpacing traditional academically-based sources of information. Given this very fast ascendance of this worldwide data bank, it can be safely assumed that the InterNet will increase its reach and importance and will probably in the long run replace written information from other general sources such as academic or professional literature.

Engineering and computer firms make also frequent use of academic and professional conferences and seminars, and, albeit to a markedly lesser extent, of university or government laboratories. By contrast, universities and their activities are of little interest to wood sector firms. It can probably be safely assumed that this difference reflects at least two factors. Firstly, the level of qualifications of the workforce in both industries. While all professional engineers have received university training and are therefore at ease with academic language and science-based approaches to problem definition and solving, there is relatively few staff in wood manufacturing firms with an academic background. As is known from other studies, this difference in professional culture and lack of familiarity tends to bar easy access to universities.

Larger firms are therefore making greater use of academic conferences, and of university or government laboratories, than smaller companies. This is partly a logical consequence of size - larger firms have more university-trained personnel than smaller ones - but it also confirms other studies, which suggest that smaller firms are generally not well served by universities, owing to the way university research is organized and the fact that applied research, which might contribute to actual problems that firms experience, is less prestigious for university researchers to engage in than is basic research (Schuetze, 1996). The case studies have shown, that contacts and cooperation with university researchers are often made directly rather than mediated through Industrial Liaison Offices or other intermediate bodies set up to enhance technology transfer. This means that there is no formal contract research agreement between a firm and the university but researchers are often approached individually by the firm (and compensated for their work as private consultants). Clearly, the most frequent mode of knowledge transfer from university laboratories to small firms remains the recruitment graduates who have a background and research experience of relevance to the firm.

In spite of the relatively low importance, compared to other external of innovation, accorded by firms to university and government research laboratories, the linkage for small firms to R&D institutions remains important. As the case studies show, technical innovations, except for minor or incremental ones, are rarely realized without resorting at some point to R&D-based knowledge generated externally. There is some evidence from the survey that reliance on external R&D related know-how might be even increasing in importance as in-house R&D is decreasing.

10.5 ORGANIZATIONAL INNOVATION

It is one of the premises of this study that organization and the capacity to innovate are closely related, and that access to, generation of, and management of information and knowledge, a flat and flexible firm structure, and the development of human resources are of specific importance. The organization of a firm comprises the internal structures and modes of operation, in conjunction with the external relationships the enterprise has developed with its broader business constituency, consisting of competitors, customers, suppliers, government, stockholders and unions.

Some of the organizational features of a firm are 'tangible' and can be described and measured, such as formal levels of decision-making, ownership, or formal partnerships or alliances with other firms. Others are intangible and cannot be easily captured by descriptive or quantitative categories (although some of the tangible data can be used as proxies), such as the 'organizational culture' or 'learning environment' within an enterprise.

One of the principal features of 'flexible' organizations is that they are 'flat', as opposed to hierarchical. While there are numerous management concepts of how flexibility can be designed and implemented (Newton 1995), one simple criterion is the authority structure, i.e. levels of supervision and participation in decision-making. According to the survey, approximately two thirds of firms classified their internal structures as flat and one third as hierarchical. Not surprisingly, the two service sectors were on the top end of those seeing their structures as flat. When this assessment by firm managers was checked against the actual levels of supervision however, it appeared that the classification was more a matter of management style and degree of formality than the actual authority structure.

An additional question about the dominant work organizational unit provided additional insights showing that the majority of the service sector firms are organizing their work in work teams (computer and engineers), while in the manufacturing sectors the majority have a departmental structure. Work teams or groups are also the dominant organizational structure in more than half of the computer and engineering firms. Working in these kinds of group situations entails the enlargement or 'enrichment' of job functions and responsibilities at the individual level, requiring team members to not only explain to the one another, but also to apply their skill base and know-how, thereby helping their colleagues to develop new know-how and understanding.

Such team-based organization is not only seen as an important feature for facilitating 'learning by interacting' but also for securing the access, dissemination and absorption of information and knowledge from external sources, a most important feature of especially small firms as pointed out already.

The sheer volume of new technical knowledge has resulted in the two high tech sectors becoming more selective in terms of access to information, with filters in place to prevent too much company time being devoted to interesting yet seemingly unproductive areas. As a result, knowledge creation through the sharing and appropriating of external technical information and knowledge tends to be less systematic and structured, and occasionally somewhat haphazard in comparison with the process of creating knowledge from internal sources through the various forms of knowledge conversion.

In high tech and engineering firms the sharing of information and knowledge from external sources occurs typically by way of presentations and discussions at internal seminars, 'brown bag lunches', by the circulation of papers and articles among colleagues, and more targeted mentoring and training activities. By contrast, secondary wood firms involved in the production of standardized goods tend to use these arenas very sparingly, depending more on middle managers such as department heads or production supervisors to disseminate relevant information.

By contrast, organizational knowledge, i.e. know-how related to how work or people are organized, tends to be concentrated at the upper levels of the organization and is often marginalized in favour of more immediate business concerns, such as product design or marketing. An example is provided by the secondary wood products sector where many companies are going through a fundamental shift in production techniques. In these companies, the emphasis is on new technology, and the ways in which they can be employed to enhance productivity and quality, but rarely is there more than a very direct and instrumental interest in the organization of the workplace. In other words, the know-how brought into the firm is almost exclusively directed at production and quality requirements, and organizational know-how is often missing. The result is that even where new technologies have been installed, work is still sometimes organized along lines more appropriate to the traditional process of design and production in manufacturing.

However, the opposite trend can sometimes be observed too. In one of the case study firms, using older and partly out-of-date production equipment, management has achieved considerable improvements both of productivity and quality, and of staff morale through a rather drastic change in management style, work organization and corporate culture by actively encouraging employees to improve both work attitudes and skills, involving them in decision making about their work, and increasing their latitude of action through the building of work teams. This particular example is illustrative as the attempt to create a more worker-centred company culture has produced tangible benefits in a relatively short period of time, not only in terms of increased productivity, quality and sales, but also in terms of worker. This was achieved not through the purchase of modern equipment but through a modern management approach which built on internal sources by using the creative potential and the tacit knowledge and 'community of practice' (Brown and Duguid 1991) of the existing workforce. The success of this particular case demonstrates the need to rethink a traditional and often narrow definition of knowledge and knowledge management, stemming from the times of Taylorism, with principles which were applicable to large scale manufacturing plants, and even then of questionable value for small firms.

10.6   HUMAN RESOURCES MANAGEMENT (HRM)

The example illustrates dichotomies of approach and practice with respect to work organization which exist in all the sectors studied. Some of the managers see knowledge as an integral part of a broader commitment to an organic learning environment, where know-how creation through skill development and learning are a genuine part of the working process. In contrast, others seem motivated primarily by the short-term 'bottom line' and direct cost-benefit efficiencies, and tend to view training primarily as a cost and time factor, to be incurred only when specific training needs emerge, often in conjunction with new technologies, products, or markets.

The picture is therefore mixed with regard to HRD. On the one hand, while the majority of firms said that HRD was a strategic part of their business plan. this strategic commitment did often not translate into policies for training and company practice.

For example, the study found an absence in most firms of designated or formally trained professionals working in HRD. Across the board, firms cited a lack of funds and work related time pressures to explain the apparent inconsistency between a perceived need to develop into learning organizations and the prevailing operational practice of marginalizing the responsibility for the planning and administration of training and professional development. Thus, senior managers typically combine their regular responsibilities with that of HRM, making strategic decisions regarding skill needs, the allocation and provision of resources for training, and operational choices about providers, programmes or courses, and participants. In practice however, many of these decisions (with the exception of the budget), are left to middle managers who often lack the experience necessary to make informed decisions. Given the group or project nature of work in engineering, telecommunications, and computer services, and the movement towards group work in the secondary wood products sector, this type of watered-down human resources management system, even if it is viewed as a more efficient use of precious executive time, is clearly problematic.

The financial aspects tend to be treated with equal low priority. Most case study firms do not have a formal training budget nor do they have a procedure in place to systematically track the outcomes of courses and programmes taken. Although there are a few notable exceptions, this picture appears fairly representative across all sectors.

This result is in line with findings from other studies. For example, a large-scale study on Canadian firms (Betcherman, 1994) found that an employers' commitment to HRD was often more 'rhetoric than reality' and that 'Canadian firms often base their competitive strategies on cutting costs and introducing new technology. Strategic approaches that place human resources front and centre are less typical and often at odds with dominant strategies.' The study found further that 'he majority of firms studied still operate under traditional models that place a low priority on human resources and involve a low degree of commitment between employer and worker. These organizations tend to integrate HRD issues into overall business plans, have Taylorist job designs and work processes, a low level of investment in training, and rarely share information or decision-making with workers or their unions.'(p.64)

On the other hand, it must be stressed that the informal yet organized learning activities typically found in smaller firms are far more numerous and varied than the incidence of formal training arrangements would suggest. Both the case studies and the survey revealed informal mentoring, often within project teams or focus groups, brainstorming sessions and other interactive mechanisms are standard practice in the majority of firms. This tends to confirm the assumption mentioned above that small firms, which are found by all surveys to provide little training for their employees, are compensating formal training arrangements by organizational arrangements that facilitate informal ways of training and learning.

When knowledge needs are of a more specialized nature and not available from within the firm, know-how must be acquired from outside. This is particularly true of technical skills related to new techniques, equipment and software etc., but also soft skills such as communication and problem-solving skills. Skill training in all of these areas is available from both education and training institutions, public and private, but also through professional associations, and in some cases, voluntary organizations and unions.

The study found that there is a perceived lack of suitable courses at universities and colleges in terms of content, mode, and scheduling. This can be partly explained with the nature and structure of supply, namely more long term and generic courses provided by post-secondary institutions, in particular universities, which are targeted to degree-seeking students rather than to professionals and other employees seeking short-term and more technically-oriented training. The apparent dismay felt by many managers over what universities have to offer is also rooted in the nature of demand. Small firms typically seek short-term but intensive training courses ('just-in-time-learning'), a demand that can be more easily met by smaller private providers or industry associations in a position to react more quickly to short-term demand than large public, and in particular academic organizations. Even if continuing education programmes and special units in universities and colleges are trying to orient their continuing education offerings more to market demand, they are often still at a disadvantage in comparison to other providers, especially when it comes to 'just-in-time' delivery, the availability of particular state-of-the-art equipment and technology, and sometimes also with regard to the familiarity and expertise of their instructors with state-of-the-art technology.

Universities and colleges are seen as doing better in providing initial education. The high-tech and engineering firms, who recruit the majority of their junior level personnel from B.C. institutions, were by-and-large content with the quality of graduates, even if occasionally complaining about the lack of practice relevance in both computer science and engineering programmes. By contrast, firms in the wood sector were generally critical about the lack of basic skills that many young people bring to the job. As most of these young people have very little in the way of vocational qualifications, some managers blame the school system, which they feel has traditionally done a poor job of preparing young people for working life or to imbue them with a sense of responsibility, willingness to learn, and a positive work ethic.

As markets, technologies, customer relations and other parts of the business environment change, most firms have become aware of the importance of 'soft' skills when working in flexible structures and required to communicate effectively with others, both within and outside the firm. Seemingly paradoxically however, such soft skills do not appear to be a priority when firms interview candidates for a job opening. When interviewing, firms regard technical or professional competence, documented by degrees, certificates, diplomas etc., as a basic requirement while candidates are expected to demonstrate, and bring to the job, additional qualifications and attitudes which make them fit into the firms' culture and organization. The importance of the 'fit' factor is evidenced by the practice of virtually all case study firms of hiring at the junior level, a practice which permits firms to acculturate new employees. Firms are also very positive about co-op programmes and like to make use of work release students when possible. Many of them go on to hire former co-op students upon graduation, attracted by their knowledge of the company, its products and culture, and confident in their ability to produce within a relatively short period of time. Companies from the two high tech sectors appear particularly interested in the potential and 'fit' of prospective employees. In line with this emphasis and as a consequence of the current shortage of skilled computer professionals, firms in both these sectors indicated they were preferred to hire new graduates with limited experience but are enthusiastic, hard working, unattached, and relatively free from the preconceived ideas and work related problems associated with more seasoned professionals.

Specific training in the areas of inter-personal. analytical and problem-solving appears to be the exception rather than the rule. It is generally expected that these can be picked up on-the-job, facilitated through team work, job-rotation and other flexible organizational structures and procedures. This practice also applies in engineering where management skills are considered to accrue over time, a result of learning by doing and assuming management responsibilities during the course of a career rather than systematic learning. However some firms are slowly beginning to address the problem by hiring professionals outside the field with backgrounds in management or other 'soft' skills areas.

10.7 SUPPORT FOR INNOVATION

The study found very different views among managers on how government and other institutions from the public sector can and should support innovation in firms. The general consensus is that government has a responsibility to create the type of framework that supports private initiatives, and to interfere where market mechanisms fail. The latter point, market failure, is the main argument in favour of a number of subsidies and other support for R&D which, it is argued, are too costly and risky for small firms and would therefore not take place if government did not pick up some of the cost.

Tax credits remain the primary source of government support for innovation, with the high tech areas making most use of this type of funding. In contrast, engineering firms make little use of the scheme since R&D expenditures are mostly accounted for under the heading of project costs, and billed to clients. In the case of large projects, involving substantial R&D, the client is often the government or a public utility, which accounts for the importance of public contracts in overall fiscal strategy rather than concerns over tax credits. In the secondary wood sector, managers seemed unfamiliar or dubious of how the scheme might be applied to this sector.

The major criticisms of the scheme centred on the credibility of the assessors to fully comprehend the dynamics of emerging technologies, the extensive paperwork and time required to navigate the process, and confusion regarding appropriate definitions for both research and development, that not only make sense but are workable within current business contexts. Many companies would ideally like to see the scheme extended to include some of the outlays for commercializing new products and support services such as training and administration, which many consider an integral part of the R&D process.

Beside tax credits, most firms surveyed have also received R&D grants and associated financial assistance for their innovative work, however, direct R&D grants are also a problem for many companies. Although the funds are welcome, there are concerns over the long delays and excessive paperwork involved, coupled with the low rates of success and need to deal with a bureaucracy with little understanding of the way small firms work. One manager complained that by the time a company has provided yet another business plan to apply for assistance, the rest of the application would probably be out of date "and the process likely to amount to no more than an exercise in frustration". Companies would like to see application processes streamlined with agencies bound by specified turnaround times.

As international markets and competition have become increasingly significant to firms in all four sectors, many firms, especially in the high-tech sectors and in engineering, think that more appropriate support is needed related to exports and other forms of international business. Increased competition in regional and national markets, coupled with improved access to international markets has many companies caused to develop overseas marketing strategies and linkages with business partners abroad they are often ill prepared to sustain. There is a consensus that more could be done to assist companies with introductions to foreign markets than the occasional high profile trade mission, and more funds need to be made available to assist firms deal with both the increased overheads and cash flow problems associated with building up business contacts and negotiating sales overseas.

10.8 CONCLUSIONS

The paper has attempted to show that in many respects standard ways of measuring innovation are ill suited the innovation process in a knowledge-based economy in general and to the situation of small firms in particular. Standard surveys as conceived by the OECD and the European Union (OECD 1997) and practices in several countries. In the light of the close and symbiotic relationship between technical and organizational innovation and the importance of human knowledge, skills and learning in the innovation process, the failure to include data in these fields is clearly problematic. Admittedly, of the more intangible elements of this process are difficult to quantify and measure, a different methodological approach is needed. In this paper, a combination of case study research and survey instruments is advocated. Rather than measuring the input and output of the process of innovation, such an approach would lead to a better understanding of the process, its main elements and their interplay. By looking more closely in the black box, namely the innovation process at the enterprise level, governments and other public bodies could better identify those areas where support through public infrastructures and targeted programs are particularly needed and effective. Such a new approach is especially necessary for understanding the process of innovation in smaller firms, in particular in the high-tech and the service sectors. Since only one in eight of the newly founded small firms survives the first two years of operation, a review of public assistance and support for small firms seems appropriate.

The approach suggested here has been applied to a recent study on innovation in small firms in British Columbia. Targeting small firms in four sectors, the study has yielded many insights into how the methodology can be improved for assessing the innovative capability and behaviour of small firms. It has also produced a picture of small firms that is, in some aspects, different from that painted by standard innovation surveys which tend to concentrate on larger firms and on enterprises in the manufacturing sector.

REFERENCES

Acs, Z., and Audretsch, D.B. (1990). *Innovations and Small Firms*. Cambridge, Mass. MIT Press.

Baldwin, J. (1994). Strategies for Success: A Profile of Growing Small and Medium-Sized Enterprises (GSMEs) in Canada. (Revised ed.). Ottawa: Statistics Canada/Industry Canada.

Betcherman, G., Leckie, N. and McMullen,K. (1997). *Developing Skills in the Canadian Workplace: The Results of the Ekos Workplace Training Survey.* CPRN Study No. W02. Canadian Policy Research Networks. Ottawa.

Betcherman, G., Leckie, N., and McMullen, K., Caron, C. (1994). "The Canadian Workplace in Transition". Kingston, Ont.: Queen's University (Industrial Relations Centre).

Betcherman, G. (1993). "Research Gaps Facing Training Policy-Makers". *Canadian Public Policy*, vol. XIX, 18-28.

Brown, J.S. and Duguid, P. (1991). "Organizational Learning and Communities in Practice: Toward a Unified View of Working, Learning and Innovating" *Organization Science*, 2, 2.1, 40-57.

Feldman, M.P. (1994). *The Geography of Innovation.* Dordrecht: Kluwer.

Gjerding, A.N. (1991). "Work organization and the innovative design dilemma". In B.A. Lundvall (ed.), *National Systems of Innovation: Towards a Theory of Innovation and Interactive Learning.* London: Pinter.

Nonaka, I. And Takeuchi, H. (1995). *The Knowledge Creating Company - How Japanese Companies Create the Dynamics of Innovation.* New York: Oxford University Press.

OECD (1997). *National Systems of Innovation.* Paris, OECD.

OECD (1996). *Technology, Productivity and Job Creation.* Paris: OECD.

OECD (1996b). *Innovation, Patents and Technological Strategies,* Paris, OECD.

OECD (1994a). *The Comparability of the Innovation Survey Questionnaires and Analytical Projects.* Paris: OECD.

OECD (1994b). *The OECD Jobs Study.* Paris: OECD: 2 vols.

OECD (1992). *Technology and the Economy - The Key Relationships.* Paris: OECD.

Senge, P.M. (1990). *The Fifth Discipline: The Art and Practice of the Learning Organization.* London: Doubleday.

Voyer, R. and Ryan, P. (1994). *The New Innovators - How Canadians are Shaping the Knowledge-Based Economy.* Toronto: James Lorimer and Co.

11 THE DYNAMICS OF REGIONAL INNOVATION IN ONTARIO

Meric S. Gertler, David A. Wolfe and David Garkut
University of Toronto

11.1 INTRODUCTION

Over the past decade, the industrial economies have witnessed a wave of economic and political change that most find difficult to comprehend. The phrase that keeps reappearing in attempts to explain this phenomenon is 'a shift in the tectonic plates' that shape our society (Stewart 1997, xvii). At the root of this change are three inter-related processes: the emergence of a new information technology paradigm that is dramatically altering the economic calculus of production and distribution throughout the industrial economies; the phenomenon of globalization which is intensifying the linkages and interdependence between the economies of Europe, North America and East Asia; and the gradual replacement of the old Taylorist and Fordist methods of mass production with a new paradigm of innovation–mediated production. A critical part of all three processes is the increasing reliance on knowledge and information in the economic activities that create value in capitalist economies.

The industrial economies are clearly entering into a new era of knowledge–based growth and innovation. While knowledge has played an ever increasing role in the process of growth since the onset of the industrial revolution, what is changing is its relative importance in relation to the other factors of production. In the eyes of commentators ranging from Bell to Drucker and Nonaka, the determining factor of production is no longer land, labour nor capital, it is knowledge (Drucker 1993, 6). This new system of knowledge–based production relies on an increasing application of the human intellect to generate new products and processes in a more productive fashion. In direct contrast to the Taylorist doctrine that utmost efficiency was generated by the separation of conception and execution, knowledge–based production relies upon a more effective integration of physical and intellectual labour. The generation and use of knowledge is no longer something that occurs in an isolated fashion in the university, government or corporate laboratory and is then transferred to the production site. Rather it is applied at every phase of the conception, design, production and distribution of goods and services to enhance both the speed and efficiency with which they are delivered to the hands of the consumer (Florida and Kenney 1993, 637-52).

The ability to survive and prosper in this new form of knowledge–based production is linked to the process of innovation — the process by which new ideas are developed and deployed for commercial purposes in the industrial economies. The concept of innovation itself is subject to a number of different interpretations and definitions. In its narrowest sense, it is limited to the creation of new *technical* innovations that lead to the introduction of new products or processes. In its wider sense, however, it refers not only to the creation of new products, but also new forms of organization that affect the production process or new ways of organizing and accessing markets (Edquist 1997, 9). The use of the concept in its broadest sense is strongly endorsed by Lundvall, who argues for an inclusive definition. For him "innovation is a ubiquitous phenomenon in the modern economy." It involves "ongoing processes of learning, searching and exploring, which result in new products, new techniques, new forms of organisation and new markets" (1992, 8). He attaches a high degree of importance to the patterns of interaction between firms as part of a collective learning process in the acquisition and use of new technical knowledge. This flows from his belief that innovation is increasingly tied to a process of interactive learning and collective entrepreneurship, especially in terms of the relationships between producers and users of new technology.

At its root, the process of innovation is grounded in the firm, but firms do not operate in isolation. In the process of innovating, they interact with a wide range of organizations that sustain this ability. Thus, the social and institutional settings within which the firm operates are critical for their innovative capacity. Recent work on this subject has focused attention on the way that the structure of firms and the supporting infrastructure of institutions cohere into different national or regional systems of innovation. Again, Lundvall maintains that "a system of innovation is constituted by elements and relationships which interact in the production, diffusion and use of new, and economically useful knowledge . . ." (1992, 2). The main elements of the system in his conception include: the internal organization of firms; the network of inter–firm relationships; the role of the public sector; the institutional set–up of the financial sector; and the degree of R&D intensity and the nature of R&D organization.

Lundvall's definition of a system of innovation emphasizes both the social and the interactive nature of the innovation process — characteristics that are stressed by a growing number of commentators. This central thesis found in this literature involves two key ideas (Storper 1992; Cooke and Morgan 1993; Saxenian 1994). First, production systems exhibit a much more finely articulated *social* division of labour, with the greater degree of specialization that is implied by this. Second, relations between firms in this more heavily externalized, transaction–intensive, production system have undergone a significant *qualitative* change. In place of earlier relations dominated by short–term, price–based considerations and arm's–length exchanges, firms are now said to be engaging in longer–term, closer, trust–based interaction, in which cooperation and exchange of proprietary information allow partners to compete more successfully together in national and international markets, as each one focuses on its 'core competencies'. In contrast to more conventional forms of inter–firm relations — markets and hierarchies — this alternative form of resource allocation is characterized by transactions that "occur neither through discrete exchanges nor by administrative fiat, but through networks of individuals engaged in reciprocal, preferential, mutually supportive actions" (Powell 1990, 303).

This emphasis on the importance of networked relationships is rooted in recent insights into the nature of the innovation process itself. Technologies tend to develop along pathways or trajectories characterized by strong irreversibilities. These trajectories are reinforced as certain choices are made and others are foreclosed; this results from the fact that technologies are the product of interdependent choices, or 'network externalities'. They are given to a variety of user–producer and user–user interactions; as the number of users of a given technology rises, its reduces the possibilities for different patterns of use by others (Arthur 1994). One reason why technologies follow these pathways is because of the effects of technological spillovers in the economy. The knowledge of how to do certain things technologically frequently derives from the knowledge of how to do other things and it, in turn, contributes to the knowledge of how to do related things. Sometimes these spillovers follow the lines of traded input–output relationships in the economy, but often they occur along lines that are not traded. These technological spillovers are tied to knowledge and practices that are not always codified or explicit. They are frequently shared among firms or transferred from firm to firm through various forms of networks. They may also derive from other institutional arrangements, such as the particular norms and rules governing the functioning of local labour markets. These forms of collaboration and networking give rise to the existence of untraded interdependencies within a local economy (Dosi 1988, 1142-47; Storper 1995a).

Another noteworthy example of the new, more social, economic relations comes in the form of increasingly common collaborative research and development (R&D) in science–based industries such as pharmaceuticals and microelectronics/computing/telecommunications (Mytelka 1991; Dodgson 1993; Hagedoorn 1993). In such sectors, collaborative relations in such forms as strategic alliances between two or more firms have become a popular organizational forms by which to spread the costs and risks of new product development, or to bring together the complementary assets (including distribution systems) of individual firms. In such arrangements, normally at least one (and frequently both) of the partners is large in size. The resulting production 'networks' and the collaborative relations between firms and the supporting infrastructure of institutions that sustain innovation, are seen as essential responses to a new set of international competitive conditions, in which quality, differentiation, responsiveness, and time–to–market with new products have joined cost as the principal factors separating winners from losers (Best, 1990).

The ability of firms to compete effectively in the emerging information technology paradigm is determined, in part, by their ability to develop a sustained capacity for innovation along the lines discussed above. Increasingly, this implies a growing investment in, and emphasis on, research and development to feed the flow of new ideas into the production process. According to the OECD, R&D is increasingly seen as being essential for enhancing a firm's capacity to absorb and make use of new knowledge of all kinds, including technological knowledge. In this sense, research is viewed not merely as a source of inventive ideas, but as an activity which contributes to the firm's absorptive capacity and its ability to solve problems. "When problems arise in the innovation process, as they are bound to do, a firm draws upon its knowledge base at that particular time, which is made up of earlier research findings and technical and practical experience" (1997, 38).

But the ability to sustain a capacity for innovation no longer depends on R&D alone. Successful firms combine a capacity for *technical* research and development with the ability to implement a wide range of new organizational practices within the firm and inter–firm practices with their network of suppliers and customers. Those who succeed are refered to as 'high performance firms'. According to Richard Florida and his collaborators, high performance firms are marked by a number of key characteristics. They rely on technical innovations to compete in export markets around the world. They effectively integrate the process of innovation between the R&D laboratory and the factory floor in a new form of innovation–mediated production. New technologies are combined effectively with new organizational designs to increase the productivity, efficiency and time–to–market of their entire operation. They make extensive use of new forms of work organization and greater degrees of employee involvement to achieve these goals. They invest heavily in the training of their employees as a key aspect of their commitment to continuous improvement and organizational learning. In addition, they also adopt newer, more closely networked relations with both suppliers and customers to optimize customer service, quality and flexibility. They exhibit an increasing reliance on just–in–time production techniques as part of a new relation between end–users and their suppliers. High performance operates as an integrated system that relies on the successful combination of all the factors outlined above. In a series of studies of firms in the Great Lakes economy, they assess the extent to which firms have made the transition to this new form of economic interaction (Heinz School of Public Policy and Management 1993, 19–27; cf. also Kochan and Osterman 1994). In the discussion which follows, we report on the results of a recent survey of manufacturing firms across five industrial sectors in Ontario. The concept of the high performance firm provides a benchmark against which to assess the innovative behaviour of these firms.

11.2 THE CHANGING GEOGRAPHY OF PRODUCTION

The increased attention paid to the social and interactive dimensions of the innovation process has also been linked to an emerging debate about the new geography of production in the knowledge–based economy. This new geography is marked by a 'paradoxical consequence of globalization' (Acs, de la Mothe and Paquet 1996, 340). As the information and communication networks created by new digital technologies link the disparate economies of the globe more closely together, they simultaneously increase the importance of space and proximity. There is some irony in the fact that the information and communication technologies simultaneously increase the flexibility of firms to operate on a global basis, yet accentuate the importance of regional concentrations of related firms and industries. The reality remains that certain types of information and knowledge exchange occur more effectively through direct face-to-face contact. "Put simply, the more tacit the knowledge involved, the more important is spatial proximity between the actors taking part in the exchange" (Maskell and Malmberg 1995, 29). The reason for this is twofold: first, it is partly a function of the economics of time and distance — it is normally less costly and easier to interact with others who are close at hand; second, it involves the question of trust and understanding — the transfer of tacit knowledge is normally facilitated by an environment or context in which the collaborators share a common set of values and culture. Both these factors are clearly facilitated by geographic proximity.

Thus, the new production paradigm is highly dependent on localized, or regionally–based, innovation. Innovative capabilities are often sustained through regional communities that share a certain base of knowledge and the increments to that knowledge base. Industrial geographers have long observed that patterns of production tend to aggregate over time among networks of firms drawing upon the distinctive skills and characteristics of local labour markets in specific regions. Geographer Michael Storper uses the term 'territorialization', to describe the range of economic activity that depends on resources which are territorially specific. The types of resources involved can include specific assets that are only available in a certain place, or more critically, assets whose real value emerges out of the context of particular inter–organizational or firm–market relations that depend upon geographic proximity. Relations based upon geographic proximity constitute valuable assets when they generate positive spillover effects in an economic system. The more grounded the economic activities of a region are to the specific assets of that region, the more fully territorialized are those activities (Storper 1995b).

The dynamism of these regional economies is tied to the totality of their industrial system and the broader social and cultural context within which they are embedded. The industrial system of a region includes three important dimensions: the indigenous mix of institutions and culture in the region; the structure of the industrial system; and the internal organization or industrial culture that prevails in firms in the region. The relevant range of institutions can include both public and private ones, such as universities, business and professional associations, local training or industrial institutes and other associations that may contribute to a dynamic local culture in the region. The industrial structure of the region refers to the inter–firm organization of its production system, especially the extent and nature of the relations between suppliers and customers within the individual sectors or networks of interrelated sectors, and the role played by the larger firms within the regional economy. Finally the internal organization or industrial culture of the firm includes the extent to which the production system is organized on traditional hierarchical lines or is more decentralized, the degree to which relations between management and the workforce are characterized by a cooperative or conflictual approach and the relative importance attached to training and the continuous upgrading of skills (Saxenian 1994, 7).

As noted above, more and more cases can be found of emerging cooperative relationships between networks of producers — between large assemblers and smaller suppliers in the auto industry, between networks of small producers, such as exists in the Emilia–Romagna industrial district of Italy, and even among large producers in the computer and telecommunications industries that make up the core of the new information technologies. The growing costs of R&D, as well as the increasing complexity and knowledge–intensity of new scientific research and product development make the challenge more forbidding for individual firms — hence the growing importance of regional agglomerations of innovative firms in the collective advance of technical knowledge. The key elements of a networked regional economy include a dense complex of public and private industrial support institutions, high–grade labour market intelligence and related vocational training mechanisms, rapid diffusion of technology transfer, a high degree of interfirm networking and receptive firms well–disposed towards innovation. The very density of these networks and institutional supports is often interpreted as a sign of the vibrancy of a regional economy (Cooke and Morgan 1993, 562).

Drawing upon this insight, a number of writers suggest that the key to economic success in the future depends upon the development of closer relations among networks of firms and a broader infrastructure at the regional level on which complexes of firms can draw for support in the innovation process. According to Florida, the economic significance of regions has been dramatically increased by the shift to a knowledge–based economy.

In this new economic environment, regions build economic advantage through their ability to mobilize and to harness knowledge and ideas. . . . In effect, regions are increasingly defined by by the same criteria and elements which comprise a knowledge–intensive firm — continuous improvement, new ideas, knowledge creation and organizational learning. Regions must adopt the principles of knowledge creation and continuous learning; they must in effect become *learning regions* (1995: 532).

One of the problems with this set of claims is that its empirical base has been rooted heavily in the experiences of a handful of colourful and much studied regional cases: notably, Emilia–Romagna in northeastern Italy, Baden–Württemberg in southwestern Germany, and California's Silicon Valley. Despite the existence of these celebrated examples, there remains considerable debate about two issues. First, the extent to which such collaborative behaviour is widely diffused to other industries and regions within the industrial world remains unclear. In particular, most research to this point has been focused on new regions of industrial growth, to the relative neglect of more mature industrial spaces, such as Ontario (Gertler 1992). In the European context, numerous efforts are underway to replicate these results. To attain this level of capability, regions are encouraged to develop a collective learning process in which the key actors — private firms, public agences and a wide array of intermediary associations — work to enhance the broad capabilities of the region's infrastructural support for the innovation process (Morgan 1995). Some other regions, such as Wales, have enjoyed success in emulating these examples, but the extent to which the model can be generalized remains an open question. Furthermore, even in those instances where collaborative relations have been documented, there is some debate over the effectiveness or benefits arising from such activity (Harrison 1994).

There is thus a critical need for further investigation of the extent to which firms and their supporting infrastructure of institutions in other regions are developing similar capabilities. In the case of Ontario, despite considerable attention to these issues over the past decade, there is still a substantial lack of evidence about the extent to which it conforms, or not, to the characteristics outlined above. In an earlier paper, we outlined the nature and characteristics of the regional innovation system in Ontario at the broad macroeconomic and sociological level (Wolfe and Gertler 1997). In this chapter we begin the process of supplementing that perspective with a more detailed examination of the nature and extent of innovation at the firm and inter–firm level.

11.3   THE SURVEY DESIGN

The survey of innovative behaviour among manufacturing firms was designed as part of a broader research effort on the regional innovation system in Ontario. The broader nature of the project and some of the issues under investigation in related parts were critical in determining the initial construction of the survey sample. The selection of firms was drawn from a list of those known to have participated in a range of Ontario government programs and research centres during the period 1990-1995 and for which we could obtain information in the 1995 Scott's Directories.2 These included firms that participated as industrial partners in the eight provincial Centres of Excellence, as well as firms who had some involvement with the formation of the sector strategies for the auto, computing, electrical, health, machine, tool, die and mold, plastics, aerospace and telecommunication sectors. It also included firms which had received assistance from the Industry Research Program and the University Research Incentive Fund (URIF) administered by Technology Ontario, firms which had been clients of the Ontario Innovation and Productivity Service (OIPS), as well as a small number of firms designated as industrial partners in the joint research projects between researchers in Ontario universities and two of the Four Motors for Europe.[3] The control group sampling frame was drawn from those firms in the Scott's Directories in the same industrial sectors, but that were not identified as members of these programs or research centres.

Both the target and control groups were then further stratified on the basis of industrial sector, ownership and size. Firms were divided into eight sectors on the basis of their first four-digit SIC codes. In terms of ownership, firms were classified as being either foreign or Canadian owned. The criteria used to stratify firms by size varied by sector. For the auto, aerospace, electrical, telecommunications, computing and health sectors, firms were divided into small (1-99 employees), medium (100-249 employees) and large (250 + employees) firms. The corresponding figures for the machine, tool die and mold and the plastics sectors were small (1-49 employees), medium (50-149 employees), and large (150+ employees). Once fully stratified, firms were then selected using a systematic sampling procedure with a random start. A sample of 444 firms was selected for the target group while 446 firms were selected for the control group, for a total sample size of 890 firms. Table 11-1 shows the sample breakdown by sector.

The survey questions drew upon elements found within a range of previous innovation surveys. A number of different surveys proved helpful in the design of our own instrument, including Statistics Canada's Survey of Growth Companies (1992), the Survey of Innovation and Advanced Technology (1993) and the Survey of Innovation (1996). In addition, we drew upon the PACE (Policies, Appropriability and Competitiveness for European Enterprises) survey (Arundel et al., 1995) of large European firms as the source for some of our questions. As indicated above, the survey of high performance firms carried out at the H. John Heinz III School of Public

Policy and Management (1993) by Richard Florida and his associates provided a useful benchmark for comparing the relative performance of Ontario firms. Finally, the survey implemented in Wales as part of the EU–sponsored project (CASS, 1996), as well as the survey of advanced technology firms in Oxfordshire carried out by Helen Lawton Smith (1990) provided additional sources of ideas.

Table 11.1 - Response Rate by Sector

Sector	Number of surveys sent	Number of surveys received	Response Rate (%)
Transportation Equipment	132	32	24.2
Computing, Electrical & Telecommunications	260	67	25.8
Machine Tool & Die	238	57	23.9
Health	130	34	26.2
Plastics	130	47	36.2
Entire sample	890	237	26.6

The survey was pretested a number of times to control for potential problems of length and complexity. At least two firm or industry association representatives in each sector were surveyed to obtain feedback about the clarity of the questions, overall organization, and potential problems that might be encountered in responding to it. As a result of the feedback obtained through this procedure, the survey was reduced in half, a number of the sections was dropped or integrated and the format of several questions was modified to reduce the possibility of response errors.

The questionnaire itself was divided into five sections. Section one asked firms to assess the competitive environments they were facing, as well the workplace technologies and innovations they were utilizing in their efforts to cope with those competitive pressures. The second section asked the firms to describe the nature of their relationships with their customers, their suppliers and other firms in their industry. Section three asked them to assess the effectiveness of various federal and Ontario government programs in assisting their firm and their industry, while the fourth section asked similar questions with regard to the role played by federal and provincial research centres. The final section asked for relevant background material such as employment levels, R&D expenditures and sales figures.

Target group firms were surveyed over the period December, 1996 through May, 1997. The control group was surveyed over the period May, 1997 to October, 1997. Firms were initially contacted by phone to assess their willingness to participate and to obtain appropriate contact names. Firms that agreed to participate received a survey, cover letter, and business reply envelope, with these materials being sent to the contact person. The cover letter explained the purpose of the survey, under whose auspices it was being carried out, the type of information that would be gathered, an estimation of the amount effort required to complete the survey, assurances regarding confidentiality protocols, contact information and a promise to provide an executive summary of the survey results that the firms could use for benchmarking pruposes.

For the purpose of this study, the sampling unit was set at the plant or establishment level, not at the level of the company as a whole. This influenced decisions regarding which individuals would be contacted within the company. In the case of larger firms, efforts were made to contact plant managers, while for small firms, the focus was on senior executives at the level of vice president or president. If a questionnaire was not received within two to three weeks of the initial mailing, follow–up phone calls were made to determine if the firm was still willing to participate and to address any concerns the firm's representative might have regarding the survey. An offer to send another survey to the firm was also made if the firm agreed to complete the survey at this point.

Reasons for refusal to participate varied. One of the most common was the length and complexity of the survey and the effort required to complete it. This clearly influenced the overall response rate. Another response provided by a number of smaller firms, especially within the control group, concerned their lack of innovative behaviour. Because they did not utilize a large number of the workplace innovations or advanced process technologies listed in the survey, they felt they could not make a meaningful contribution.[4] At the time of writing, the overall number of respondents stood at 237, for a response rate of 27%. Though these rates may appear low, they are probably respectable given the issues of length and complexity noted above.

11.4 HIGH PERFORMANCE IN ONTARIO: HOW INNOVATIVE ARE ITS MANUFACTURERS?

As the preceding discussion makes clear, the concept of 'high performance' can be disaggregated into a number of distinct dimensions or elements. For our analysis of the current picture within Ontario manufacturing, we have employed a set of indicators which, taken together, provide a well–rounded and holistic assessment of the degree to which innovative practices — broadly defined — have been adopted to transform the workplace and the social organization of production. Our indicators fall neatly into two complementary groups: those that reflect the use of innovative practices *inside* individual establishments and those that indicate the development of innovative practices in *inter–firm relations*. Our approach here follows the recent literature on networks and regional development (Cooke and Morgan, 1993), which argues that the adoption of 'network relations' is transforming practices not only between firms but also in–house, through the adoption of progressive forms of workplace reorganization. Our specific indicators in these two categories are listed below.

Innovative Practices Internal to the Establishment:
- Research and development expenditures
- Training practices
- Adoption and use of advanced process technologies
- Adoption and use of innovative workplace organization

Innovative Practices in Inter-Firm Relations:
* Collaborative relations with suppliers
* Collaborative relations with customers
* Collaborative relations with other firms in the same industry

In the discussion that follows, we investigate each of these dimensions individually. For the sake of this analysis, we combine both the 'target' group and 'control' establishments to construct an aggregate picture. Where appropriate, we disaggregate the sample according to principal dimensions such as establishment size, sector, and ownership in order to sketch a more detailed picture of how the adoption of these practices might vary across different subsets of the firms in the study.

11.4.1. Transforming the Workplace: How Innovative Are Ontario Manufacturers' Internal Practices?

One of the most widely used indicators of innovative activity by firms is the amount of money spent on the performance of research and development. In Table 11-2, we adopt the standard classifications used by the OECD to assess the intensity of R&D expenditures by our sample firms and to investigate how this has changed between 1989 and 1995. At the start of this period, the sample of firms appears to be somewhat bifurcated between low and high R&D performers: nearly one–half of all establishments fell into the low category (devoting less than one percent of their sales to R&D expenditures), with just over one–third in the high category. Over time, the proportion in the low category declines steadily, with most firms apparently shifting into the medium category of R&D intensity.   However, the proportion of establishments spending more than three percent of sales on R&D (and therefore falling into the high category) has not risen significantly over the six years. This result is broadly similar to that reported in the most recent Statistics Canada service bulletin. R&D intensity was higher for small firms and decreased as the size of the firm's employment increased (1997, 5).

Based on previous research, we would expect the degree of R&D intensity to vary strongly by sector (Wolfe and Gertler, 1997). For example, aggregate statistics indicate that Ontario's most prominent R&D performers (in both absolute and relative terms) are found in sectors such as telecommunications equipment, computing, software, and pharmaceuticals. At the same time, important sectors such as automotive assembly and parts have been shown to perform notoriously low levels of R&D in Ontario.   The findings from our survey (Table 11-3) confirm this pattern.   The computing, telecom, and electrical/electronic equipment group has by far the largest share of establishments in the high category, followed by the health products group (which includes both pharmaceuticals firms and producers of medical instruments and products). The transportation equipment sector (made up of automotive and aerospace producers) performs according to form, with a relatively small proportion of its establishments in the high category and a large share in the low category.   Only the plastics industry fares less well on this indicator, with nearly 60 per cent of establishments spending less than one percent of their sales on R&D. Furthermore, the changes in shares between 1989 and 1995 appear to accentuate the inter–sectoral differences just summarized. Hence, the computing/telecom/electrical group leads all other sectors with the best improvement in the high category. The transportation equipment group actually saw the share of its establishments in the high category fall by more than two percentage points over this period, although the medium category did undergo a significant expansion of its share (up nearly 13 points).

Table 11.2 - R&D Intensity Over Time

Level of R&D expenditures as a percentage of sales*	1989	1992	1995
	(valid percentage)		
Low	46.7	43.1	38.5
Medium	17.0	17.0	23.0
High	36.3	39.9	38.5
Valid N	135	153	161

* The low, medium and high categories correspond to the OECD classifications for this ratio of less than 1% for low, 1 to 2.999% for medium and 3% or greater for high. See Anthony Arundel, Gert van de Paal and Luc Soete, *Pace Report: Innovation Strategies of Europe's Largest Industrial Firms* (MERIT: June 1995), p. 7.

Table 11.3 - R&D Intensity by Industrial Sector, 1995

Percentage distribution of establishments (with change in percentage share from 1989 in parentheses)

R&D intensity level	Transportation Equipment*	Computing, Telecom & Electrical	Health	Machine Tool & Die	Plastics	Row Total
Low	54.5 (-10.5)	25.5 (-14.5)	16.0 (-12.6)	46.2 (+2.4)	57.1 (-6.5)	62 38.5%
Medium	22.7 (+12.7)	12.8 (+5.3)	36.0 (+17)	23.1 (-1.9)	28.6 (+1.3)	37 23%
High	22.7 (-2.3)	61.7 (+9.2)	48.0 (-4.4)	30.8 (-0.5)	14.3 (+5.2)	62 38.5%
Column Total	22 99.9%	47 100%	25 100%	39 100.1%	28 100%	161 100%

Note: R&D intensity is calculated in the same manner as in Table 2. Figures in the column total may not come to 100 due to rounding errors.

*The transportation equipment sector includes firms from the automotive and aerospace sectors

There has also been a strong contention over the years that R&D performance varies considerably by nationality of ownership. In particular, a vivid debate has raged over the extent to which foreign–owned firms perform their share of R&D in Canada. More recent evidence has suggested that apparent differences between Canadian and foreign–owned firms may instead result from systematic differences in establishment size between these two groups. One study suggests that, on balance, "it is probably the smaller size of Canadian firms that accounts for their observed higher R&D intensity" (Holbrook and Squires 1996, 373).

In Table 11-4, we disaggregate R&D performance by ownership and establishment size (measured in terms of employment) in order to examine the validity of these two perspectives. Some fascinating patterns emerge. For small and medium–sized enterprises (SMEs), the differences between Canadian and foreign–owned firms is striking and verges on statistically significant at the p = .05 level. Almost one–half of Canadian establishments fall into the high R&D intensity category, compared to a share of less than one–quarter for foreign–owned establishments. For the low intensity category, the shares are almost reversed. At the upper end of the size distribution (establishments employing 250+ workers), similar differences emerge when comparing Canadian and foreign–owned establishments in the high R&D intensity category. However, this difference all but disappears when comparing shares in the medium and low categories. Overall, SMEs emerge as a more R&D intensive group than do larger establishments. Moreover, this size difference is especially marked within the Canadian–owned firms. The results obtained in this part of the survey are quite close to those reported by Statistics Canada in its latest review of the effect of foreign ownership on R&D performance. Since 1990, the proportion of revenue targetted to R&D by Canadian firms has risen, while that of foreign–controlled firms has declined (1997).[5]

If there is a central core to the high–performance firm envisioned by Florida and others, it is in the transformation occurring in the way work is organized and performed. A key aspect of this is the effort devoted to training by employers. Training expenditures expressed as a share of annual sales is one simple measure of this effort (see Table 11-5). What is clear from this table is that the extent of training effort by firms in our sample has increased quite markedly over the 1989–1995 period. The share of establishments within the low category declined steadily from 56 to 36 per cent. This decline of 20 points was equally shared as gains by the medium and high categories.[6]

When one disaggregates this information by sector, the pattern that emerges is strikingly different from the sectoral patterns for R&D intensity. Virtually across the board, changes since 1989 show that all sectors have seen their distributions shift upwards in favour of the medium and high categories. Moreover, the computing/ telecom/electrical group (a strong R&D performer) also scores well on training intensity, but the plastics sector actually registers a larger proportion of establishments in the high category. In general, there is much less differentiation between sectors on the dimension of training intensity than was seen in the case of R&D intensity, suggesting that the move toward greater training effort is widespread and pervasive. Only when we disaggregate the analysis by size of establishment (Table 11-6) do we see any real variation. Here, large plants emerge as relative underperformers when

Table 11.4 - R&D Intensity Incidence (with Column Percentages Shown in Parentheses) by Ownership, Controlling for Firm Size. (1995).

R&D Intensity	Ownership			
	Canadian	Foreign	Row Total	Chi-Square (P)
Small and medium-sized enterprises (1-249 employees)				
Low	27 (27.6)	11 (50.0)	38 (31.7)	
Medium	23 (23.5)	6 (27.3)	29 (24.2)	
High	48 (49.0)	5 (22.7)	53 (44.2)	5.77 (0.056)
Column total	98 (100.0)	22 (100.0)	120 (100.0)	
Large enterprises (250+ employees)				
Low	6 (42.9)	6 (54.5)	12 (48.0)	
Medium	3 (21.4)	3 (27.3)	6 (24.0)	
High	5 (35.7)	2 (18.2)	7 (28.0)	0.939 (0.625)*
Column total	14 (100.0)	11 (100.0)	25 (100.0)	

Note: R&D intensity is calculated in the same manner as in Table 2.
*In this case, more than 20 percent of the cells in the crosstabulation had expected frequencies less than five

Table 11.5 - Training Intensity Over Time

Level of training expenditures as a percentage of sales*	1989	1992	1995
	(valid percentage)		
Low	56.2	42.6	36.2
Medium	21.5	27.0	31.6
High	22.3	30.5	32.2
Valid N	130	141	152

* The distribution of training intensity for 1995 was divided into three equal segments to create the definitions for the low, medium and high categories for this table. The actual cut-off points were less than 0.2% for low, 0.2-0.6% for medium and greater than 0.6% for high.

Table 11.6 - Training Intensity by Industrial Sector, 1995

Percentage distribution of establishments (with change in percentage share from 1989)

Training intensity level	Transportation Equipment	Computing, Telecom & Electrical	Health	Machine Tool & Die	Plastics	Row Total
Low	30.0 (-28.8)	36.2 (-25.3)	34.8 (-17.8)	30.6 (-13.5)	50.0 (-16.7)	55 36.2%
Medium	45.0 (+21.5)	27.7 (+14.9)	39.1 (-3.0)	38.9 (+12.4)	11.5 (+2.0)	48 31.6%
High	25.0 (+7.4)	36.2 (+10.6)	26.1 (+20.8)	30.6 (+1.2)	38.5 (+14.7)	49 32.2%
Column Total	20 100%	47 100%	23 100%	36 100%	26 100%	152 100%

Note: Training intensity is calculated in the same manner as in Table 5.

compared to small and medium establishments. And a further important distinction becomes evident when training effort is compared for Canadian and foreign–owned establishments (while again controlling for size) (Table 11-7). Within the SME category, Canadian–owned plants are significantly more likely to engage in a high training effort than are foreign–owned establishments (of which nearly two–thirds are found in the low category). Although a similar pattern emerges for large enterprises, this difference is most evident within the high category but is greatly diminished for the lowest category (Table 11-8).

Table 11.7 - Training Intensity Incidence (with Percentages Shown In Parentheses) by Establishment Size, 1995

Size category

Level of training expenditures as a percentage of sales	Small (1-49 employees)	Medium (50-249 employees)	Large (250+ employees)	Row Total
Low	17 (31.5)	22 (31.4)	13 (52.0)	52 (34.9)
Medium	16 (29.6)	24 (34.3)	8 (32.0)	48 (32.2)
High	21 (38.9)	24 (34.3)	4 (16.0)	49 (32.9)
Column Total	54 (100.0)	70 (100.0)	25 (100.8)	149 (100.0)

Note: Training intensity is calculated in the same manner as in Table 5. Chi-Square=5.52, *p*=.24

Table 11.8 - Training Intensity Incidence (with Column Percentages Shown in Parentheses) by Ownership, Controlling for Firm Size. (1995).

Training Intensity	Ownership Canadian	Foreign	Row Total	Chi-Square (P)
Small and medium-sized enterprises (1-249 employees)				
Low	21 (23.6)	16 (64.0)	37 (32.5)	
Medium	32 (36.0)	5 (20.0)	37 (32.5)	
High	36 (40.4)	4 (16.0)	40 (35.1)	14.67 (0.00065)
Column total	89 (100.0)	25 (100.0)	114 (100.0)	
Large enterprises (250+ employees)				
Low	5 (50.0)	8 (57.1)	13 (54.2)	
Medium	2 (20.0)	5 (35.7)	7 (29.2)	
High	3 (30.0)	1 (7.1)	4 (16.7)	2.38 (0.300)*
Column total	10 (100.0)	14 (100.0)	24 (100.0)	

Note: Training intensity is calculated in the same manner as in Table 5
*In this case, more than 20 percent of the cells in the crosstabulation had expected frequencies less than five

One of the main forces underlying the drive to increase training efforts is, according to the literature on 'intelligent manufacturing', the growing use of advanced manufacturing technologies that make greater demands on the skills of operators. Tables 11-9 and 11-10 provide a window on the extent to which these advanced process technologies (APTs) have been implemented by firms in our sample. The first of these shows clearly that the likelihood of adopting at least one APT (from a list of twelve different types) rises consistently with establishment size. Of the large plants, fully 100 per cent indicate that they are making use of at least one APT. Our survey also requested that respondents indicate how extensively such technologies were used in their operations (see Table 11-10). Here, we see that the most extensively adopted technologies are (in order) computer–aided design (CAD), materials requirements or manufacturing resource planning (MRP), programmable controllers, computer numerically controlled (CNC) machines, and technical data or factory networks. By way of comparison, these results are quite similar to those reported in Baldwin, Sabourin and Rafiquzzaman (1996, Table 5, p. 18), based on Statistics Canada's 1993 Survey of Innovation and Advanced Technology. This earlier study reported that the most frequently adopted technologies (weighted by the shipments produced by adopting firms) included CAD, programmable controllers, computers for factory control, and technical data or factory networks. Table 11-10 also allows us to examine the extent of use of each technology by sector. For nine out of twelve technologies, the transportation equipment group exhibits the highest average extent of use, suggesting that the stringent technical requirements of the automotive and aerospace assemblers has compelled suppliers to upgrade their process techologies quite extensively, even if these firms are not intensive performers of R&D.

Table 11.9 - Incidence (with Percentages Shown in Parentheses) of Advanced Process Technology (APT) Usage by Establishment Size.

Size category APT usage	Small (1-49 employees)	Medium (50-249 employees)	Large (250+ employees)	Row Total
Not used	16 (26.2)	6 (7.6)	0 (0.0)	22 (13.3)
Used	45 (73.8)	73 (92.4)	26 (100.0)	144 (86.7)
Column Total	61 (100.0)	79 (100.0)	26 (100.0)	166 (100.0)

Note: Chi-Square = 15.11 $p=.00052$

Table 11.10 - Advanced Process Technology (APT) Usage: Mean Scores

Sector

Type of APT*	Transportation Equipment	Computing, Telecom & Electrical	Health	Machine Tool & Die	Plastics	Entire sample
AI	1.0	1.1	1.2	1.0	1.2	1.1
AGVS	1.3	1.1	1.0	1.2	1.1	1.1
CAD	3.8	3.7	3.0	3.3	2.9	3.4
CAM	2.4	2.2	1.9	2.2	2.0	2.1
CIM	1.7	1.3	1.2	1.5	1.3	1.4
CNC Machines	3.1	2.1	1.6	2.9	2.0	2.4
Programmable Controller	3.3	2.4	2.1	2.6	3.0	2.7
FMC	2.2	1.5	1.5	1.4	1.6	1.6
Material Working Lasers	1.1	1.3	1.2	1.3	1.2	1.2
MRP	3.3	3.3	2.6	2.5	2.6	2.9
Robots	1.8	1.7	1.4	1.5	1.7	1.6
Technical Data Network	2.1	2.7	2.5	2.3	1.9	2.3

Note: Extent of use was measured on a scale running from "1=not used" through "5=extensive usage in all product lines".

*APT abbreviations:

AI=Artificial Intelligence
AGVS=Automoated Guided Vehicle Systems or Automated Storage and Retrieval Systems
CAD=Computer-Aided Design
CAM=Computer-Aided Manufacturing
CIM=Computer Integrated Manufacturing
CNC=Computer Numerically Controlled Machines
FMC=Flexible Manufacturing Cells or Flexible Manufacturing Systems
MRP=Materials Requirement Planning or Manufacturing Resource Planning
Robots=Robots, including Pick and Place Robots
TDN=Technical Data Network or Factory Network

It is important to go beyond the adoption of advanced process technologies and to consider other, more holistic approaches to transforming the workplace from within. As Kochan and Osterman (1994) argue, these complementary changes are considerably more profound — and more challenging to implement successfully — than the simple adoption of new machinery, since they imply the introduction of fundamentally different relations between workers and managers, and new ways of organizing the work process itself, in order to achieve the 'high performance' or 'mutual-gains enterprise'. Our survey collected information on the extent of use of nine distinct forms of workplace innovation (see Table 11-11). The most extensively adopted innovative practices were the involvement of shop floor employees in production planning, profit–sharing and other forms of incentive–based pay, and ISO 9000 certification. However, concurrent/simultaneous engineering (in which product innovations are developed by multi–function task forces integrating research and development, production, and marketing), and quality circles/total quality management were close behind the leaders. In every case, large establishments were the most extensive adopters of these innovative practices.

Table 11.11 - Usage of Workplace Innovations by Establishment Size: Mean Scores, 1995

Size category

Type of workplace innovation	Small (1-49 employees)	Medium (50-249 employees)	Large (250+ employees)	Entire Sample
Involving shop-floor employees in production planning	2.9	2.6	3.3	2.9
Concurrent engineering	2.7	2.6	3.1	2.7
ISO 9000	2.0	3.2	3.7	2.9
Frequent job rotation	2.0	2.2	2.7	2.3
Labour-management committees	2.5	2.6	3.5	2.8
Profit-sharing	2.6	3.0	3.1	2.9
Self-directed work groups	2.5	2.3	3.0	2.5
Statistical process control	1.7	2.4	2.8	2.3
Total quality management	2.4	2.6	2.9	2.6

Note: Extent of use was measured on a scale running from "1=not used" to "5=used in all departments or divisions."

Investigating the sectoral incidence of these workplace innovations (Table 11-12), there appears to be little obvious correspondence between the patterns evident here and those evident in Table 11-10 reporting APT use. Where the transportation group stood out in the earlier table, this is no longer the case in the current analysis. Here the most prominent sector (measured by mean scores on extent of use) is plastic products, which is the most extensive adopter of joint labour–management committees, profit–sharing/incentive–based pay, frequent job rotation, and statistical process control.

Table 11.12 - Usage of Workplace Innovations by Industrial Sector: Mean Scores, 1995

Sector

Type of workplace innovation	Transportation Equipment	Computing, Telecom & Electrical	Health	Machine Tool & Die	Plastics	Entire sample
Involving shop-floor employees in production planning	2.7	2.6	2.7	3.1	3.0	2.9
Concurrent engineering	2.6	2.8	2.4	2.9	2.5	2.7
ISO 9000	3.3	3.2	2.2	2.7	3.1	2.9
Frequent job rotation	2.3	2.2	2.1	2.1	2.5	2.2
Labour-management committees	2.7	2.9	2.7	2.4	3.1	2.8
Profit-sharing	2.4	3.1	2.5	2.9	3.1	2.9
Self-directed work groups	2.3	2.9	3.0	2.1	2.5	2.5
Statistical process control	2.4	2.3	2.0	2.1	2.5	2.3
Total quality management	3.0	2.5	2.3	2.5	2.6	2.6

Note: Extent of use was measured on a scale running from "1=not used" through "5=used in all departments or divisions".

Having examined a variety of indicators of innovative practice internal to the firm, and having found a number of encouraging signs, one is naturally led to question the extent to which the use of such practices might have brought about tangible improvements in the adopting firms' competitive performance. While this relationship is undoubtedly a complex one requiring an extensive analysis on its own, a preliminary examination is provided in Table 11-13. Here we find that strong R&D performers in 1989 were also, on the whole, those establishments most likely to generate the highest growth rates in employment in the subsequent 1989–1995 period. We also see that the most training-intensive workplaces in 1989 were also strong job creators in the ensuing six years of recession and recovery. Hence, on the basis of this very preliminary examination, there appears to be tentative support for the argument that adoption of progressive internal practices leads to superior economic performance subsequently.

Table 11.13 - Relationship between Employment Growth and Selected Indicators of Innovative Internal Practice.

A. Employment growth (1989-95) by R&D intensity (1989)

R&D intensity (column percentages in parentheses)

Employment growth *	Low	Medium	High	Row Total
Low	23 (37.7)	7 (31.8)	14 (28.6)	44 (33.3)
Medium	27 (44.3)	7 (31.8)	11 (22.4)	45 (34.1)
High	11 (18.0)	8 (36.4)	24 (49.0)	43 (32.6)
Column Total	61 (100.0)	22 (100.0)	49 (100.0)	132 (100.0)

Note: Chi-Square = 12.64, p= .01320.

B. Employment growth (1989-95) by training intensity (1989)

Training intensity (column percentages in parentheses)

Employment growth *	Low	Medium	High	Row Total
Low	19 (34.5)	9 (34.6)	11 (23.9)	39 (30.7)
Medium	21 (38.2)	15(57.7)	10 (21.7)	46 (36.2)
High	15 (27.3)	2 (7.7)	25 (54.3)	42 (33.1)
Column Total	55 (100.0)	26 (100.0)	46 (100.0)	127 (100.0)

Note: Chi-Square = 19.04, p= .00077.

*Categories for employment growth and training intensity were created by dividing establishments into three equal segments. R&D intensity is calculated in the same manner as in Table 2.

11.4.2 Transforming Inter–Firm Relations: How Innovative Are Ontario Manufacturers' External Practices?

As our introduction to this chapter makes clear, the shift toward more knowledge–based production depends not only on the introduction of new work methods inside the firm, but is increasingly predicated on a reorganization of the social division of labour in production systems. Industrial organization is being restructured in order to foster the interaction of individual firms with one another and to support the creation of *collective learning processes*. As a way of tracking the extent to which such inter–firm learning processes are taking hold in Ontario manufacturing, we asked our respondents a series of questions about their current inter–firm practices and how these are changing. The results provide a fascinating perspective on these developments in Ontario for the first time.

Establishments surveyed were asked to indicate the extent to which they have developed close, collaborative interaction with their customers, suppliers and other firms in the same industry (i.e. potential competitors) with respect to six distinct activities (see Table 11-14). The findings paint an interesting picture of the extent and precise ways in which collaborative practices are being adopted. First, it is immediately evident that manufacturers in Ontario are adopting collaborative interaction for *vertical* relationships with their customers and suppliers up and down the supply chain. These appear to be notably more prevalent than *horizontal* collaboration with potential competitor firms in the same industry. Second, among the vertical relationships, firms are more likely to characterize their relations with customers as collaborative, compared to their relations with suppliers.   Third, the tendency to move towards more collaborative inter–firm relations appears to be strongest (for the most part) amongst the larger establishments in our sample. Interesting exceptions to this pattern include research and joint production scheduling with customers, where small establishments show mean scores equal to those of large establishments, and in process/product design and product development, where mean scores for small plants are only marginally lower than those for large ones. These findings seem to suggest that co–operative relationships are easier to establish and maintain when the firms involved are not in direct competition with one another. This result is quite logical and predictable within an Anglo–American framework of macro–regulation, in which there remain systemic disincentives against inter–firm co–operation (Gertler, 1997).

When one approaches the same issue from the perspective of individual sectors (Table 11-15), further interesting patterns emerge. When it comes to collaborating with customers, the most active sectors appear to be transportation equipment (judging by the number of times that this sector shows the highest mean scores for a given form of activity). As for relations with suppliers, the plastics sector registers more top mean scores than any other sector, followed by the transportation equipment group. As for relations with other firms in the same industry, while again these scores are generally much lower, the sector emerging as most collaborative in a horizontal sense is health products.

Table 11.14 - Interfirm Relations by Establishment Size: Mean Scores.

Sector

Interfirm activity*	Small (1-49 employees)	Medium (50-249 employees)	Large (250+employees)	Entire Sample
A. Relations with customers				
Joint production	2.0	1.8	2.3	1.9
Joint production scheduling	2.9	2.8	2.9	2.9
Joint marketing/ export promotion	2.0	2.2	2.4	2.1
Process or product design	3.6	3.2	3.7	3.4
Product development	3.7	3.6	3.9	3.7
Research	3.4	3.1	3.4	3.3
B. Relations with suppliers				
Joint production	1.8	2.1	2.7	2.1
Joint production scheduling	2.3	2.8	3.1	2.7
Joint marketing/ export promotion	1.7	1.9	1.8	1.8
Process or product design	2.9	3.1	3.3	3.1
Product development	2.8	3.3	3.2	3.1
Research	2.5	2.8	2.8	2.7
C. Relations with other firms in the industry				
Joint production	1.3	1.4	1.5	1.4
Joint production scheduling	1.2	1.3	1.5	1.3
Joint marketing/ export promotion	1.5	1.6	1.5	1.5
Process or product design	1.7	1.7	2.0	1.7
Product development	1.7	1.8	2.0	1.8
Research	1.7	1.8	1.9	1.8

Note: The strength of interfirm relationships was measured on a scale ranging from "1=no relationship through to "5=close collaborative interaction".

Table 11.15 - Interfirm Relations by Industrial Sector: Mean Scores.

Sector

Interfirm activity*	Transportation Equipment	Computing, Telecommunications & Electrical	Health	Machine Tool & Die	Plastics	Entire Sample
A. Relations with customers						
Joint production	2.5	1.8	1.2	2.1	2	1.9
Joint production scheduling	3.3	2.7	2.2	3.1	3.1	2.9
Joint marketing/ export promotion	2.5	2.3	1.9	2.0	2.1	2.2
Process or product design	3.4	3.4	3.3	3.6	3.3	3.4
Product development	3.6	3.9	3.5	3.6	3.9	3.7
Research	3.0	3.4	3.3	3.3	3.3	3.3
B. Relations with suppliers						
Joint production	2.7	2.1	1.8	2.1	1.9	2.1
Joint production scheduling	3.0	2.8	2.6	2.6	2.3	2.7
Joint marketing/ export promotion	1.9	1.6	2.0	1.8	1.8	1.8
Process or product design	2.9	3.1	3.1	3.0	3.4	3.1
Product development	3.2	3.0	3.1	2.8	3.5	3.1
Research	2.3	2.5	2.8	2.8	3.2	2.7
C. Relations with other firms in the industry						
Joint production	1.6	1.4	1.3	1.2	1.3	1.4
Joint production scheduling	1.5	1.5	1.3	1.1	1.2	1.3
Joint marketing/ export promotion	1.6	1.6	1.8	1.4	1.3	1.5
Process or product design	1.5	2.0	2.0	1.7	1.5	1.7
Product development	1.6	2.0	2.2	1.7	1.6	1.8
Research	1.7	1.9	2.0	1.7	1.6	1.8

Note: The strength of interfirm relationships was measured on a scale ranging from "1=no relationship through to "5=close collaborative interaction".

Our survey also gathered information about the prevalence of specific innovative inter–firm practices. Up until now, most of the analysis of such developments has remained anecdotal and piecemeal, focussing on individual case studies. The importance of these data is that they allow us to draw some *general* conclusions about the degree to which such practices have in fact diffused throughout the Ontario economy. From Table 11-16, we see that just–in–time (JIT) delivery relationships with customers and suppliers are the most prevalent form of innovative inter–firm practice, followed closely by the outsourcing of parts and components production. No other innovation on this list produces a score higher than the midpoint on the scale (3.0), although the mean score for strategic alliances and joint ventures is at this level. The other noteworthy pattern concerns establishment size. With two exceptions (strategic alliances/joint ventures and EDI), small plants record the highest mean scores, indicating the most extensive use of these specific forms of inter–firm interaction. Especially prevalent here are outsourcing (all types) and JIT (with both customers and suppliers). While the results reported in Table 11-14 suggested that, in many cases, large establishments were most likely to engage in innovative inter–firm relations, the balance seems to shift clearly in favour of small enterprises when the analysis focusses on these specific forms and practices.

One way to reconcile these two sets of findings is to infer that, while small firms are most likely to engage in specific practices such as outsourcing and JIT, they do so in the context of a relationship (possibly with larger firms) which is not overwhelmingly collaborative in nature. In other words, they could be subject to exploitive or 'unenlightened' treatment by such partner firms — again, a finding that is wholly consistent with earlier analyses of the difficulty of superimposing collaborative practices on top of a deeply engrained business culture which privileges aggressive competition and rugged individualism (Wolfe and Gertler, 1997; cf. also Harrison, 1994).

As with the measures of internal innovative practices examined in section 4.1, it makes sense to examine, at least in a preliminary way, the extent to which the adoption of innovative inter–firm practices is associated with better subsequent establishment performance. For this we turn to Table 11-17, which provides the analysis for two specific dimensions of inter–firm practice. This table suggests that those establishments that have moved farthest to adopt practices such as joint research and just–in–time delivery relationships with their customers have also generated higher than average employment growth over the 1989 to 1995 period. Once again, it is reassuring to find at least tentative evidence that such practices might lead to tangibly better overall results in terms of key indicators, such as job creation.

Table 11.16 - Usage of Interfirm Practices by Establishment Size: Mean Score (1995)

Size category

Interfirm practice	Small (1-49 employees)	Medium (50-249 employees)	Large (250+ employees)	Entire Sample
Outsourcing R&D	2.8	2.2	2.1	2.3
Outsourcing production of parts	3.6	2.9	3.0	3.2
Outsourcing production of complete product	3.0	2.1	2.7	2.5
Strategic alliance or joint venture	2.9	2.9	3.1	3.0
Just-in-Time with customers	3.6	3.5	3.4	3.5
Just-in-time with suppliers	3.5	3.3	3.3	3.3
Electronic data interchange	2.5	3.0	3.2	2.9

Note: Extent of use was measured on a scale running from "1=limited" to "5=extensive".

Table 11.17 - Relationship between Employment Growth and Selected Indicators of Innovative Inter-Firm Practice.

A. Employment growth (1989-95) by research relationships with customers

Strength of relationship (column percentages in parentheses)

Employment growth *	No relationship	Limited relationship	Close relationship	Row Total
Low	4 (23.5)	30 (35.7)	27 (31.4)	61 (32.6)
Medium	10 (58.8)	29 (34.5)	25 (29.1)	64 (34.2)
High	3 (17.6)	25 (29.8)	34 (39.5)	62 (33.2)
Column Total	17 (100.0)	84 (100.0)	86 (100.0)	187 (100.0)

Note: Chi-Square = 6.97, p= .13732.

B. Employment growth (1989-95) by just-in-time with customers

Extent of use (column percentages in parentheses)

Employment growth *	Limited use	Moderate use	Extensive use	Row Total
Low	13 (54.2)	7 (26.9)	14 (25.0)	34 (32.1)
Medium	6 (25.0)	12 (46.2)	15 (26.8)	33 (31.1)
High	5 (20.8)	7 (26.9)	27 (48.2)	39 (36.8)
Column Total	24 (100.0)	26 (100.0)	56 (100)	106 (100.0)

Note: Chi-Square = 11.59, p= .02067

*Categories for employment growth were created by dividing establishments into three equal segments.

11.5 CONCLUSION

The evidence adduced from this preliminary analysis of our survey results provides limited support for the hypotheses sketched in the first part of the paper. While there does appear to be some indication that firms in Ontario's manufacturing sector are moving in the direction of the 'high performance' model outlined by Florida, there is less substantial evidence that they are also building the kind of densely networked regional economy described by Cooke and Morgan. In part, these results are quite consistent with those of our earlier macroeconomic and sociological examination of the regional innovation system where we concluded that many features of the regulatory context surrounding the workplace in Ontario discourage and militate against the types of meaningful inter–firm cooperation found in the European regional economies. Cooperation between firms is more difficult to achieve when, as a result of a highly decentralized training regime and a system of labour market regulation that encourages frequent turnover and instability, potential cooperators undermine a sense of mutual trust by poaching each other's skilled workers. Similarly, the flexibility and emphasis on short–term results engendered by the structure of Ontario's capital markets and system of industrial finance also militates against the establishment of the long–term, cooperative relations found among firms in some European regions that operate under a different financial regime. It may prove difficult to achieve more meaningful levels of cooperation among networks of firms in Ontario unless greater attention is paid to these broader regulatory factors.

On the more positive side, however, the survey provides strong confirmation for the view that firms in Ontario are responding to the challenge of globalization and closer integration into a continental market under the conditions of the North American Free Trade Agreement. Along most of the dimensions identified as elements of the high performance model — growing R&D intensity, increased training effort, relatively more extensive adoption and use of advanced process technologies, the shift to more innovative workplace practices and closer relations with buyers and suppliers — firms in Ontario display strong evidence of moving towards the positive end of the spectrum. Thus they seem to be matching the benchmark set by their counterparts in the proximate states south of the Great Lakes. This augurs well for the ability of Ontario to compete in the new continental economy emerging in the wake of the free trade agreements.

REFERENCES

Acs, Z., de la Mothe, J., and Paquet, G. (1996). "Local Systems of Innovation: In Search of an Enabling Strategy". In P. Howett, *The Implications of Knowledge-Based Growth for Micro-Economic Policies*, (pp. 339-360). Calgary: University of Calgary Press.

Arthur, W.B. (1994). *Increasing Returns and Path Dependence in the Economy*, The University of Michigan Press. Ann Arbor.

Arundel, A., van de Paal, G. and Soete, L. (1995). *Innovation Strategies of Europe's Largest Industrial Firms: Results of the PACE Survey for Information Sources, Public Research, Protection of Innovations and Government Programmes*. Final Report. MERIT, University of Limberg. Maastricht.

Baldwin, J., Sabourin, D. and Rafiquzzaman, M. (1996). *Benefits and Problems Associated with Technology Adoption in Canadian Manufacturing.* Catalogue 88-514E, Statistics Canada, Ottawa.

Best, M.H. (1990). *The New Competition: Institutions of Industrial Restructuring.* Polity Press. Cambridge, UK.

Centre for Advanced Studies in the Social Sciences (1996). *Welsh Innovation Survey.* University of Wales. Cardiff.

Cooke, P. and Morgan, K. (1993). "The Network Paradigm: New Departures in Corporate and Regional Development", *Environment and Planning D: Society and Space,* 11, pp. 543-564.

Dodgson, M. (1993). *Technological Collaboration in Industry: Strategy, Policy and Internationalization in Industry.* London, Routledge.

Dosi, G. (1988). "Sources, Procedures and Microeconomic Effects of Innovation." *Journal of Economic Literature* 26 (September):1120-71.

Drucker, P.F. (1993). *Post-Capitalist Society.* HarperBusiness. New York.

Edquist, C. (1997). "Introduction: Systems of Innovation Approaches - Their Emergence and Characteristics," In *Systems of Innovation: Technologies, Institutions and Organizations.* Pinter. C. Edquist (ed.), London. 1-35.

Edquist, C. (ed.) (1997). *Systems of Innovation: Technologies, Institutions and Organizations.* London: Pinter.

Florida, R. and Kenney, M. (1993). "The New Age of Capitalism: Innovation–Mediated Production," *Futures* 25 (July/August): 637-651.

Gertler, M.S. (1992). "Flexibility Revisited: Districts, Nation-States and the Forces of Production," *Transactions of the Institute of British Geographers, New Series* 17: 259-78.

Hagedoorn, J. (1993). "Understanding the Rationale of Strategic Technology Partnering: Interorganizational Modes of Cooperation and Sectoral Differences," *Strategic Management Journal* 14: 371-85.

Harrison, B. (1994). *Lean and Mean: The Changing Landscape of Corporate Power in the Age of Flexibility.* Basic Books. New York.

Heinz, H. John III School of Public Policy and Management (1993). *Reinventing the Heartland: A High Performance Strategy for the Great Lakes Region.* Carnegie Mellon University. Pittsburgh.

Holbrook, J.A.D. and Squires, R.J. (1996). "Firm-level Analysis of Determinants of Canadian Industrial R&D Performance," *Science and Public Policy 23*: 6 (December): 369-74.

Kochan, T.A. and Osterman, P. (1994). *The Mutual Gains Enterprise.* Harvard Business School Press. Cambridge, MA.

Lawton-Smith, H. (1990). *The Location and Development of Advanced Technology Industry in Oxfordshire in the Context of the Research Environment,* Unpublished DPhil thesis, University of Oxford.

Lawton-Smith, H. (1990). *The Location of Innovative Industry: The Case of Advanced Technology Industry in Oxfordshire.* School of Geography, Oxford University. Oxford.

Morgan, K. (1995). *The Learning Region: Institutions, Innovation and Regional Renewal,* Papers in Planning Research, No. 157. Department of City and Regional Planning, University of Wales. Cardiff.

Mytelka, L.K.(ed.) (1991). *Strategic Partnerships: States, Firms and International Competition.* Pinter Publishers. London.

Powell, W.W. (1990). "Neither Market nor Hierarchy: Network Forms of Organization," in B.M. Staw and L.L. Cummings (eds.), *Research in Organization Behaviour*, Vol. 12, JAI Press. Greenwhich CT.

Saxenian, A. (1994). *Regional Advantage: Culture and Competition in Silicon Valley and Route 128,* Cambridge: Harvard University Press.

Statistics Canada (1992). *Survey of Growth Companies,* reproduced in John R. Baldwin, *Innovation: The Key to Success in Small Firms,* Analytical Studies Branch, Research Paper No. 76. Ottawa. 1995.

Stewart, T.A. (1997). *Intellectual Capital: The New Wealth of Organizations.* Currency Doubleday. New York.

Storper, M. (1995a). "The Resurgence of Regional Economies, Ten Years Later," *European Urban and Regional Studies* 2(3).

Storper, M. (1995b). "Territories, Flows and Hierarchies in the Global Economy," *The Swiss Review of International Economic Relations (Aussenwirtschaft).* (June).

Storper, M. (1992). "The Limits to Globalization: Technology Districts and International Trade," *Economic Geography* 68.

Wolfe, D.A. and Gertler, M.S. (1997). "The Regional Innovation System in Ontario." in P. Cooke, H.-J. Braczyk and M. Heidenreich (eds), *Regional Systems of Innovation: Designing for the Future.* London, Taylor and Francis.

12 CANADA'S TECHNOLOGY TRIANGLE

Jeffrey Roy
University of Ottawa

12.1 INTRODUCTION

A short distance west of Toronto, Ontario, one finds what has come to be called 'Canada's Technology Triangle' (CTT). The Triangle is an established system of four networked municipalities - Cambridge, Kitchener, Waterloo and Guelph which together comprise one of Canada's most thriving socio-economic regions as supported by a range of indicators such as regional employment, new firm creation and regional growth rates (Smith 1996).

While much of the literature on local and regional innovation emphasises both government (structures, institutions, rules) and governance (culture, trust, etc.), North American socio-economic competition continues to privilege the separateness of socio-economic sectors and actors (Fukuyama 1995). This creates a tension in the innovation and adjustment processes as more-or-less creative and adaptive locales and regions struggle to engage the global economy. But in successful areas such as CTT, the successful evolution and successful of the region is driven in part by its capacity to integrate all socio-economic agents into an adaptive, flexible and synergistic structure. This set of dynamics sets out our basic interests in Canada's Technology Triangle.

In approaching the CTT, however, it is worth noting the guidelines provided by Bryant and Preston who provide ten points to evaluate the efforts of Canadian municipalities: i) local participation as necessary; ii) inclusive institutional frameworks; iii) promotion of territorial identity; iv) promotion of regional economic and financial circuits; v) stimulation of local entrepreneurship; vi) initiatives to complement market mechanisms; vii) innovation as central; viii) understanding the broader regional economic system; ix) need for diversified activities; and x) strategies are varied and not homogenous (1987). This list is worthy of review for two reasons. First, many of these points emerge as part of CTT's explicit efforts to further its own development strategies. Secondly, and equally important, is the notion that many of these points cannot be undertaken by government alone without a fundamental shift towards being part of a governance alliance, network or system.

The evolution of economic geography is one which builds on Marshall's cluster-based industrial districts to today's cartography of local innovation systems emphasising a co-evolution of socio-economic sectors (Saxenian 1994; Kanter 1995; de la Mothe and Paquet 1996; Moore 1997). Local embeddedness has been stressed particularly by the "milieu" approach (Aydalot 1986; Aydalot & Keeble 1988; Mailit 1991; Todtling 1994; Paquet 1996) driven by the collective benefits derived from the interactions and interdependencies amongst actors within a specific regions: what results are processes of agglomeration where territorial-based innovation systems harness the practical, technical and tacit benefits of proximity.

This link between proximity, flexibility and learning becomes a self-enforcing process within the prism of governance systems such as Richard Florida's "learning region" (1996) captures knowledge flows, information-sharing, and a basis of shared externalities which provide shared benefits within the community if the (cultural) presence of social capital acts as an enabling force for collective learning and action to take hold. Sub-nationally-defined entities are now the key shells within which such dynamics are understood (Ohmae 1995; Paquet and Roy 1996). The result are "distributive" compacts of socio-economic processes (Paquet 1997). Wolfe invokes Cooke's "associative governance to signify the growing shift from hierarchical forms of organisation to more heterarchical ones in which network relations are based on conditions of trust, reciprocity, reputation, openness to learning" (Hirst 1994; Cohen and Rogers 1995; Amin 1996; Wolfe 1997).

There is little question as to whether or not local governments are being asked to play a more important role in today's society. There is nevertheless considerable debate as to whether or not local government is up to the challenge of subsidiarity or devolution (Blais 1995; de la Mothe and Paquet 1994). Proximity in industrial clustering and innovation also extends to public sector reform: the demise of the Keynesian Welfare State, with its emphasis on national-level processes is explained by the simultaneous forces of globalization on the one hand, and the need for concertation locally on the other; the resulting (albeit awkwordly phrased) "globalization" tendencies have impacted discussions on state reform and embedded this interest in local state structures within the same parameters more broadly-defined models of local governance (Jessop 1993; Naisbitt 1994; Courchene 1996).

The resulting push to empowerment of local government is also supported by some subsidiarity-driven arguments in favour of proximity - government to the people; more open systems than national institutions; improved efficiency and effectiveness through greater public input and participation - of a direct rather than representational nature; and enhanced degrees of flexibility for innovation in "re-inventing government" (Osborne and Gaebler 1993), either through the re-creation of state agencies or in partnership with other sectors (Carr 1996; Barnett 1997).

This latter point links subsidiarity and flexibility. In his examination of the link between "enabling government" and "urban governance" Barnett puts forward the following issues as the nexus between reforming traditional state structures towards alternatively construed arrangements inspired by subsidiarity: i) "what size, level and structure of government is best suited to perform an enabling role; ii) how can government facilitate activity by the private and voluntary sectors in order to bring about pluralistic governance; and iii) how can higher levels of government facilitate lower levels government in the discharge of their duties" (Barnett 1997)? In addition, and following a logic built from the potential benefits of proximity and flexibility, the application of subsidiarity to government is also based on the view that organisational learning can best be nurtured within discursive forums of interchange and interaction requiring a loosely-defined and decentralised setting (EU 1994).

Some mistakenly view a diminishment of the government's role. (Ohmae 1995) Yet this is to overstate. The modest successes of Canada's Technology Triangle originate with the actions and persistence of local governments and ligatures.

Local governance models capturing processes of innovation go beyond the nexus of proximity, flexibility and learning. While these elements reveal much, there remains an under-appreciation of the importance of other socio-economic partners in contributing to the competitive and adaptive capacities of groups of firms. To better tease-out these multi-stakeholder connections which lead to the convergence of governance and innovation, we can underline three synergistic dynamics, each representing a manner by which socio-economic actors inter-relate and co-evolve in an interdependent fashion: they are the nutrients and lifelines of a "complex adaptive system or ecological system encompassing an array of inter-dependent socio-economic elements (Paquet 1992; Moore 1996). They are: i) competitive advantage - determined through the strength of market-driven clusters; ii) intelligent advantage - shaped by the tacit and technical knowledge transfers between individuals, firms (and networks of firms) and research and education institutions; and iii) collaborative advantage - nurtured through the collective capacities within the region allowing individuals and organisations to both recognise and harness synergistic interdependencies emerging, in part from civic forums and community-based identities.

i)   Competitive Advantage: Market Clusters

Models of industrial competitiveness point to the strength of market forces within industry (or regional) clusters (Porter 1989): strong market forces lead to demanding customers who in turn, promote both high quality standards and continual improvement and innovation. Similarly, the presence of several successful firms in related industries can promote a healthy dose of competition between them. With both competition and co-operation both playing roles in economic governance today (Best 1991; Rugman and D'Cruz 1992; Moore 1996; Paquet and Roy 1997) market linkages stress the former.

Such linkages include customers, suppliers and competitors. An important variable in the strength of a region for sustaining high-technology businesses is its ability to provide a window, and even a spring-board towards global markets. What this means for business firms is that a strong local base of competing firms, sophisticated customers and reliable suppliers can foster a healthy climate for innovation strategies leading to competitive advantage. Such market linkages can be considered from multiple viewpoints. Though a regionally-defined socio-economy may represent only a small portion of most firm's total sales there, other regional attributes are key determinants of competitive advantage, such as: i) the region's suitability to supplier-chain networking and outsourcing arrangements; ii) the local customer base for technology-intensive products and services, and the presence (or not) of sophisticated regions of lead-product technologies; and iii) the strength of the "cluster" of intense competitors are located in close proximity, strengthening the local market.

ii)   Intelligent Advantage: Knowledge Infrastructure

The role of knowledge transfers is a key determinant of regional governance and firm performance (and the connection between the two). This infrastructure includes knowledge-generating institutions such as universities, colleges, research institutes and laboratories, and various consortia/hybrid arrangements among them. The infrastructure also includes the region's supply of human resources and skills, a critical determinant for technology-intensive industries.

Local R&D partnerships link firms with many post-secondary institutions. As "supporting institutions", universities are integrated as key factors for a region's economic competitiveness (Porter 1989; Gulbrandson 1995). Yet firms engagements with universities and/or other organisations also impact the firm's internal organisation. The challenge facing R&D-driven firms and regions is the ability to effectively co-ordinate and nurture these knowledge systems in order to ensure value for investment: such is the nurturing of a firm's "absorptive capacity" through ongoing strategies of institutionalising networks of learning from existing resources and generating new knowledge, both explicit and tacit (Cohen and Levinthal 1989; Tassey 1992; Nonaka and Takeuchi 1997).

Knowledge infrastructure thus includes all potential sources of new ideas available for the firm to innovate. We can summarise three main elements attaching firms to its knowledge infrastructure: i) external partnering within the region's research environment for product or process-based technology transfers and innovations; ii) the availability of tacit knowledge and specialised skills in human resources; and iii) the capacity for continual organisational learning.

iii)   Collaborative Advantage: Community Culture

Social capital is an indicator of collaborative capacities: it reflects the strength of the underlying "civic community", the propensity of people to form organisations not bed on kinship, and the "spontaneous sociability" within a region. (Putnam 1993; Fukuyama 1995). Thus, the "new competition" actually emerges as a simultaneous set of competitive and co-operative relations, the latter shaped by cultures of trust (Best 1992; Sabel 1993; Fukuyama 1995). For Saxenian (1994) social capital appears as local culture, shaped by the region's formal and informal institutions, similar to Moss Kanter's (1995) linking of the efficacy of market-based organisations (firms) and their surrounding community within increasingly "distributed" governance (Paquet 1997).

This socialisation of competitive practises underscores the interdependence and connectivity between individual entrepreneurs and sectoral and regional institutions.

Accordingly, civic entrepreneurs emerge as catalysts for social capital, both in forging alternative (non-market, non-state) organisations and shaping local culture in ways that aim at explicitly at the generation of social capital. The linking of social capital with civic entrepreneurialism is a direct response to the challenge of sparking and nurturing local innovation to enhance global competitiveness: the recent Silicon Valley-based vision of "grassroots leaders" typifies this movement shaped by the presence and formation of local culture (Granovetter 1985; Lundvall 1988; Saxenian 1994; Gertler 1996; Henton, Melville & Walesh 1997). Such attention also focuses on the meaning and types of culture at work within regional governance and innovation systems, as well as the vagueness by which it is often invoked. "We seem to arrive at the conclusion that regional (economic) development presupposes - yet is not reduced to - cultural change; the centrality of the need to understand various aspects of culture and its change becomes evident" (Oinas 1995).

The map below sketches out the set of forces shaping the performance capacities of the region: the dynamic processes on the right, most often characterised as components of new governance systems reflect various networks and hybrids of private sector networks, the community and social sectors and the extended public sector - through research and knowledge institutions; the traditional processes on the left reflect those state structures impacting, in various forms, the evolution of the governance system denoted by the three dynamic processes.

From the above loose map, we may derive a number of initial propositions on the nature of the government's role - at each of the three levels and relative to each of the three governance processes:

i) *the federal government* - indirectly impacting the region, with its policies linked to its responsibilities for the national socio-economy as a whole;

ii) *provincial governments* - indirectly impacting specific clusters and community-based cultures, but directly connected to the knowledge infrastructure with resources and jurisdictions in research and education; and

iii) *local governments* - situated in close proximity to strategically partner with local-area clusters, knowledge and research institutions, and civic and community organisations.

The point of these initial directions is not to suggest that in Canada, governments are embracing such roles as dictated by these contours, but rather to provide a starting point for the assessment of government action within our case, Canada's Technology Triangle.

Nevertheless, the resulting emphasis on local actions can be underlined as an important premise of this exercise: local government is being called upon to play a widened and more strategic set of roles - aimed at each of the synergistic governance processes, and its ability to do so is a key source of advantage (or handicap) for the region.

With respect to CTT, acknowledged as region already noted for a strong socio-economic performance and local governments responsible for the creation of CTT, an assessment of government within such a region should yield important insights transferable elsewhere, pending further research and investigation.

12.2 OVERVIEWING AND ASSESSING CTT

As noted at the outset, Canada's Technology Triangle incorporates the four cities of Kitchener, Waterloo, Cambridge, and Guelph. Initially, 10 years ago, its purpose was to collectively pool resources and efforts from the four urban governments to create a quasi-regional governance entity capable of servicing the CTT as a whole: the CTT "grew out of the mutual awareness that the four communities could band together and draw on their complementary strengths"(Lindsay 1994).

The focal point of the CTT initiative was the nurturing, promotion, and expansion of technology-driven industrial activities: they include telecommunications, computer software and hardware, environmental technologies, advanced manufacturing, the automotive sector, agriculture and others. Along with local government and businesses, CTT's evolution has also been based on other local actors such as universities and colleges, research and innovation centres, and community groups servicing the infrastructural needs of the region.

12.2.1 The Cities and CTT Government

CTT is what some refer to as a "virtual network" between the economic development officers (one from each) of the member-cities: until now there have been no formal organisations in addition to this four-node policy network. The initiative's modest origins were mainly to achieve economies of scale in areas such as international marketing and investment attraction schemes (clearly there was a much greater incentive for the smaller regions to participate here). Over time, however, the focus of CTT's purpose and activities has shifted from an external orientation to one emphasising the internal aspects of CTT's regional economy.

Such a shift came about from a consensus among the city development officers that spending designed to attract new investment had become less "important" than priority policy areas of internal concern. These internal concerns include four key areas: i) strategic planning for growth and sustainable economic development; ii) institutional infrastructure; iii) business promotion and support for existing industries; and iv) entrepreneurship and new business (and jobs) generation with an according emphasis on small firms. Transforming such a consensus into concrete measures and underpinning governance is not a given.

12.2.2 Recent Initiatives in CTT Governance

CTT city-members turned to horizontal alliances as a tentative first step towards alternative governance, attempting to overcome their perceptions of barriers that had been internalised within traditional, and functionally-based bureaucracies of local economic development. The initiative is a response to the "local-clustering phenomenon challenging governments to develop the programs needed to service these knowledge-based centres or technopolies" (Coffey; Blais; Voyer & Ryan 1994).

CTT's Steering Committee (the four municipal EDOs) subsequently decided to commission a strategic review of CTT's options available for its future evolution. Given its ambition, the process is worth reviewing in some detail, including the mission statement, guiding principles and options considered for a renewed CTT governance system. Beginning with the former, the revised mission statement for CTT as a "partnership of the municipalities of Cambridge, Guelph, Kitchener and Waterloo committed, in co-operation with the private sector, other levels of government, and the education and training sectors, to encouraging economic development and facilitating technological innovation in a cost-effective manner" (CTT 1996). The following set of CTT objectives were then developed to guide the strategic review process of CTT structures and future options:

i)   *to design and implement internal and external <u>marketing programmes</u> for the area's overall economic development;*

ii)   *to facilitate the establishment and maintenance of <u>networks and partnerships</u> among the business, technological, education and other communities, through promotion, participation, information dissemination, and other supportive activities;*

iii)   *to design, establish, and manage an area economic development <u>information system;</u>*

iv)   *to facilitate <u>innovation</u>, through the design, development, application and commercialisation phases in all technologies;*

v)   *to promote a <u>climate</u> conducive to innovation, investment and enterprise;*

vi)   *to complement and augment individual <u>municipal economic development priorities</u>; and*

vii)   *to continue and improve upon CTT's diverse activities in <u>enquiries management</u> (CTT 1996).*

Finally, within these parameters the Steering Committee considered a wide range of options for a revision of CTT governance: they included refinements to the present structure, the creation of a new private-public sector partnership, and CTT's abolishment (in its present form) in favour of a more centralised economic development efforts within the confines of regional government. The latter option, running counter to the logic of subsidiarity, can be viewed as a poor response to the challenges of proximity, flexibility and learning; the need to better connect to the private sector, however, was widely embraced as a central challenge.

Thus the first option emerges as the guidepost for change, the details of which are presented in the CTT's internal review study:

This alternative is characterised by increased formality, the implementation of CTT's first Strategic Workplan, and a volunteer CTT Board made up of representatives from the private sector, the four Municipal Councils, and the higher education sectors.

This alternative would involve a very close and expanded working relationship between CTT and the private sector, through the CTT Board, and through ongoing participation in investment recruitment, marketing, feasibility assessments and other tasks central to the operations of the organisation (Douglas 1996).

As indicated, the importance attached to partnering with the private sector is reflective of the government-governance logic developed earlier on. The failings to engineer such a meaningful partnership, a laudable goal, are testament to the difficulties of their integration. Despite wide spread support for the philosophy underwriting the CTT initiative, there has never been a mechanism to bring together CTT-area clusters into a complementing governance structure. As such, fruitful partnerships remain elusive.

One consequence of this difficulty, however, has been an attempt by local area technology firms to act pro-actively: the creation of "Communitech - CTT's Technology Association" is the first effort to create a network of businesses defined by CTT boundaries (as opposed to municipal bodies such as Chambers of Commerce). This new organisation, accountable to its membership and based on a similar model currently operating on Ottawa-Carleton, will pay particular attention to nurturing linkages between business and education sectors (the knowledge infrastructure), the other objective highlighted in CTT's strategic review. Briefly, where governments have been unable to do so, alternative governance vehicles are forged.

A final element of CTT's shifting governance portrait for 1996 was the awarding of federal government funding to a CTT-sponsored initiative, Canada's Technology Triangle Accelerator Network (CTTAN) designed to accelerate the financing and growth of small technology firms in the region (the funding is through the Liberal government's Canada Community Investment Program). This latest development is of great interest for two reasons: i) the federal involvement, the first of its kind for anything related to CTT brings to the forefront questions bottom-up dynamics converging with top-down (albeit limited) intervention; and ii) the person recently named to act as the Head of CTTAN was, until recently the EDO of Cambridge (and thus one of four members of the CTT steering committee).

12.2.3   Rating the Synergistic Capacities

Much like other Regions, the regional advantage of CTT is continually determined by the relative performance of its three dynamic sub-processes of governance: market linkages, the knowledge infrastructure and community culture. In order to assess CTT's success as an initiative, we can examine the extent to which the network of local state actors has contributed either positively or negatively to each.

i)   *Market Linkages:*

With respect to the clustering dynamics of CTT-based industries, then, the contribution of local governments has been tentative and mixed. The preceding description of CTT's inception underlined its emphasis on working to enhance existing clusters across the four municipalities rather than creating new ones; though knowledge-intensive areas such as telecommunications, software and environmental technologies may appear as recent exceptions - given their relative importance in CTT's new initiatives, the broad emphasis has been inclusionary rather than exclusionary. Cambridge's view of the automotive sector as "high-technology a well" is supported by the global evolution of this once, traditionally-characterised industry (Morales 1995).

While some attempts have been made to co-ordinate a better profiling of area clusters, the CTT network has done little to improve market linkages; instead, such efforts remain functional jurisdictions of each municipality. While the area is recognised and marketed as a highly-competitive region, local development officers express a sense of frustration that instruments and benchmarks remain locally-fixated, rather than collectively expanded. Their sentiment would seem to be reflected in the tensions of the latest "Team Kitchener" initiative: politically driven by the Mayor to focus on the city's needs and priorities, the report articulated the need for a stronger CTT role. In other words, local government lost out to the growing acceptance of region governance.

Reflecting the latter point, and a benchmark of the contribution of CTT's existence has been the growing base of economic and statistical data on CTT - collectively as opposed to rough aggregates of the four municipalities. The point, driven by difference between state and region boundaries, is common to policy-makers within many local systems: the lack of quantitative information on Silicon Valley's region (comprising over twenty municipalities) was an important impetus for creating JVSN. Similarly, support from CTT-based governments (including the Region) has allowed for a modest beginning in providing a pool of information conducive to socio-economic realities (and region governance) as opposed to dated political boundaries (and their local governments).

ii) Knowledge Infrastructure:

CTT is rightly promoted as a region emphasising human capital, as the notable lack of substantial natural resources has always been countered with an emphasis on training, education and the capacities of individuals to demonstrate adaptiveness and resilience (as already noted, this historical legacy implies that CTT is but the latest phase of this evolution). Today the three local area universities solidifies the region's reputation, as does the recognised success of Conestoga College as the provincial leader in placing local graduates in the labour market.

Yet there is also a growing view from within CTT that the "intelligent advantage" fostered from the region's strong base of post-secondary training and research institutions is under-performing, relative to its potential. In this regard, there is a notable void in looking to CTT's role, other than promoting the area's three universities under a common flag (all three are located in Guelph and Waterloo). No formal linkages have evolved between the CTT network and the universities.

Developing meaningful ties to the education sector, it should be recalled, was a central element of CTT's recent strategic review. The notion of a local innovation system applies closely here, as such linkages are deemed necessary to nurture the sorts of tacit, technical and human knowledge transfers required to sustain the learning, adaptive and commercialisation capacities of local firms. Moreover, the region's strong emphasis on co-operative education acts as a powerful incentive to both lure students and further linkages to other socio-economic centres, both nationally and internationally. Yet the inability of CTT's economic development committee to parlay the network into a more broad and encompassing system is without question a missed opportunity.

Such a view, increasingly acknowledged, partly explains the recent inception of Communitech (designed to foster connections and better synergies between its private sector membership and the educational sector). Priority areas are better access to research and knowledge, better support for entrepreneurs and stronger mechanisms with respect to retaining local university graduates, is partly an impetus for creating Communitech (Smith in Kubasta 1997). Clearly, the next stage of CTT should place education and human capital at the forefront of the region's strategy, not only for attracting ideas and talent but also to improve the retention capacities (at times, paradoxically, at odds with the dynamics of co-op which facilitates studying in the region and working elsewhere).

Thus a key challenge for the CTT initiative is the necessity of a mechanism to link local, business, government and education sectors: step one could be co-operative ties between Communitech and the CTT Steering Committee, but such action is a first step of what holds considerable potential for advancing region governance. Here as well, local governments - working through their CTT alliance have an important role to play in partnering with the province: the latter's jurisdiction is central here, both for education at all levels of study and for R&D resources such as Centres of Excellence. One CTT member points to an important success of the alliance as provincial recognition of three technopoles rather than two (the Greater Toronto Area, Ottawa-Carleton plus CTT); more concerted efforts are needed to strengthen the region's knowledge infrastructure.

iii)   Community Culture:

An important component of CTT's governance capacities is the area's community culture: the region has a long history of a strong community spirit supporting its economic resilience (Walker 1987), though consensus as to why is elusive. From an ethnic-orientation, it may be problematic to the extent that German culture and historical influences underpins it since this view is widely accepted across the local population (diffusing itself in the local culture accordingly).

A key engine of entrepreneurship is the educational institutions, including both The University of Waterloo and Wilfrid Laurier University. The success of Waterloo's linking of learning and research with business is well-documented (Fournier 1991; Roy 1996; Kubasta 1997) and is not our prime focus here (the university's primary role is clearly connected to knowledge infrastructure. Yet the socialisation of an entrepreneurial culture from university-based spin-offs provides a partnering capacity to strengthen social capital. The further institutionalisation of such vibrancy, and its nurturing into a more mature set of clusters is directly associated with the perceived need of Communitech and CTTAN respectively.

Mapping the Governance System

Levels of Government

Yet it would be mistaken to over-emphasise the importance of CTT's local network in contributing to, and fostering the sense of local culture permeating the region. While there is evidence that the CTT network is an important factor in contributing to the region's resilience and prosperity (Roy 1996), the point here is that its existence cannot be entirely accredited to this partnership of local governments. Much like the historically-nurtured presence of social capital in Northern Italy documented by Putnam, CTT is but the most recent, and perhaps most advanced example of the region's resilience, made possible by the enabling force of a supportive community culture (Walker 1987).

This foundation - of a local identity and supportive civic ties provides a basis to reinvest in its stock of social capital: in contrast to other regions (such as Silicon Valley in Califorinia, attempting to recognise social capital as a crisis remedy (JVSN 1992) CTT appears more enviably situated as a region looking to create mechanisms to utilise and expand an already existing pool. Nevertheless the creation of new collaborative mechanisms has not materialised, the current gridlock within CTT governments is worrisome and for these reasons one can only conclude that the region, as a region is being poorly serviced relative to its potential (we return to what should be done below).

12.3 SCENARIOS FOR INTEGRATIVE EVOLUTION

Based on the preceding assessment, it is clear that traditional government structures are not smoothly integrated into regional governance, the result being that potential synergies from the three processes of the latter are under-utilised. The purpose of this section is to envision some possible avenues for forging a better degree of connectivity, with the view that stronger linkages, based on the underpinning and overwhelming rationale for their interdependencies, lead to better local performance capacities for sustaining prosperity through regional advantage.

12.3.1 Strategic States

The point has already been made that the local, municipal governments, in particular the economic development officers, deserve credit for launching and sustaining the CTT initiative. The fact that the recent "Team Kitchener" economic development strategy stressed the importance of CTT to Kitchener's future economic fortunes provides testament to its importance (especially given its city-centric mandate).

Yet at the same time, recent experiences and developments suggest more ominous signs for the capacities of local governments to become key catalysts in such governance arrangements. There are two reasons for such an assessment: first, the political structures and processes often act as formidable barriers to making the transitions to local governance systems based on subsidiarity and multi-stakeholderism; and secondly, while the economic development function remains an important aspect of local government it may be the case that other venues - non-state forums are providing greater degrees of freedom and thus incentives for innovation in governance arrangements to occur.

On the first point, the current "CTT Start-up Board" is, on paper, a high-powered group key public officials including the Regional Chair and local Mayors and CAOs; in practise such representation is a recipe for gridlock through crowded and complex agendas. Equally important are the notable differences between political jurisdictions (and thus incentives) and socio-economic boundaries. Thus, although ideally it should be the role of local governments - acting as strategic states - to nurture the synergies between market, knowledge and community components of the region the evidence from within CTT suggests that they cannot be counted on, within he current structures to act as key agents of change in engineering new governance arrangements conducive to current socio-economic challenges.

The second point, the notion of alternative venues, ties itself to the importance of not only a vibrant and supportive community but also a requisite degree of civic entrepreneurialism to nurture a set of non-market, non-state actors capable to fulfil governance roles within the region.

12.3.2 Civic Entrepreneurialism

The preceding assessment of CTT suggests that the governance alliance has not, up until the present time played an explicitly active role in shaping the market competitiveness of the region. In fact, with the disconnectedness between the private sector and the CTT initiative the sectoral separation of state and market has resulted in an under-utilisation of region capacities. While this is not to say that industry has been performing poorly, it does suggest that CTT has done little to contribute to market-based activities of the region.

Such a scenario has been partly responsible for the formation of a rift between various local actors of the future evolution of CTT. The Kitchener-Waterloo in particular has been perhaps the most radical proponent of a new CTT entity based on a new agency which would be close to the privatised CTT option considered among the options. The Chamber's plan, however, does not benefit from widespread support among the region's businesses, nor does the Chamber speak for the business sectors of Guelph and Cambridge. Thus a key barrier to a more effective and integrative role for the private sector has been the lack of CTT-based mechanisms to facilitate such participation.

The implications for the performance of the region is an underdevelopment of integrative synergies between the market, knowledge and community components: such separateness reinforces distinctions between industries, governments and communities in each of the municipalities despite the growing, and largely undisputed evidence pointing to their economic interdependence.

Clearly, this point is well-recognised, and it is partly the impetus for Communitech to service technology firms on a CTT-wide basis. This fledging organisation could provide crucial to the future governance capacities of this region: as a civic catalyst, it is well placed to mobilise the energies of the private sector (in turn responsible for its creation), to raise the profile of local area technology firms (as flag carriers for the new economy) and to better connect these firms to other sectors such as educational institutions and local governments.

On the latter, Communitech should make every effort to emerge as an agent of positive reform, becoming vigilant allies of local EDOs in their own struggles with traditional structures of the local public sector (a positive relationship, for which the seeds clearly exist would be in notable contrast to relations between most EDOs and their local Chamber of Commerce). An interim but equally important step for Communitech would be a solidification of its ties with CTTAN: both are new actors, acting as civic entrepreneurs with their credibility and success tied to their impacts locally, particularly with the private sector (the logic here on CTTAN is that its continuing existence will be dependent on local recognition and funding support which, in turn, will be decided by is measurable impacts). With CTTAN focusing on the needs of entrepreneurs and fledging firms, Communitech's concern lies in mobilising the efforts of established local enterprise.

12.3.3   A CTT Action Plan

CTT's tenth anniversary should be seized not as a moment of celebration, but rather as a time for renewal. Local municipalities deserve credit for introducing and sustaining the concept (as it has since attached itself to others) but as the local EDOs themselves recognise, the current system is inadequate and attempts at reform have been tedious and continuously stalled. If CTT is to emerge as a recognised centre for high-technology excellence, an impetus for a more aggressive forms of local governance must be fostered. CTT must be transformed into a basis for continual adaptation, encouraging and accelerating growth in the region rather than merely managing it.

The first step is a greater presence. CTT's impetus has been for external marketing, minimising local recognition in the process. Local governments, the region, Communitech and CTTAN are now, collectively a critical mass which can encourage others, such as the educational institutions to network the CTT identify intensively across the region.

On CTT government, local municipalities and the region will have to find the resolve to move forward. CTT must become a para-public form of organisation with its own leader, working on broad-based development strategies for the region, while leaving local planning and management issues to the city EDOs. As such, it may be inevitable that the Region of Waterloo become more aggressive in promoting such an endeavour: given the devolution of powers to the Region, its growing power policy capacities will provide it with a greater voice. While the local consensus to maintain separate economic development units should be respected, the Region's clout and resources will prove pivotal in the future.

A new CTT Development Corporation, funded primarily by governments could then interact with the private sector through Communitech (and perhaps CTTAN). The educational sector could clearly serve as a partner to both sides, and one could envision a form of CTT Steering Body comprising these different stakeholders, acting collectively in a periodic fashion to elevate government and governance to the needs of the region. The genreaizable challenge facing CTT governments is the extent to which they can integrate themselves in a dynamic governance context and work in concert with the complementing efforts of civic entrepreneurs (a particular area of strength for CTT).

To return to the three synergistic dynamics, municipal economic development units would continue to service their local clusters, and co-ordinate with others when warranted. Cities should also partner vigorously with all local groups to utilise civic ties and social capital, but attaching themselves to the CTT identity in the process. The new CTT partnership between the an autonomous development corporation and Comunitech could then focus on the broad-based market linkages of the region, as well as strategies to forge a like-minded approach to the knowledge infrastructure (in concert with the educational sector). This CTT Development Corporation could then consolidate the external marketing efforts and strengthen them through a better usage of local resources and a united voice (for future Team Canada adventures as just one example). The Corporation's Head, as the recognised spokesperson the CTT economic development could then better service the region on provincial and federal forums, the latter particularly crucial to human capital factors within the knowledge infrastructure.

A private-public sector partnership, construed on working relationship between Communitech and a CTT Development Corporation must then, as its first priority mobilise local stakeholders into a co-ordinated learning strategy to strengthen the region's knowledge infrastructure. The "intelligent advantage" of CTT should no longer be taken for granted; instead it can be the basis for an integrated approach to utilising local human, tacit and R&D resources.

12.4 CONCLUSIONS

In terms of the lessons learned from CTT as a regional case study, there are five points to underline. First, a new template of government action is required, sensitive to horizontal governance networks within the region and the difficulties facing local public agencies is responding accordingly. Secondly, local government action is necessary but insufficient (as the initial CTT creation must now be transformed) without effective ties to other governance mechanisms. Thirdly, within the new template of government action the space for strategic co-ordination between local and provincial authorities may be emerging as key for local prospects (with the emphasis on local actors managing these relations). Fourthly, there is a growing need for information and data for the region, as opposed to traditional political units. Fifthly, the emergence of alternative, civic-based venues for intermediation and innovation in governance should be recognised and supported as a healthy means to bettering local capacities (conducive to the new realities of associational or distributed governance).

At the same time, the preceding discussion has revealed that the prospects for a region are determined by three forms of capital to nurture local economic activity: social, human and financial (investment and venture) capitals are equally central. We can postulate then that local governments are catalysts and points of connection for social capital; provincial governments are key engineers for education and skills training; and the pool of financial, investment and venture capital is a more complex area for co-ordinated action at all levels.

Finally, speculations on the future of CTT depend very much on one's perspective. As a region, CTT is well-diversified and dynamic with solid growth prospects. Yet the creation of a more effective system of local governance is central to CTT's continued success, particularly with respect to knowledge and technology-intensive sectors. Other regions provide lessons of reactive governance solutions: CTT still has a window of opportunity, albeit a limited one, to develop proactive measures.

REFERENCES

Cohen, W.M. and Levinthal, D.A. (1989). "Innovation and Learning: The Two Faces of R&D" *The Economic Journal*, 99, pp. 569-596.

de la Mothe, J. and Paquet, G. (eds) (1996). *Evolutionary Economics and the New International Political Economy,* London: Pinter.

de la Mothe, J. and Paquet, G. (1994). "The Dispersive Revolution" *Optimum* 25,1,pp.42-48.

de la Mothe, J. and Paquet, G. (1994). "The Technology-Trade Nexus: Liberalization, Warring Blocs, or Negotiated Access?" *Technology in Society*, 16, 1, pp. 97-118.

European Commission (1994a). *The Community Innovation Survey: Status and Perspectives.*

European Commission (1994b). *Evidence from Europe and North America on "Intangible" Factors Behind Growth, Competitiveness and Jobs*, Industry Panorama, August.

European Commission (1994c) *The European Report on Science and Technology Indicators* EUR 15897, Luxembourg

Fukuyama, F.(1995). *Trust: The Social Virtues and The Creation of Prosperity,* New York: The Free Press.

Gulbrandsen, M. (1995). *University and Region - Cooperation Between Universities and Regional Industry in the Nordic Countries,* Copenhagen: Nordic Council of Ministers.

Henton, D., Melville, J. and Walesh, K. (1997). *Grassroots Leaders for A New Economy,* San Fransisco: Jossey-Bass Publishers.

Moore, J.F. (1996). *The Death of Competition - Leadership and Strategy in the Age of Business Ecosystems,* New York: HarperCollins.

Morales, R. (1995). *Flexible Production*, Oxford, Polity.

Moss Kanter, R. (1995). *World Class - Thriving Locally in The Global Economy,* New York: Simon and Schuster.

Naisbitt, J. (1994). *Global Paradox,* New York: William Morrow and Company.

Ohmae, K. (1995). *The End of the Nation State,* New York, Harper Collins.

Paquet, G. (1997a). "States, Communities and Markets: The Distrbuted Governance Scenario" in *The Nation-State in a Global Information Era: Policy Challenges,* Queens University Bell Canada Conference.

Paquet, G. (1997b). "Alterative Service Delivery: Transforming the Practices of Governance" in Ford, R. and Zussman, D. (eds.) (1997), *Alternative Service Delivery: Sharing Governance in Canada*, Toronto, KPMG and IPAC.

Paquet, G. and Roy, J. (1996). *Governance in Canada: Competition, Cooperation and Co-evolution in Business-Government-Society Relations,* (unpublished mimeo, the University of Ottawa, pp. 368).

Paquet, G. (1992). "The Strategic State" in J. Chrétien (ed.), *Finding Common Ground - Proceedings of the Aylmer Conference,* Ottawa: Voyageur.

Putnam, R.D. (1993). *Making Democracy Work: Civic Traditions in Modern Italy,* Princeton University Press.

Roy, J. (1996). "Regional Advantage and Local Economic Development: Government and Governance in Canada's Technology Triangle" in *Colloqui,* Cornell.

Sabel, C.F. (1993). "Studied Trust: Building New Forms of Cooperation in a Volatile Economy" in D. Foray and C. Freeman (eds), *Technology and The Wealth of Nations,* London: Pinter.

Saxenian, A. (1994). *Regional Advantage: Culture and Competition in Silicon Valley and Route 128,* Cambridge: Harvard University Press.

Tassey, G. (1992). *Technology Infrastructure and Competitive Position,* Boston: Kluwer.

Todtling, F. (1994). 'The Uneven Landscape of Innovation Poles: Local Embeddedness and Global Networks' in A. Amin and N. Thrift (eds), *Globalization, Institutions and Regional Development in Europe,* Oxford, Oxford University Press.

Voyer, R. and Ryan, P. (1994). *The New Innovators - How Canadians are Shaping the Knowledge-Based Economy.* Toronto: James Lorimer and Co.

Walker, D.F. (ed.), (1987). *Manufacturing in Kitchener-Waterloo: A Long-Term Perspective,* University of Waterloo - Department of Geography Publication Regionies.

13 THE CHAUDIÈRE-APPALACHES SYSTEM OF INDUSTRIAL INNOVATIONS

Réjean Landry and Nabil Amara
Laval University

13.1 INTRODUCTION

One should not foretell, as some have done, the end of the importance of regions from the observation that innovation takes place in a context of globalization. Indeed, the flows of exchanges in ideas, information, goods and technologies that take place at the local territorial level of the regions are based on institutional arrangements that permit the integration of explicit codified knowledge circulating at the global level with the tacit contextual knowledge mastered at the regions level. The identification of the institutional arrangements and determinants that explain these arrangements singularizes the studies on local innovation systems.

This approach to innovation differs from previous approaches, first, by stressing the importance of the flows of ideas, information and technology between firms, and second, by emphasizing the importance of regional institutional arrangements that support these flows or exchanges.

In terms of research design, this paper deviates from prior studies on local innovation systems by using data from a firm level survey on innovation instead of using aggregated data.

The paper is structured as follows: Section 1 reviews the analytical issues regarding the innovation systems approaches and the research designs developed to investigate these systems. Section 2 provides empirical evidence on the Chaudière-Appalaches innovation system(s). This section is subdivided into five main parts: a) overview of the region; b) descriptive attributes about innovation and interactions between firms; c) determinants of innovation; d) variations between firms in matter of interactions; e) differences between the sub-regions within the region. Section 3 discusses the findings of the study and points to the high complexity of the flows that singularize regional innovation systems.

13.2   ANALYTICAL ISSUES REGARDING INNOVATION SYSTEMS

In a world of perfect information with no transaction costs, where buyers and sellers interact anonymously, institutional arrangements and social norms do not explain differences of innovative activities between countries and regions. However, as pointed out by Lundvall (1992: 45-46): "What makes national systems of innovation important, is that markets are organized differently in different national systems, and that the behaviour of agents belonging to different systems is governed by different rules and norms reflecting differences in the institutional set-up". As a result, an explanation of differences in innovative activities should focus on two basic questions: what are the institutional arrangements that explain variations in innovative activities?; and what factors explain these institutional variations?

13.2.1   Why Do Institutions Matter?

Recent literature on innovation claims that innovation depends on institutional factors (Le Bas,1995; Dodgson and Rothwell,1994) and that these institutional factors are localized either at the national (Lundvall,1992; Nelson,1993; Niosi and al,1993; Edquist,1997) or at the local level (Antonelli,1986; Lecoq,1991; Maillet,1992; Pyke and Sengenberger,1992; Benko and Lipietz,1992; Rallet and Torre,1995). In a world where information is not perfect, where transaction costs exist, and for which markets are not anonymous, institutions facilitate transactions by co-ordinating the use of information between buyers and sellers, by providing incentive and monitoring systems and finally, by reducing uncertainty.

This literature assumes that innovation does not depend on isolated individual agents, but instead, on how agents interact with others as elements of a system. Following the supporters of the various innovation systems approaches, we will assume that innovation depends on the flows (interactions, transactions, exchanges) of information and technology that circulate among people and institutions.

In this perspective, the innovation performances of a firm, a region or a nation depend to a large extent on how people and institutions interact in a systemic manner to create knowledge and to use it.

13.2.2   Where Are the Boundaries of the Innovation Systems?

The focus here is on interactions within those systems that produce innovations. Any interactions that are essential to the explanation of innovation must be included inside a system's boundary. Therefore, the study of an innovation system should start with questions like: "where is the boundary that encompasses the smallest number of components whose interactions explain the innovative activities?", and does the smallest number of components which explain innovation correspond to the national, regional, or sub-regional boundaries? The studies of national systems of innovation have assumed that the national components structure interactions between people and institutions sufficiently strongly to explain adequately the innovative performances at the national level. Alternatively, studies of local systems of innovation assume that the components of the local milieu are the most significant explanatory factors of innovation and growth performances.

This issue cannot be easily resolved. At the theoretical level, one should recall that if there are no clear boundaries, one has an open system: that is a system characterized by outputs that respond to inputs, in a context where the outputs are isolated from and exert no impact on the inputs. From the perspective of systems theory, an open innovative system would not assess and adjust to its own past innovative performances. Clearly, the systems that the students of innovation systems have in mind are not open systems. Therefore, if the systems approach is to provide new insights, the boundaries must be localized somewhere.

The localization of the boundaries of innovation systems appears even more paradoxical if one attempts to tackle it at the empirical level. Empirical evidence shows that innovative firms do not only depend on exchanges with people and institutions located within their country or region, but that they also heavily depend on exchanges with people and institutions outside their country or region. Indeed, given that the most frequent and the most important sources of information and technology to develop innovations are the clients and suppliers of innovative firms, and given that the most innovative firms, regions and countries have the highest levels of exchanges with clients and suppliers located either outside their region or their country, one must conclude that the localization of the boundaries becomes more and more difficult as innovative performances increase. Hence, we are confronted with a paradox: the innovation systems approaches would be more valid to explain innovation performances of systems where innovation performances are low. We do not think that this issue can be resolved only at the theoretical level. It has to be resolved at the empirical level based on solid empirical evidence.

13.2.3 How Do the Analytic Approaches to Innovation Systems Differ from Previous Approaches?

The systems approaches to innovation depart from previous approaches in laying stress on two neglected factors: the impact of interactions of firms with people and institutions outside the firms, and the impact of spatial factors. The analytic approaches to innovation can be classified by their explanatory factors: innovation depends either on sources of ideas and technologies that are endogenous to the firm, or alternatively, on sources of ideas and technologies that are exogenous to the firm. Additionally, innovation depends on sources of ideas and technologies that may or may not have spatial specificities. These explanatory factors may be used to distinguish and compare four analytical approaches to innovation (Table 13-1).

Table 13.1 - Analytical Approaches to Innovation

Sources of ideas and technology are:	Not spatialized	Spatialized
Endogenous	(1) neo-classical models	(2)
Exogenous	(3) evolutionary models	(4) models of innovative milieux: a) national milieux b) local milieux

Let us consider in turn each of the four cells. The approaches corresponding to cell 1 assume that the firm is self sufficient in terms of technologies and ideas. Therefore, the sources of technologies and ideas used to develop innovations are entirely endogenous to the firm in a context where spatial factors do not impact on the innovative performances of the firm. In this explanation, the emblematic firm is the large firms that, until the 1960's, developed their technologies and ideas of innovation through their in-house R&D laboratories. These assumptions correspond to the explanation provided by the neo-classical model of innovation.

Alternatively, innovation can be explained by sources of ideas and technology that are located outside the firm in a context where, again, spatial factors do not contribute to explain innovation performances. As manufacturing firms assess their own needs and requirements, they must necessarily decide whether to develop innovation in-house or to seek external sources. Existing studies (McFetridge, 1985; Teece, 1986; Sen and Rubenstein, 1990; Steele, 1990; Hagedoorn, 1990; Hull and Slowinski, 1990; Rothwell, 1991; Brockhoff, Gupta and Totering, 1991; Reddy et al., 1991; Robertson, 1992; Chatterji and Manuel, 1993; Palda, 1993) demonstrate that firms currently face certain incentives to seek external sources of technologies and ideas of innovations. Acquisition of ideas and technologies from external sources can serve to complement in-house technical capabilities but cannot substitute for it (Faulkner and Senker, 1994). Firms must absorb the know-how acquired from outside sources and transform it into commercially competitive products and processes. Successful absorption and transformation of external ideas and technology require the development of organizational learning capabilities within the firm. Failure to do so may generate institutional failures accounting for differences in competitiveness amongst industrial firms. This explanatory model (cell 3) corresponds to the evolutionary models of innovation.

There is, for the time being, no explanatory models that explain innovation from sources of ideas and technology that are endogenous to the firm and influenced by spatial specificities.

While the models of cell 3 attempt to integrate socio-institutional factors in their explanations of innovation, the models of cell 4 assume, in addition, that innovation also depends on local socio-institutional factors. In these models, firms are not considered as isolated agents, but as part of a milieu, which incubates innovation. This type of explanatory models refers to the innovation systems approaches that have emerged during the last decade. As mentioned earlier, the supporters of this new type of explanatory model can be classified according to their tendency to assume that the boundaries of the innovation milieux correspond either to national boundaries or regional boundaries.

13.2.4   Some Research Designs

The innovation systems approaches differ from prior analytical approaches to innovation in assuming that innovation depends primarily on interactions between people and institutions. In these approaches, the units of analysis are the flows of ideas and technology that circulate among people and institutions. Innovation, it is assumed, depends on the variety and intensity of these interactions. In recent literature, knowledge flows among people and institutions have been measured primarily by using five techniques:

1. Literature based-survey of joint research activities. Although generating new insights on the flows of exchanges, the main shortcoming of this technique is that it does not capture the variety of the flows of interactions that actually take place within an innovation system. Based on this technique, we would have found that 40% of the Chaudière-Appalaches firms having R&D activities had also undertaken joint research activities with other partners. This clearly underestimates both the variety and the level of interactions of the firms.

2. Co-publications and co-patents developed in collaboration by people from different institutions. The main shortcoming of this technique, again, is that it measures only one type of flows in contexts where a large variety of flows of exchanges take place. Based on this technique, we would have found that 4% of the Chaudière-Appalaches firms interviewed in our study have relied on patents as sources of ideas and information, whereas only 2 have used patents as outside sources of technology to develop innovation.

3. Citation analyses to assess the degree to which enterprises use patents and publications developed by outside institutions. Once more, the main shortcoming of this technique is to stress only one type of flows. In addition, this technique does not capture the interactions developed by SMEs because SMEs do not usually publish in scientific journals.

4. Qualitative case studies of innovation systems. The main strength of these studies is to stress the importance of social norms and institutional arrangements as mechanisms to foster exchanges of technologies and ideas of innovations. Qualitative case studies permit intensive examination of particular innovation systems but opportunities for systematically testing hypotheses are more limited than with other techniques. It is the technique that has been used most frequently to study the local innovation systems.

5. Firm level surveys of innovation. This technique permits to measure a wide variety of flows while it also permits, more easily than qualitative case studies, to systematically test hypotheses. It is the technique that has been selected to collect the data used in this paper.

13.3   THE STUDY ON THE CHAUDIÈRE-APPALACHES INNOVATION SYSTEM(S)

13.3.1   The Region Under Study

The Chaudière-Appalaches region covers 15,000 square km. sub-divided into 11 regional municipalities. The region is located between Québec City and the U.S border. The Chaudière River valley has always constituted a natural link with the United States. Half of the 380,000 inhabitants of the region live in six cities. The population of the regions totals 5% of the population of the Province of Québec. On the whole, the population of the region is less educated than that of the province, and the regional unemployment rate tends to be about 2% lower than the provincial average. The average income derived from employment is also lower than the provincial average. In 1994, the region had 1269 manufacturing firms providing jobs to 37,422 people. Thirty percent of these jobs were in firms with 49 employees or less, 36% in firms of 50 to 199 employees, and 34% in firms of 200 employees or more. The most important industrial sectors are, by decreasing order of importance, food, wood, transportation equipments, clothing, furniture, metal products and plastic products.

13.3.2   The Chaudière-Appalaches Survey on Innovation

On the whole, the diagnosis made by the business and political leaders of the region focuses around two inter-related issues: 1) the manufacturing firms of the region have massively renewed their technologies over the last few years, and 2) one needs to improve the technical competence of the regional manpower.

This is the context in which we were asked to conduct a survey on the innovation activities of the firms of the region in order to specify more precisely this diagnosis, and especially, to identify leverages for intervention. To achieve this objective, we have worked in very close collaboration with representatives from regional business organizations, regional municipalities, as well as with representatives from the Departments of Industry of the provincial and federal governments.

As a starting point, we have used the Oslo Manual (OECD,1992) and the questionnaire used by Statistics Canada in the survey on innovation. We adapted this latter instrument to meet the demands of the various representatives involved .The short time frame of the study led us to administer the questionnaire by telephone, to a sample of 265 enterprises drawn by the provincial Department of Industry. The interviews were completed in February 1996 with 193 enterprises, for an overall response rate of 82%. Compared to the population, the enterprises with more than 50 employees and the enterprises conducting R&D are slightly over-represented in the completed questionnaires.

13.3.3   Some Descriptive Attributes About Innovation and Interactions in the Region

A system approach to innovation requires that we tackle simultaneously two dimensions for each and every component of the innovation system: the levels of the variables and the variety of the linkages between the variables. We will hereafter lay all the emphasis on two main variables especially relevant for a system analysis of innovation: innovation and interactions between firms and other institutions. Let us consider these two variables in turn.

13.3.3.1 Levels and Variety of Innovation

In our study, as in the Oslo manual, product refers to goods and services, and innovation refers to the practical use of an invention to produce new products, to improve existing ones, or to improve the way in which they are produced or distributed. With this context in mind, a product innovation refers to the commercial adoption of a new good or service, whereas a process innovation refers to the adoption of new or significantly improved production methods. When applied to measure innovation, these definitions permit to assess both the level and the variety of the innovations developed by firms.

As indicated in Table 13-2, variety can take four values: 4% of the firms surveyed that have developed innovations during 1993-1995 have offered to their customers product innovations with changes in the production process. These firms can be considered as the most innovative firms of those interviewed. One might consider that the 55% of the firms that have introduced product innovation without changes in the production process are slightly less innovative than those that have introduced product innovations with changes in the production process. In the same vein, one might also consider the firms (12%) that have offered process innovation without changes in the product as slightly less innovative than those that introduced product innovations on the market. Finally, one might consider that the firms interviewed that offered no innovation (29%) to their customers during 1993-1995 are less innovative than the three other categories of firms. On the whole, one must point out that the introduction of product and process innovations on the market is much more widespread than in-house R&D and claims of patents.

Table 13.2 - Level and Variety of the Innovations Developed in the Chaudière-Appalaches Region between 1993 and 1995

Variety of innovation		Level	
		Number of firms	% of firms
(1)	Product innovation with change in the process	7	3.65
(2)	Product innovation without change in the process	106	54.90
(3)	Process innovation without change in the product	24	12.45
(4)	No innovation	56	29.00
Total		193	100

The most important sources of ideas, information and technology to develop innovations are presented in Table 13-3. Nearly one fifth of the firms that offered product or process innovations during 1993-1995 relied exclusively on their own internal sources of ideas and information to create and develop innovation, whereas nearly half of the firms relied exclusively on their internal technology to develop new products and process innovations. Eighty percent (123/153) of the firms that introduced innovation during 1993-1995 relied on external sources of ideas and information to develop their innovations whereas 52% (80/153) used technology from outside sources to develop their innovations during the same period. As indicated in Table 13-3, the external sources of ideas and information that firms rely the most frequently on are, by decreasing order of importance, clients, fairs-exhibitions-conferences-and-meetings, federal and provincial government laboratories. With respect to external sources of technology, the most frequent sources that firms tend to go to are, by decreasing order of importance, suppliers, associated firms, and fairs-exhibitions-conferences- and meetings. Local institutions are among the external sources that are the least frequently used by the innovative firms, although, as indicated in Table 13-3, they play a more important role to provide ideas and information than to provide technology. This short description points to the importance of clients, government laboratories and fairs-exhibitions-conferences-and-meetings as external sources of ideas and information, and to the primary importance of suppliers as external sources of technology.

Table 13.3 - Most Important Sources of Ideas and Technology to Develop Innovation in Chaudière-Appalaches Region during 1993-95

Variety	Level	Level
of Sources	of ideas and information	of technology
	N = 153 % / 153	N = 153 % / 153
Internal sources only	**19.6**	**7.7**
External sources	**% / 123**	**% / 80**
Market sources		
- Clients	45.7	12.5
- Competitors	11.8	0
- Suppliers	13.7	48.7
- Consultants	5.2	7.5
Education and research		
- Universities	8.5	2.5
- Colleges	4.6	2.5
- Fed. and prov. Labs	20.9	7.5
Local institutions	3.9	0
Generally available information		
- Fairs, exhibitions, conferences and meetings	25.5	8.7
- Journals	2.0	3.7
- Patent disclosure	2.6	2.5
Others	17.6	12.5

13.3.4. The Determinants of Innovation

Innovation is studied using nine explanatory variables. The dependent variable refers to different types of innovation. A categorical variable, it is divided into four categories identified previously: product innovation with change in the production process, production innovation without change in the production process, process innovation without change in the product, and finally, no innovation at all. To study the impact of the explanatory variables on such a qualitative dependent variable, we have developed the following multinomial logit model:

$$Log(P_i/P_j)=$$
$$b_o+b1_{ij}LNPR\&D+b2_{ij}LNPET+b3_{ij}LNPSAO+b4_{ij}LNNADT+5b_{ij}LNTNE$$
$$+b_{ij}6LNM\$OP+b_{ij}7LNPOF+b_{ij}8LNNESI+b_{ij}9LNNEST$$

Where

Log(Pi/Pj) = Log of the ratio of the probability of having an innovation from the i th category relative to that of having one from the j th category when a firm has offered innovation to its customers during 1993-1995;

LNPR&D = Log of expenditures in R&D/sales

LNPET = Log of number of engineers and technicians/total number of employees

LNPSAO = Log of percentage of sales shipped outside the Chaudière-Appalaches region

LNNADT = Log of number of advanced technologies used by a firm

LNTNE = Log of total number of employees

LNM$OP = Log of millions of dollars of operating revenues

LNPOF = Log of percentage of financing of innovation by outside sources

LNNESI = Log of number of external sources of ideas and information used to develop innovation

LNNEST = Log of number of external sources of technology used to develop innovation

We have eliminated from the regression the category ''product innovation with change in the production process'' because the number of observations (n=7) was too small in this category of innovation. Results of the regression are presented in Table 13-4. The percentage of correct predictions of the model is 80.10%. The McFadden R2 is .23, which is quite acceptable for such models. The likelihood ratio is 22.80, which is higher than the critical value of the chi-square statistics (20.09) with 8 degrees of freedom. The overall model is thus significant at the 1% level.

The regression results indicate that the firms that have not offered innovations to their customers during 1993-1995 have a larger number of employees than the innovative firms, and that the firms that have produced no innovation during the same period invest a lower percentage of their sales in R&D activities compared to the firms that innovate. Likewise, the firms that have offered no innovation to their customers have smaller operating revenues than the innovative firms do. Finally, as expected, the firms that do not innovate use a lower number of external sources of ideas and technology than the innovative firms do.

Results of Table 13-4 indicate that only two detectable differences exist between the firms that have offered product innovations without changes in the production process and the firms that have offered process innovations without changes in the production process: the firms that have developed product innovations during 1993-1995 show a higher percentage of financing of their innovation by outside sources, whereas the firms that have developed process innovation during the same period have relied on a larger number of outside sources of ideas and information to develop their innovations.

Table 13.4 - Estimated Mutinomial Logit Model Results of Factors Affecting Firms' Innovation

	Dependent variables[a]		
	ProcessVs Products	**No innovation Vs Products**	**No innovation Vs Process**
Intercept	-1.288 (-.123)	2.393 (.771)	6.505 (1.789)**
% R&D/ sales (LNPR&D)[b]	-.078 (.138)	-.219 (-3.437)***	-.401 (-1.850)**
% Eng. Techn. /N. of employees (LNPET)	.126 (.197)	.036 (.027)	-.002 (-.0001)
% of sales outside region (LNPSAO)	-.089 (.354)	.056 (.094)	.168 (.476)
N. of advanced technologies (LNNADT)	.130 (.871)	.010 (.014)	.020 (.013)
Total N. of employees (LNTNE)	.110 (.144)	.342 (2.584)***	.579 (2.33)**
Operating revenue (LNM$OP)	-.093 (.119)	-.339 (-3.117)***	-.473 (-1.808)**
% of outside financing (LNPOF)	-.164 (7.36)***	.022 (-.331)	.237 (.389)
N. of E.S. of ideas (LNNESI)	.010 (1.38)*	-.191 (-12.051)***	-.161 (-2.90)***
N. of E.S. of technology (LNNEST)	-.049 (.467)	-.099 (-3.156)***	-.067 (1.523)*

*** implies that the variable is significant at 1% level
** implies that the variable is significant at 5% level
* implies that the variable is significant at 10% level
Percentage of correct predictions = 80.10 %
Overall Chi-square statistics (DF)= 22.80 (9); (Critical value = 20.09) at 1% level
McFadden R^2 = .23

[a] the T-ratios are in parentheses
[b] All independent variables are expressed in Logarithmic form

13.3.5 Variations Between Firms in Matters of Interactions

The system approach to innovation departs from the previous approaches primarily in developing two hypotheses: 1) innovation depends on interactions with people and institutions external to the firms, and 2) the level and variety of interactions vary between innovative milieux. This section pays attention to the first hypothesis. More precisely, we answer the following question: "Are the firms really different in matters of interactions with outside sources of ideas and technology?"

The results of the analysis of variance for the external sources of ideas and information used to develop innovation are reported in Table 13-5. There are detectable differences between the firms that use market sources of ideas and information and the firms that do not. Hence, the firms that rely on clients as external sources of ideas and information to develop innovation invest more in R&D, have a higher percentage of engineers and technicians relative to their total number of employees, a smaller number of employees, and they use a greater number of advanced technologies than the firms that do not use clients as sources of ideas and information. On the whole, there are no significant differences between the firms that rely on ideas and information from their competitors and those that do not, except that the firms that do not rely on their competitors invest more in R&D than the others. The firms that do not use ideas and information from their suppliers differ from the others by having a higher percentage of their sales shipped outside the Chaudière-Appalaches region. Finally, the firms that use consultants as external sources of ideas and information use a greater number of advanced technologies and have a larger number of employees than the rest of the firms.

Noticeable differences between the firms are also observed regarding the external sources of ideas and information coming from education and research institutions. The firms that use ideas and information from universities have a higher ratio of expenditures in R&D, a higher percentage of sales shipped outside the region, and rely more heavily on external sources of financing to develop innovations than the firms that do not use this source of ideas and information. As for the firms that use community colleges as sources of ideas and information, they differ from the others by having a greater number of advanced technologies and a greater number of employees. Finally, the firms that use federal and provincial research laboratories differ from the others by using a smaller number of advanced technologies, having a smaller percentage of sales shipped outside the Chaudière-Appalaches region, and smaller operating revenues. We have also detected noticeable differences between the firms with respect to the use of generally available information. More specifically, the firms that go to fairs, exhibitions, conferences and meetings differ from the others by using a greater number of advanced technologies and by financing a lower percentage of innovation with funds from external sources. As can be seen in Table 13-5, the firms differ remarkably concerning the use of scientific publications as external sources of ideas and information. Indeed, the firms that use publications differ from the others by having a lower ratio of engineers and technicians, a lower percentage of sales shipped outside the Chaudière-Appalaches region, a lower percentage of self-sufficiency regarding the financing of innovation, a greater number of employees, and finally, larger operating revenues.

Table 13.5 - Differences Between Firms that Use Outside Sources of Ideas and Information to Develop Innovation

Variables	$R&D/Sales (%)		N. Eng.tech/N employees (%)		Sales outside region (%)		Advanced technologies (N)		Employees (N)		Operating revenue (M$)		Outside funding (%)	
	Used	No	Used	No	Used	No	Used	No	Used	No	Used	No	Used	No
Sources of ideas														
Market sources														
Clients	14.01*	6.86	50.88*	36.18	84.29*	82.79	3.64*	3.19	69.35	99.41*	10.32	10.51	13.89	15.35
Competitors	3.26	10.25*	51.28	40.17	81.94	83.50	3.89	3.30	128.22	84.43	25.06	8.88	9.83	15.37
Suppliers	9.70	8.10	40.65	41.42	74.38	84.51*	3.47	3.34	61.19	92.02	6.01	11.05	16.19	14.53
Consultants	4.81	9.77	27.80	42.23	89.87	83.05	4.12*	3.32	157.9*	85.53	17.02	10.11	10.62	14.97
Education and research														
Universities	18.76*	8.73	38.27	41.51	92.64*	82.75	2.92	3.38	69.61	90.0	9.45	10.52	28.46*	13.55
Community colleges	6.85	9.65	35.04	41.63	89.28	83.11	4.85*	3.29	167.3*	85.57	15.53	10.22	10.86	14.93
Public laboratories	6.51	9.99	40.44	45.34	81.80	93.62*	1.10	3.18*	86.39	101.85	9.57	16.14*	14.18	17.99
Generally available information														
Fairs, exhibitions, conferences, meetings	8.80	9.77	39.10	41.91	84.71	82.99	3.92*	3.21	107.76	83.77	11.61	10.14	6.82	17.10*
Journals	5.32	9.88	21.55	42.25*	63.50	84.50*	3.08	3.37	180.1*	81.84	28.77*	9.14	5.53	15.56*
Others	8.10	9.82	39.89	48.59	81.30	83.75	1.14	2.73*	75.39	158.6*	9.02	17.74*	13.03	22.92*

* indicates that the differences between the means are statistically significant ($p < .1$) between the firms that have used external sources of ideas and information and the firms that have not used external sources. When « * » are indicated in the columns «used», it signifies that the mean of all the firms that have used external sources of ideas and information is, for the source examined, higher than the mean of the firms that have not used this external source. In the same manner, when « * » are indicated in the columns «no», it signifies that the mean of all the firms that have not used external sources is, for the source examined, higher than the mean of the firms that have used this external source.

The results of the analysis of variance for the external sources of technology used to develop innovation are reported in Table 13-6. Given that a smaller proportion of firms has used outside sources of technology rather than outside sources of ideas and information, the number of observations regarding particular sources was too small to conduct the analysis of variance. Therefore, we have had to aggregate the sub-categories regarding "education and research" and those falling under the heading "generally available information". The firms that rely on clients as external sources of technology differ from the others in having a higher ratio of expenditures in R&D, a higher ratio of engineers and technicians, a lower total number of employees, and a lower percentage of self-financing of innovation. As for the firms that do not use suppliers as sources of technology to develop innovation, they differ from the other firms in two ways: they have a higher ratio of R&D expenditures and a higher ratio of engineers and technicians. No other differences between firms were statistically significant regarding the market sources of technology.

As for the external sources of technology regarding universities, community colleges and government laboratories, they were aggregated under the general heading "education and research". The firms that have used education and research institutions as external sources of technology differ from the others in two ways: they have a greater number of advanced technologies and a lower ratio of expenditures in R&D.

Finally, even after aggregating the sub-categories falling under the heading "generally available information", no significant differences were detected between the firms that use and those that do not use these external sources of technology.

13.3.6   Differences Between the Sub Regions Within the Chaudière-Appalaches Region

As mentioned in the literature review above, one of the most difficult and persistent issues of the innovation systems approach is the localization of the boundaries of the system. This issue is inasmuch difficult to settle as regions tend to be defined in terms of geographical and administrative milieux that do not necessarily coincide with the idea of industrial innovation milieux. In order to determine whether or not the Chaudière-Appalaches region is a system, we have desegregated the region into sub-regions corresponding to much smaller innovative milieux. We had a sufficient number of observations to conduct an analysis of variance between five sub-regions. They are Beauce-Sartigan, Bellechasse, Les Chutes-de-la-Chaudière, L'Amiante, and the aggregation of the other sub regions, which are Desjardins, La Nouvelle-Beauce, Les Etchemins, L'Islet, Lotbinière, Montmagny, and Robert-Cliche. The smallest sub region, Les Etchemins, has 82 firms providing jobs to 1,190 people whereas the biggest sub-region, Beauce-Sartigan, has 215 firms providing jobs to 7,167 people. The analysis of variance was made in taking into account all the nine explanatory variables used to analyze the innovation system of the Chaudière-Appalaches region. Indeed, are the sub-regions different or alike?

The results of the analysis of variance presented in Table 13-7 indicate that there are no discernible differences between Beauce-Sartigan, Bellechasse and l'Amiante regarding the number of external sources of ideas and information used to develop innovation. However, the firms of Les Chutes-de-la-Chaudière sub-region use a smaller number of external sources of ideas and information than the firms of the other sub-regions. No significant statistical differences exist between the sub-regions regarding the number of external sources of technology used to develop innovation.

The average percentage of sales dedicated to R&D expenditures are significantly lower in Beauce-Sartigan and Bellechasse than in the other sub-regions. There are no discernible differences between the three other sub-regions.

As for the percentage of engineers and technicians relative to the total number of employees, the figures of Table 13-7 show that this average percentage is lower in Beauce-Sartigan than in Bellechasse and Les Chutes-de-la-Chaudière, and that there are no significant differences between Bellechasse, L'Amiante, les Chutes-de-la-Chaudière and the other sub-regions.

The percentage of sales shipped outside the region is higher in Beauce-Sartigan than in Bellechasse. There are no statistically significant differences in this matter between the other sub-regions.

The number of different advanced technologies used by the firms is higher in L'Amiante than in Beauce-Sartigan and Les Chutes-de-la-Chaudière. In addition, the firms located in Les Chutes-de-la-Chaudière use a lower number of advanced technologies than is the case for the firms included under the heading "other sub-regions".

On the whole, the firms of Beauce-Sartigan have a larger number of employees than the firms of Bellechasse and les Chutes-de-la-Chaudière. No statistically significant differences were detected in this matter between the firms located in Beauce-Sartigan and L'Amiante, as well as between L'Amiante and the three other sub-regions.

The mean operating revenues of the firms located in Beauce-Sartigan and Bellechasse are alike but bigger than those of the firms located in les Chutes-de-la-Chaudière and L'Amiante.

Finally, as can be seen in Table 13-7, there are not many statistically significant differences between the sub-regions with respect to the percentage of external financing of innovation. However, there are, in this respect, significant differences between the firms of Les Chutes-de-la-Chaudière and those falling under the category "other sub-regions."

Table 13.6 - Differences Between Firms That Use and Do Not Use Outside Sources of Technology to Develop Innovation

Variables	$R&D/Sales (%)		N. Eng.tech/N employees (%)		Sales outside region (%)		Advanced technologies (N)		Employees (N)		Operating revenue (M$)		Outside funding (%)	
	Used	No	Used	No	Used	No	Used	No	Used	No	Used	No	Used	No
Sources of technology														
Market sources														
- Clients	16.08*	9.22	59.57*	40.15	73.0	83.95	3.50	3.34	33.60	91.67*	5.39	10.73	14.09	24.70*
- Suppliers	4.66	10.78*	44.69	40.30	77.13	84.98*	3.59	3.29	86.29	89.18	11.03	10.30	17.50	13.90
Laboratories of research	2.09	9.80*	52.49	40.50	87.22	83.15	4.22*	3.31	58.78	90.09	11.00	10.42	5.12	15.25
Generally available information	5.15	9.93	44.33	41.15	89.08	82.91	2.25	2.66	97.21	87.91	11.55	10.35	17.84	26.06
Others	10.40	9.48	42.73	41.21	88.50	82.92	3.34	3.50	126.36	85.58	13.47	10.21	28.64*	13.43

* indicates that the differences between the means are statistically significant (p < .1) between the firms that have used external sources of ideas and information and the firms that have not used external sources. When « * » are indicated in the columns « used », it signifies that the mean of all the firms that have used external sources of ideas and information is, for the source examined, higher than the mean of the firms that have not used this external source. In the same manner, when « * » are indicated in the columns « no », it signifies that the mean of all the firms that have not used external sources is, for the source examined, higher than the mean of the firms that have used this external source.

Table 13.7 - Differences Within the Sub-Regions of the Chaudière-Appalaches Region

Variables / Sub-region	E.S. Ideas (N)	E.S. Techno (N)	$R&D/Sales (%)	N. Eng.tech/N employees (%)	Sales outside region (%)	Advanced technologies (N)	Employees (N)	Operating revenue (M$)	Outside funding (%)
Beauce-Sartigan (a)	1.41=ab+c=d=e	0.54=b=c=d=e	4.95=b-c-d=e	33.85-b=c=d=e	86.78+b=c=d=e	3.13=b=c=d=e	115.0+b=c=d=e	16.27=b+c+d+e	13.60=b=c=d=e
Bellechasse (b)	1.26=a=c=d-e	0.67=a=c=d=e	4.92=a-c-d=e	47.30+=a=c=d=e	74.28=a=d=e	3.37=a=c=d=e	63.58-a=c=d-e	11.11=a+c+d-e	16.67=a=c=d=e
Chutes de la Chaudière (c)	0.88=a-b-d-e	0.46=a=b=d=e	14.64+++b=d=e	53.66+=a=b=d+e	80.83=a=b=d=e	2.77=a=b-d-e	42.44=a=b=d-e	5.04-a+b-d-e	21.55=a=b=d+e
L'Amiante (d)	1.33=a=b+c=e	0.63=a=b=c=e	13.33++b=c=e	42.83=a=b=c=e	88.25+b+c=e	3.80+=a=b+c=e	76.73=a=b=c=e	5.10-a+b-c=e	15.62=a=b=c=e
Other sub-regions (e)	1.66=a+b+c=d	0.48=a=b=c=d	11.18+++b=c=d	38.08=a-b-d	83.22=a=b-c=d	3.53=a=b=c=d	103.85=a+b+c=d	10.50=a=b+c+d	12.29=a=b=c=d

Note : « _a_ », « _b_ », « _c_ », « _d_ », « _e_ » refer to the different sub-regions. The signs « _+_ » and « _»_ indicate that, for the variable considered, the sub-regional mean is statistically significantly (p < .1) greater or smaller for the sub-region considered in the row than for the other sub-regions considered. The sign « _=_ » indicates that no significant statistical differences exist between the sub-regions considered for the variable examined.

13.4 DISCUSSION AND CONCLUSION

Nearly 70% of the firms studied have offered product or process innovations to their customers during 1993-1995. This figure demonstrates that the firms are much more innovative than indicated by the studies based on indicators relative to R&D and patents.

Likewise, the analysis of the most important sources of ideas and technology to develop innovation demonstrates that the firms rely on outside sources much more frequently than predicted by the studies based on indicators relative to R&D and patents. The firms of the Chaudière-Appalaches region resort to outside sources 80% of the time to have access to ideas and information, and slightly more than half of the time to have access to technology. Clients, government laboratories and fairs are clearly the most important external sources of ideas whereas suppliers are clearly the most important external sources of technology. The level and variety of interactions between the firms and the outside sources of ideas, information and technology clearly suggest that the firms interact with other people and institutions as elements of a system.

Furthermore, the analysis of the determinants of innovation clearly demonstrates, as predicted by the innovation systems approach, that the innovative firms develop more interactions with outside sources of ideas, information and technology than the non innovative firms do, thus confirming again the hypothesis that the firms interact with other people and institutions as elements of a system.

The analysis of the variations between the firms in matters of interactions clearly demonstrates that the firms really differ regarding their interactions with outside sources of ideas, information and technology. Likewise, it clearly demonstrates that the firms are elements of a very complex network of interactions, where the firms and outside sources interact following a pattern of division of labour that suggests that complex patterns of interdependent interactions do exist between the firms and their outside sources of ideas, information and technology.

Finally, when attention is turned to the differences that are discernible between the sub-regions within the Chaudière-Appalaches region, one finds that three sub-regions, Beauce-Sartigan, Bellechasse and L'Amiante stand apart on many determinants of innovation, thus suggesting that one does not have a single regional innovation system, but instead, many sub-regional innovation systems that are significantly different. The figures reported in Table 13-6 suggest, as also supported by many business and political leaders, that these three sub-regions constitute innovative systems having their own idiosyncrasies. However, additional empirical evidence would be required to provide a definite answer regarding the exact localization of the boundaries of the innovation system(s) of the Chaudière-Appalaches region.

The findings of this chapter demonstrate that firm level surveys on innovation constitute a productive research design to better understand the flows of ideas, information and technology used to develop innovation. In order to deepen our understanding of these flows at the regional level, future firm level surveys on innovation should spatialize more precisely the answers to the questions regarding external sources of ideas, information and technology. Additionally, future research on regional innovation systems based on firm level surveys on innovation should pay attention to new explanatory variables such as regional social norms and regional social capital.

REFERENCES

Benko, G. and Lipietz, A. (1992). *Les régions qui gagnent: districts et réseaux: les nouveaux paradigmes de la géographie économique*, Presses universitaires de France, Paris.

Brockhoff, K., Gupta, A.K., and Rotering, C. (1991). "Inter-Firm R&D Co-operations in Germany,", *Technovation*, 11, 4:219-229.

Chatterji, D. and Manuel,T.A. (1993). "*Benefitting from External Sources of Technology*", *Research-Technology Management*, 36, Nov/Dec,21-26.

Dodgson, M. and Rothwell, R. (eds). (1994). *The Handbook of Industrial Innovation*, Brookfield, Edward Elgar.

Edquist, C. (1997). "Introduction: Systems of Innovation Approaches - Their Emergence and Characteristics," In *Systems of Innovation: Technologies, Institutions and Organizations*. Pinter. C. Edquist (ed.), London. 1-35.

Edquist, C. (ed.) (1997). *Systems of Innovation: Technologies, Institutions and Organizations*. London: Pinter.

Faulkner, W. and Senker, J. (1994). "Making Sense of Diversity, Public-Private Sector Research Linkage in Three Technologies", *Research Policy*, 23,6:673-695.

Hagedoorn, J. (1990). "Organizational Modes of Inter-Firm Cooperation and Technology Transfer", *Technovation*, 10,1:17-30.

Hull, F. and Slowinski, E. (1990). "Partnering with Technology Entrepreneurs", *Research-Technology Management*, 33,6:16-20

Le Bas, C. (1995). *Economie de l'innovation*, Paris, Economica.

Lundvall, B.-Å. (ed.) (1992). *National Innovation Systems: Towards a Theory of Innovation and Interactive Learning*, Pinter, London.

Maillet, D. (1992). "Milieux et dynamique territoriale de l'innovation", *Canadian Journal of Regional Science*, 15,2:199-218.

McFetridge, D.G., (ed.) (1985). *Technological Change in Canadian Industry*, University of Toronto Press, Toronto.

Nelson, R. (ed.) (1993). *National Innovation Systems. A Comparative Analysis*, New York. Oxford University Press.

Niosi, J., Saviotti, P., Bellon, B. and Crow, M. (1993). "National Systems of Innovation: In Search of a Workable Concept", *Technology in Society, 15*, 1:207-227.

OECD (1992). *Technology and the Economy - The Key Relationships*. Paris: OECD.

Palda, K. (1993). *Innovation Policy and Canada's Competitiveness*, Vancouver, The Fraser Institute.

Pyke, F. and Sengenberger, W. (1992). *Industrial Districts and Local Economic Regeneration*, International Institute for Labour Studies, Geneva.

Rallet, A. and Torre, A. (1995). *Economie industrielle et économie spatiale*, Economica, Paris.

Reddy, N.M., Aram, J.D. and Lynn, L.H. (1991). "The Institutional Domain of Technology Diffusion", *Journal of Product Innovation Management, 8*, 4.

Robertson, N.C. (1992). "Technology Acquisition for Corporate Growth", *Research-Technology Management*, 35,2.

Sen, F. and Rubenstein, A.H. (1990). "An Exploration of Factors Affecting the Integration of In-House R&D with External Technology Acquisition Strategies of a Firm", *IEEE Transactions on Engineering Management*, *37*, 4.

Steele, L.W. (1990). "Managing Joint International Development", *Research-Technology Management*, 33,4:16-26.

14 SAINT JOHN, NB. AS AN EMERGING LOCAL SYSTEM OF INNOVATION

Richard Nimijean
University of Ottawa

14.1 INTRODUCTION

The literature on innovation systems points to the need to build up our inventory of case studies in order to understand the processes of innovation and regional clustering (de la Mothe and Paquet, 1997; Nordicity Group, 1996; Nelson, 1994). Such an exercise is required in order to capture the complexities of the economic growth process; understand discrepancies in economic performance across time and space, particularly given the impact of globalization on national and sub-national economies (on this issue, see Boyer and Drache, 1996); and help inform policy-makers in their endeavours to promote economic growth in general, and the process of economic diversification to a knowledge-based economy in particular. This is important in Canada given the nature of its political economy: a large, regionalised country with a relatively small population and whose economic development was based on staples exploitation. This has produced uneven economic growth and settlement patterns across space and over time.[1]

This poses special challenges in the efforts to transform the Canadian economy into one based on ideas and innovation. Apart from the major metropolitan areas of Toronto, Vancouver, Montreal and Ottawa-Hull, the Canadian landscape is marked by relatively small regional urban centres, each possessing a somewhat unique blend of history, politics and economic development. Most share, however, a common feature: their initial growth was based on staples-exploitation.

Saint John, New Brunswick[2] is a prime example of a mid-sized Canadian city seeking to transform itself so that it can be competitive in the new economy. Long seen as a manufacturing centre and a transportation hub dependent on the processing and export of natural resources, it is also a centre that has been in decline, as increasingly scarce public sector dollars can no longer support manufacturing activities.[3] Moreover, the local economy is dominated by the presence of the Irving family companies which, some contend, has historically made it difficult to develop a competitive, entrepreneurial climate. Finally, it should be noted that despite its perception, the Saint John economy is actually dominated by service industries.[4] This rather unique set of circumstances presents government, business and university leaders with a great challenge in their efforts to diversify the local economy.[5]

Building on the sophisticated telecommunications infrastructure of the province, the presence of NBTel and Fundy Communications, and the provincial government's emphasis on the information highway and economic diversification, changes are under way. These include niche programming at the Saint John campus of the University of New Brunswick (UNBSJ), and the work of city and regional economic development bodies, Enterprise Saint John (ESJ) and the Fundy Region Development Commission (FRDC).

This paper will document the nature of Saint John's regional system of innovation and efforts at transformation; offer an assessment of these efforts; and propose recommendations as part of Statistic Canada's efforts at developing new statistical indicators of innovation and economic growth. Subsequent work could build upon this paper, for example through the development of an inventory of knowledge and data, similar to those performed by Holbrook (1997) on British Columbia and Landry (1997) on the Beauce region of Quebec. Indeed, these three cases illustrate perfectly the radically different nature of Canadian regional innovation systems and demonstrate the need for the development of new indicators.

14.2 LOCAL SYSTEMS OF INNOVATION

In its international comparison of twenty regional and local systems of innovation, the Nordicity Group (1996) concluded that several points help us understand the nature of local innovation and the development of industrial clustering. These lessons, which can presumably inform local decision-makers in their efforts to promote economic growth and adaptation to a knowledge-based economy, include:

- champions are instrumental to cluster development;
- a science and technology infrastructure is a necessary but insufficient condition; linkages with the business community are required;
- information networks, both informal and formal, are required;
- pro-business government support is helpful;
- cluster development is a long-term process (decades); and
- planning cluster development is feasible.

The key is to understand the interactions of key actors in the local system and to assess the stage at which a local industry or sector finds itself: "proximity is more important when firms are at the innovative stage of the industry's development where skilled labour and innovative centres (e.g. advanced infrastructure) are especially significant. As firms mature proximity becomes less important. There is a cluster life-cycle in that a cluster emerges, grows, matures and either renews itself or declines." (Nordicity Group, 1996, p. 7)

This life-cycle is clearly the result of the combination of the factors provided above. Strategic clusters in the regional economy, which could or should be the focus of local development strategies, should thus be identifiable. In the words of Rugman and D'Cruz, "a strategic cluster is... a network of businesses and supporting activities located in a specific region, where the leading flagship firms compete globally and the supporting activities perform to international standards." (Rugman and D'Cruz, 1995, p. 54) In the Canadian case, they suggest, many clusters will be resource-based, pointing to the need to continue promoting value-added resource exploitation. (Rugman and D'Cruz, 1995, p. 55)

These definitions provide us with a suitable framework for documenting and assessing a local system of innovation, although one weakness remains: it is not clear when clustering occurs. Is it self-defined? Is it measured in relation to population or economic activity? Is it simply a subjective perception that clustering is taking place (e.g. a lot of activity is taking place in sector *x*)?

14.3 THE SAINT JOHN ECONOMY

Using the definitions and criteria provided above, it is possible to assess the evolution of the Saint John economy. As will be seen, the emerging clusters in Saint John are telecommunications and tourism. They are the focus of economic development planning; a critical mass of firms are or potentially can be engaged in these sectors; and there is support for cluster development through post-secondary programs and networking. In the case of telecommunications, it is due to the presence of strong entrepreneurial firms, an energetic champion in Premier Frank McKenna,[6] and a strengthening network between business and post-secondary institutions. In the case of tourism, it is due to the combination of an extensive planning process, a comparative advantage allowing for the exploitation of the "Fundy Experience," emerging markets, and university-business links. Other sectors show promise for cluster development, notably the health sciences. Historically significant sectors will remain important, yet they are either maturing as clusters or will continue to decline in importance relative to the transformation of the economy. Most notably, this involves a reduced significance for manufacturing, particularly shipbuilding.

The Provincial Environment

Before proceeding with a direct overview of the Saint John economy, it is first necessary to examine recent developments in the provincial economy. In this case, not only has its historic structure affected the economy in Saint John; the tone and thrust of provincial initiatives have also greatly influenced local transformation efforts.

Like other western governments, the New Brunswick government has redefined its role in economic growth, emphasizing debt management and private sector economic development. As a result, its role is increasingly restricted to identifying areas for economic growth, promoting partnerships, and attracting investors and firms to the province. This is a major turnabout for the province for it, like all Atlantic Canadian provinces, has been relatively more dependent on government expenditures as a stabilizer in the economy than have the other provinces. (Grant, 1992, p. 77)

The provincial economic strategy emphasizes the diversification of the economy, value-added resource exports, and tourism. They are deemed necessary to transform New Brunswick into a "have" province, due to the loss of tens of thousands of "old economy" jobs: for example, the completion of the naval frigate program in Saint John, military base closures, and federal and provincial government downsizing.[7] The province's goals in economic strategy include promoting entrepreneurship; preparing people for the new economy; strengthening infrastructure and the investment climate; preserving and enhancing environmental and natural resources; regional development; and promoting the New Brunswick identity. (New Brunswick, 1989). To this end, the government has adopted a series of pro-business policies focussing on knowledge industries and diversification of the economic base away from natural resources, all the while recognizing the importance of the latter. Thus it is also promoting value-added exploitation of natural resources.

Information technology is seen as a key to lessening its traditional dependence on the resource industries. It is estimated that this sector is growing 25% to 30% annually, with 2,000 of the 3,000 jobs in the sector created in the last three years.[8] The province is committed to creating 3,500 to 5,000 new jobs in this sector over the next five years. As a sign of its commitment, the government was the first province to set up an information highway secretariat. As well, it recently created the New Brunswick Information Technology Council, composed of representatives from the private sector, government and academia, the mandate of which is to recommend actions to government on issues of content, usage, and infrastructure of the information highway. The province is focussing on attracting investment and preparing people for the new economy through the "Job Ready Workforce Initiative," which includes provincially-sponsored training and a national and international recruitment program.[9]

IT clusters are primarily found in Fredericton, Saint John and Moncton. In 1994, NBTel built Canada's first fibre optic telecommunications system, and Fundy Communications also recently completed a fibre optic ring surrounding the province. The province highlights these efforts when promoting the province as a leading edge environment for business in the new economy. As well, it claims that New Brunswick is the most cost effective Canadian location for call centres and data centres and one of the most cost effective in North America. (FRDC, 1996, p. 25) Government has complemented these private sector efforts: initiatives include, for example, connecting all schools to the Internet ($20 million in federal, provincial and private sector funding); a successful "Get Connected" program offering a sales tax rebate (up to $250) to consumers purchasing home computers;[10] and the *Itjobnet*, which provides a database on job opportunities in the IT sector, a c.v. service, and a training database.[11]

Resources will continue to be important for the provincial economy, but the emphasis is on value-added exploitation and the development of quality products.[12] For example, Premier McKenna foresees forestry royalties increasing from $50 million annually to $500 million annually in the next few decades. (Llewellyn, 1997, p. 16) As well, aquaculture is an important new industry, already worth $125 million annually, which combines significant new revenues and employment (including some skilled employment). (Llewellyn, 1997, pp. 18-20)

Tourism is another element of the strategy, given that it is worth $725 million annually with 25,000 jobs. For example, eco-tourism is being promoted in the south through the "Fundy Experience" in an effort to expand the traditional tourism industry. (Balcomb, 1997; Llewellyn, 1997, pp. 24-25)

Finally, the province is committed to developing sectors which promise potential growth. Thus it is supporting the proposed Institute of Health Sciences in Saint John (discussed below) in order to gain a foothold in the pharmaceuticals industry. As well, a provincially-commissioned study indicated that the province is the most competitive location in North America for plastics manufacturing; as a result, the province is aggressively pursuing investors and firms in this industry. The industry is desirable because of the use of plastics in a variety of economic activities and industries; a good supply of cheap labour in the province; proximity to the northeastern United States, and low energy costs: all combine to give the province the lowest operating costs for the industry, the study claims.[13]

It remains to be seen whether the provincial government will be successful in its efforts to rejuvenate the economy. On the one hand, growth of the provincial economy continues. One bank survey suggests that the provincial economy will grow by 2.2% in 1997, due largely to the province's efforts at diversifying the industrial base.[14] Exports have also boomed in recent years, increasing by 79% from 1992 to 1995 ($3 billion to $5.4 billion) before dropping by $57 million in 1996. (APEC, 1997)

On the other hand, unemployment remains stubbornly high. Despite government's efforts, the unemployment rate has remained level during the past decade, even with relatively low participation rates in the labour force. Moreover, there were only 2,800 net jobs gained in the province from June 1996 to June 1997. This will make it difficult for the government to attain its goal of creating 25,000 jobs by 1999 (promised in the last provincial election in 1995), a target estimated to be 20,000 jobs away.[15] This situation of perceived dynamism yet sluggish economic performance continues to confound economic analysts.[16]

Table 14.1 - Selected New Brunswick Statistics

Population (1996):	738,000
Labour Force (July 1997):	363,500
Employment (July 1997):	317,400
Participation Rate (July 1997):	60%
Unemployment (July 1997):	12.3%
Gross Domestic Product (1996):	$16.1 billion (current dollars)

Source: Government of New Brunswick, The New Brunswick Economy 1997; Government of New Brunswick, Ministry of Advanced Education and Labour.

The Economic History of Saint John

The economic history of Saint John is not a happy one: indeed, since the prosperity of the 1850's associated with wooden shipbuilding, the city's economic performance can best be characterized as stagnant, except for brief periods of prosperity. The slow growth is due to the narrowness of the economic base, which has led to an out-migration of young people, especially the educated and the skilled. Discontinuous growth is due to an emphasis, historically, on large scale projects which, while creating jobs in the short-term, did not generally produce lasting employment. (Ridler, 1977, pp. 24-25; Dick and Hood, 1978, p. 143)

The nature of Saint John's economic development has been attributed to a failure to master new technologies (such as the conversion to steel steamships); confederation and the completion of the intercolonial railway; the merger process which led to the creation of large business organizations in the late 19th and early 20th centuries, which in turn transferred control of production out of Atlantic Canada and into central Canada; and a continued focus on large-scale industrial developments and transportation industries as a means to promote growth. (See Dick and Hood, 1978, pp. 10-14; City of Saint John, 1987a, pp. 1-2)

In other words, economic fortunes in many ways appeared to be dependent on factors *seemingly* beyond the control of the city. Consequently, extended poor economic conditions contributed to the development of a pessimistic outlook of residents and a conservatism hindering proposals for change, an attitude which has only recently been overcome. For example, as far back as 1946 a city planning document noted: "skepticism of proposals of change is one of the characteristics of the Saint John Public; and general conservatism of thought in the city will prove to be one of the greatest obstacles in popular acceptance of some items of the plan." (Cited in City of Saint John, 1987a, p. 2)

Moreover, some have contended that the dominance of the Irving family companies, which are active and important players in most sectors of the economy,[17] has potentially dampened growth prospects in the region, hindered the development of an indigenous entrepreneurial attitude and hence stifled competitiveness. Indeed, De Benedetti refers to an "entrepreneurial chill" in Saint John that might encourage firms to set up in Moncton or Fredericton rather than Saint John. (De Benedetti, 1994, pp. 221-222)

However, it has also been suggested that in recent years the Irvings have opened up to local competition and other suppliers.[18] (Report on Business, 1994) As well, their role in philanthropy, while often not recognized, is growing. The various Irving companies as well as the individual family members have increased their contributions to the community, such as supporting the annual Rally of Hope, sponsoring the restoration of the Loyalist Burial Grounds or contributing $1.5 million to UNBSJ's capital campaign.

Their sheer presence in the local and provincial economy will always make the Irvings a source of controversy. For some, they are a strength, injecting money into the economy, creating jobs, and actively participating in the community; for others, they only see the potential take over or crowding out of the market of smaller businesses.

Economic Performance

One of the major influences on the Saint John economy has been its ice-free port. It spawned a shipbuilding industry which, while less vibrant today, remains an important component of the local economy. As well, this industry has been symptomatic of the fragile state of the economy and the public mind set: for example, the completion of the federal frigate program (see note 3) sparked a major debate on the future of the Saint John economy. While many simply longed for the glory days of the past, hoping for new megaprojects or federal dollars, others saw it as the need to diversify the economy and become less reliant on one major sector.[19]

The port has also made Saint John the province's major export and transportation hub. Forestry and associated paper products, for example, are central to the port's well-being, as it takes advantage of the significant New Brunswick forestry sector (including value-added sectors such as pulp and paper) and the port's strength in the shipment of natural resources. (De Benedetti, 1994, pp. 213-215) Irving Paper and Irving Pulp and Paper are significant forces in the economy, and are in the midst of a $200 million modernization and environmental upgrade. Moreover, Saint John is home to the Irving Oil refinery, Canada's largest.

In recent years, however, the greatest surge in economic activity has been in the service sector. In employment terms, trade, manufacturing and health remain the most important sectors, but from 1981 to 1991, hotels and restaurants accounted for the greatest growth (up 84% to now account in 1991 for over 10% of total employment), while business and other services experienced the greatest drop (25%, accounting for 6.2% of total employment). (De Benedetti, 1994, p. 209) Such figures must, however, be seen in context. The rise in service employment in hotels, restaurants and so forth is associated with a dramatic renewal of the downtown core in terms of accommodations, restaurants, and lifestyle services such as athletic and sporting facilities, which have made Saint John a more attractive location for business. (De Benedetti, 1994, p. 211)

More recent analyses confirm this trend. According to UNB economist David Murrell, almost all of the jobs created in the province in the last three years have been in the south (which includes Saint John); however, they are largely low-paying service sector jobs. Job losses in health and education have led to a drop in after tax income over the last three years.[20] Yet the city remains dependent, to an extent, on manufacturing, even with the completion of the frigate program, so there is some reason for analysts to be gloomy.[21]

The recent performance of the Saint John economy has been mixed. For example, while the city's unemployment rate has fluctuated in recent years, it remains high. There are approximately 9,000 people in the Fundy region who are unemployed and another 14,000 who are dependent on social assistance. (FRDC, 1996, p. 1) More worrying is the possibility that drops in the unemployment rate have not always been positive. For example, when the unemployment rate dropped by over 2% from December 1994 to 1995, the size of the labour force dropped by 3,000 people. This was attributed to people who left the city, particularly those with skills, and to those who were simply discouraged and stopped looking for work.[22]

City planners are, however, becoming more optimistic about the labour force of the future. For example, the labour force is becoming more educated, heightening prospects for a successful transformation of the economy. As the FRDC notes (FRDC, 1996, pp. 17-18), 65,000 members of the Fundy Region labour force (total 81,900) in 1991 had either completed a post-secondary education or had some postsecondary training. More importantly, this education is concentrated in the younger members of the labour force. Of particular note is that this education is overwhelmingly in the applied sciences and engineering.

Table 14.2 - Selected Saint John Statistics

Population, Saint John CMA:	129,380
Labour Force (July 1997):	65,000
Employment (July 1997):	57,000
Unemployment (July 1997):	8,000
Participation Rate (July 1997):	61.4%
Unemployment Rate (July 1997):	12.5%

Source: Statistics Canada

Emerging Cluster: Telecommunications

Broadly defined, telecommunications is the key emergent cluster for the Saint John (and greater New Brunswick) economy. Saint John is home to NBTel (2300 employees) and Fundy Communications (500 employees), which are both regarded as being amongst the most dynamic telecommunications organizations in North America. The FRDC estimates that as of early 1996, over 4,000 people were employed in the telecommunications sector in the Fundy region. (FRDC, 1996, p. 15).Many of these workers are based in Saint John performing high-skill and/or "knowledge" jobs. While NBTel's roots are in the telephone industry and Fundy Communications roots are in the cable industry, they are now engaged in full-scale competition in each other's former areas of expertise, given that each has fibre-optic rings around the province. This mirrors national trends towards convergence in the telecommunications industry.[23]

For example, NBTel is in the midst of a planned seven year, $350 million capital expenditure program. By the year 2004, its network will make use of broadband technology. Its Vibe initiative should make communications 200 times faster than fibre-optic systems. With such a system in sight, NBTel foresees offering integrated interactive services combining voice, e-mail, pagers, fax and videoconferencing. (See Llewellyn, 1997, p. 11-13)

Fundy Communications is also planning to offer full multimedia services such as long distance service, video conferencing and video on demand. It intends to invest $70 million in capital projects by 1999. Fundy plans to extend service to Nova Scotia and Prince Edward Island, enhancing its ability to compete with NBTel in areas such as cable television delivery, Internet access, data and video markets, and the long distance market. It is in the process of implementing its Internet service in Saint John, Moncton and Fredericton, which it claims is 30 times faster than regular commercial dial-up services. This alone has created 42 jobs.

Other organizations are helping to consolidate this cluster. For example, the DMR Group, part of the Amdahl Company, has 125 consultants in Saint John, making the Saint John office the largest of its three Atlantic Canadian offices. Much of its work is devoted to providing information system services to NBTel.

MIT Information Technology, headquartered in Saint John with an estimated 125 employees, is part of the J.D. Irving Group of companies. While initially its activities were focussed on providing services for Irving companies, it has now begun to export services, through its offices in Halifax, Fredericton, Moncton, Ottawa and Toronto. Among its services are facilities management, network services, systems integration, business process engineering and application development.

New Brunswick has been recognized as the first province to aggressively develop the call centre industry, building upon provincial government initiatives described above and the presence of a strong private sector in the telecommunications field, spearheaded by NBTel and Fundy Cable. Competitiveness has been increased due to NBTel's "Connectivity" software service, which allows call centre productivity to increase six-fold while reducing capital expenditures, and the province's decision to remove the sales tax on "1-800" services. Job growth in Saint John's call centre industry has been significant. For example, a recent Enterprise Saint John survey of the industry revealed that there are 13 call centres in the city employing over 1,500 people. This figure is projected to grow to 2,250 by the end of 1998.

In order to consolidate the call centre industry in Saint John, ESJ established a call centre management working group, composed of representatives from the provincial government, the local employment centre, the Saint John campus of the New Brunswick Community College, and senior management representatives from all firms with call centre operations in the city. Meeting monthly, the group promotes growth in the industry, shares knowledge about developments in the industry, and promotes the positive benefits of the industry. ESJ and the firms involved are in a campaign stressing the many high skill and higher wage jobs in the industry, in order to overcome the criticism that such jobs are low-skill, low-wage and non-unionized. As well, for ESJ the hope is that this industry will lead to greater diversification within the telecommunications field once the call centres have become entrenched in the city.

Post-secondary initiatives in the region seek to ensure that this cluster will have a steady stream of skilled workers and entrepreneurs. This will encourage sustained dynamism in the region and help keep them in Saint John, as employers will have ready access to trained graduates. As well, the nature of these initiatives cements connections between the private sector and the educational sector.

For example, in 1996 UNBSJ established Canada's first B.A. program with a major in electronic commerce. This important new field of inquiry builds upon local strengths in telecommunications and has generated strong links with the business community, both in terms of sponsorship and organizational linkages with firms such as Nortel, NBTel, DMR Consulting and Sun Microsystems. This led to the creation of the Electronic Commerce Centre, with the collaboration of ESJ and the FRDC as well as the private sector. This program thus serves as a key resource for the telecommunications sector, offering access to the latest developments in electronic commerce and preparing students to become more employable in the sector, thus enhancing business competitiveness. For example, the Centre uses its knowledge of the field to assist small and medium-sized firms in their applications for grants.

The Electronic Commerce Centre is complemented by UNBSJ's NSERC-SSHRC-NB Power-Xerox Chair in the Management of Technological Change, a 5 year, $1 million program. The chair will examine issues such as the management of human resources in new work environments created by the new information and communication technologies; the management of electronic document systems; and electronic commerce.

Educational support is not restricted to the university, however. For example, the Saint John campus of the New Brunswick Community College recently launched its BITS program (Business Information Technology Specialists), developed in partnership with NBTel, Maritime Information Technology, the DMR Group and Fundy Computer Services. The program is designed to produce graduates to run a firm's information management system by retraining them as software developers or network administrators. 10 of the 24 spots are reserved for NBTel employees, while the remaining spots, reserved for the public, will be sponsored by the other corporations. It is a 62 week program, which includes a 6 month paid work term.[24] Demand for the program has exceeded expectations, so a second intake of students has been launched with courses to begin in February 1998. At the request of employers, the college has also offered call centre training programs. With the cooperation of the call centre working group, it is expected that the campus will offer in 1998 a full-fledged call centre program as part of its regular curriculum This program will be developed in consultation with the industry, and will be geared towards filling the ever-growing number of call centre jobs in the Saint John area and meeting the demand for such training from local residents.[25]

Thus it appears that telecommunications qualifies as a cluster as defined above: there is the presence of major industry firms, competitive at the national level. Competition is developing within the sector. Services are being exported. Consequently, the cluster is serving as a magnet for other firms. For example, the software company Genesys Labs of California started a joint venture with Bruncor Inc. (the parent of NBTel) to develop computer telephony integration, deemed to be central to the continued growth of the call centre industry. This venture is expected to employ 20 people in product development, sales, marketing and customer and technical support by the end of 1997. Clustering will likely be centred on two areas: systems development, given the presence of the head offices of NBTel and Fundy Communications and the fact that their operations are centred in Saint John; and the call centre industry.[26]

The software industry, however, is not likely to cluster in Saint John, as it is already concentrated in Fredericton. This is to take advantage of graduate students in engineering and computer science at the main campus of the University of New Brunswick in Fredericton. (De Benedetti, 1994, p. 220) Thus we see one of the downfalls of having a small university campus in this instance. However, telecommunications firms in Saint John will have access to the latest developments, given that Fredericton is just 100 kilometres away; it is simply that the direct economic benefits of that industry will not be captured in Saint John.

Concern has also been expressed about the emphasis on call centre jobs. While the boost to employment and the economy has been welcomed,[27] there are fears that these jobs are often low wage and low skill jobs, and do not represent the ideal of state of a vibrant telecommunications sector.[28] On the other hand, as De Benedetti notes, the steady stream of call centres to Saint John helps create the impression that Saint John and indeed New Brunswick is a good place to do business if access to an advanced telecommunications infrastructure is required. (De Benedetti, 1994, p, 219) Yet as the AT&T example described in note 27 suggests, other jurisdictions are becoming competitive in this sector as well.

Emerging Cluster: Tourism

As Foster has demonstrated in the case of Cape Breton, Nova Scotia, it is possible to conceive of tourism as a strategic cluster, for in this case, unlike in Porter's model of competitiveness, you do not need strong home market or domestic demand. He writes that the tourism cluster:

> focusses on three nodes. The economic base consists of world class tourist attractions and strong markets in emerging markets such as eco-tourism and cruise ships. To be successful, the sector needs local university support in activities such as tourism training, multi-media and cultural/heritage research. A source of innovation is the demand expectations of visitors. With the increased global competition for the tourist dollar, visitors are increasing their expectations in quality of service and valuer for money. (Foster, 1995, p. 118)

Building upon targeted marketing and advertising undertaken by the provincial government, the FRDC will be devoting much of its "proactive" energies to the development of the eco-tourism industry, particularly marine eco-tourism.[29] External demand is significant for the development of this cluster, for it is believed that the market is changing, given a wealthier, aging and more educated population in a time in which people often take several shorter vacations a year. (Balcomb, 1997) Thus, the FRDC has identified eco-tourism as a major growth area, given the presence of the Bay of Fundy and its diversity of environmental attractions, and the fact that people are more interested in this type of tourist attraction.

The uniqueness of the Fundy Experience has led to a significant increase in cruise ship visits: beginning with one cruise ship visit in 1989, it is now expected that there will be 15 visits in 1997, bringing in 20,000 visitors who wish to experience this environment. As a result, the Saint John Port Corporation has begun aggressively promoting Saint John as a port destination.[30]

Not only does this new dimension of the tourism industry attract outside revenues into the area; eco-tourism is also being used to develop spin-offs in the tourist area. For example, a major dimension of the proposed IMAX theatre in Saint John is to help keep tourists here an extra day, either before or after a whale-watching expedition. As well, the FRDC is spearheading efforts to develop an IMAX film on the Bay of Fundy as a means to promote the Bay of Fundy experience to the potential eco-tourism market. (Balcomb, 1997) It is also hoped that this industry will promote the area as a potential location for visitors interested in establishing new residences and businesses.

As a result of its strategic planning exercise, the FRDC decided that it could best fulfil its mission by helping entrepreneurs develop and promote the Fundy Experience. This is consistent with the provincial emphasis on the development of tourism and with the desire of local entrepreneurs. The latter was determined through an extensive consultation process with regional politicians, planners and entrepreneurs. A SWOT analysis conducted for the FRDC by UNBSJ revealed that development of the Fundy Experience would best capitalize on regional strengths and current opportunities for growth, taking into account regional weaknesses and threats from abroad. It also builds upon targeted marketing and advertising about the Fundy Experience undertaken by the province. (Balcomb, 1997) The process thus ensures that the various actors are likely to work together to develop this cluster.

The Fundy Experience is now in its infancy. While the economic impact has yet to be determined, there are now over 40 firms offering "Fundy Day Adventures," ranging from whale watching to hiking. Most of them are new firms, less than two years old. These "adventures" are promoted through the province's advertising and marketing campaign.

Support for this industry will be provided by post-secondary institutions. UNBSJ and the St. Andrews campus of the New Brunswick Community College recently announced the creation of a joint program, the Bachelor in Applied Management in Hospitality and Tourism. This will support the region's efforts at promoting value-added tourism as part of its growth plan by ensuring that there exists a highly qualified managerial and entrepreneurial class to develop growth in the industry. The program involves a co-op component. Moreover, the program is likely to expand to Nova Scotia and Prince Edward Island. (Miner, 1997)

This cluster shows great potential for development. There is support at the provincial level, for example, as the government invested $1.7 million in 1996 in the continuing development of the Fundy Trail network. (FRDC, 1996, p. 25) Yet it also illustrates the need to broaden our conception of clustering. As much as we tend to be enamoured with high tech industrial development, there are non-high tech sectors in our economy which will continue to be economically important and which can serve to attract other sectors to a region by promoting environmental diversity and a high quality of life.

Other Sectors

While telecommunications and eco-tourism may be the sectors which are prominent in Saint John's development plans, it is recognized that past strengths are not unimportant. For example, resources will remain important for the local and provincial economy for some time to come: the key is to increase the value of these exports through modernization, as has occurred in the forestry industry. (Carson, 1997)

Manufacturing is likely to continue its relative decline as other sectors grow in importance. In its own right, this may be a positive experience. As several commentators have suggested, Saint John's over reliance on manufacturing has actually been a source of weakness in the economy, preventing diversification as well as adaptation to a service economy. (De Benedetti, 1994; Ridler, 1977; Dick and Hood, 1978) Indeed, De Benedetti suggests that the manufacturing tradition, long the strength of the Saint John economy, could prove to be the factor undermining its performance in the new economy. (De Benedetti, 1994) As noted above, Saint John's history is dotted with short-term, large-scale industrial projects which provided a quick boost to the local economy but often did not have lasting effect. Such has been the case both, for example, with the federal frigate program as well as the construction of the Confederation Bridge connecting New Brunswick and Prince Edward Island.

One area which is receiving more attention is health sciences. UNBSJ is looking to create an Institute of Health Sciences with the Atlantic Health Sciences Corporation at the Saint John Regional Hospital (which is a neighbour of the university). (Miner, 1997) This complements the campus' focus on health sciences, which has seen the recent development of three new programs in conjunction with the Saint John Regional Hospital.[31] An illustration of partnering and utilizing the strengths of each institution, it is also an example of developing a new niche. The university would be able to use its researchers in nursing, data analysis, and a planned Institute of Polling and Policy Analysis, among others, to examine the social dimensions of health and to work in tandem with the clinical trials conducted by the hospital's medical researchers. (Miner, 1997) The process has already begun, with the creation of a clinical trial division at the hospital sponsored by ACOA, the provincial government and six pharmaceutical companies.[32] However, potential development is jeopardized by the simple fact that Saint John has been suffering a loss of medical specialists in recent years. Funding constraints in the health care system have made it difficult to hire specialists from abroad. Moreover, researchers are deterred by a serious aging of the medical equipment.[33]

There remains a need for continued economic activity in areas which are not part of the emphasized clusters. The primary clusters are likely to continue to generate economic spin-offs, for example through the continued development of businesses in the downtown core. However, there is still a need for work for the 9,000 unemployed in the city, particularly since they are the people most likely to be without the skills or education necessary to be a knowledge worker in the new economy, or do not have access to capital to start their own ventures.

As such, concern has been expressed about those workers who are not easily retrained. Given that the prime manufacturing concern in the region, Saint John Shipbuilding, is not likely to employ as many people as it has in the past, new opportunities will have to be discovered. To this end, ESJ will begin efforts to attract American firms in the "metal bashing" industries, such as pipe-fitting and welding, who are back ordered and are experiencing labour shortages. It is believed that Saint John could offer a low-cost alternative that is close to the market. (Carson, 1997) This may be seen as a traditional or "old economy" solution, but as Cohen and Zysman remind us, "manufacturing matters." (Cohen and Zysman, 1987) Moreover, as Carson duly notes, this is a city "that has always built things," which is central to the city's history and identity. (Carson, 1997)

14.4 ASSESSMENT

The preceding section provided an overview of the Saint John economy, highlighting the emergent sectors which hold promise for future economic growth. It suggested that there has been a concerted effort to overcome the historic pessimism and conservatism held by many in the city and to transform the economy, seen as necessary to prevent further economic decline. This section, using the criteria of the Nordicity Group outlined above, will assess the efforts of the city to date.

Champions

The Nordicity Group points to the importance of champions, be they individuals, organizations, government or firms, in stimulating the development of local industrial clusters. In the case of Saint John, the renewed effort at economic transformation coincides with the arrival into power of Frank McKenna's Liberal provincial government in 1987.

McKenna's role as champion[34] has several dimensions: committed to making the provincial economy less prone to cyclical fluctuations because of its dependence on resources, he and his government have focussed on structural transformation. This involves not simply a focus on information technology and taking advantage of the information highway. It also involves changing the culture of the province and promoting entrepreneurialism. Stopping the chronic brain drain experienced by the province and even encouraging New Brunswickers abroad with skills to return home are key, as is increasing the skills base of the population.[35] As well, the Premier is renowned for his constant travelling abroad, encouraging firms to come to New Brunswick to take advantage of low business costs and the sophisticated telecommunications network. ESJ and FRDC have both pointed to the Premier's instrumental role in encouraging development at the local level, and his office's habit of providing leads on investment to local development commissions. For example, the Premier was instrumental in attracting the technology centres of both IBM Canada and XEROX Canada to Saint John. (Balcomb, 1997; Carson, 1997)

Moreover, as Milne (1996, p. 106) points out, while McKenna's efforts have improved the perception of the province held by the business community outside of the province,[36] the confidence of the population as a whole has increased, stimulating an entrepreneurial spirit. As well, morale within the public service has improved.

Thus, McKenna's role as champion relates not to the development of particular clusters. Rather, it involves the instillation and refinement of an entrepreneurial attitude aimed at transforming the economy and overcoming a negative attitude which hindered development and diversification historically.

Science And Technology Infrastructure

The role of post-secondary institutions in the growth of the new economy is critical. Performing R&D, training HQP, and forming partnerships with government and business are necessary ingredients for economic transformation and renewal. Indeed, where linkages between post-secondary institutions and business are strong, clustering tends to be dynamic. (Nordicity Group, p. 39)

At first glance, New Brunswick in general, and Saint John in particular, appear to be at a disadvantage in this important area. For example, R&D expenditures in New Brunswick in the higher education sector are the second lowest of any province (only Prince Edward Island is lower), at $41.8 million in 1994-95, or 1.46% of the Canadian total.[37] Moreover, federal expenditures on R&D performed in New Brunswick in 1994-95 were only $8 million, or 0.98% of the Canadian total.[38] There were only $1 million in federal expenditures on extramural S&T in the Saint John CMA, or 0.05% of the federal total.[39] Business does not alleviate this weak showing, performing only 0.7% of the Canadian BERD in New Brunswick in 1994, and funding only 0.6% of the Canadian total. These figures are consistent with New Brunswick's historic share of GERD, which has, since 1979, always only been 1-2% of the Canadian total.[40]

The above are "traditional" indicators of the success of a local or regional economy, which might suggest that prospects for transformation are low. The political and economic reality is that such a distribution will not change. However, this does not mean that provincial post-secondary institutions cannot play a meaningful role in the transformation of the economy or that innovation at the local or regional level will not occur. As Statistics Canada itself acknowledges, investments of money and human resources in R&D present " a limited picture of science and technology."[41] Rather, it points to the need to stress partnerships, identify niches for activity and research, and highlight the new ways of looking at their role in the economy.[42] This involves exploring new possibilities and opportunities in the innovation and economic growth process. (de la Mothe and Paquet, 1997, p. 2).

An outsider might initially suggest that UNBSJ might not be of great assistance in the efforts to transform the local economy. Critics and cynics would point to the small student population (approximately 3,000 full and part-time students), a small faculty (approximately 100 full-time faculty members), a small range of graduate programs, limited undergraduate programs in the physical and natural sciences and engineering (in some programs, students complete the third and fourth years of their programs at the Fredericton campus) and a relative isolation from the mainstream university community.

Contrary to this rather pessimistic outlook, however, the university is indeed a major partner in local transformation efforts. While the campus is relatively small, in comparison to other Canadian universities, it actually is a significant and visible actor in the local system of innovation, for it is a major player in a mid-sized city. As Dr. Rick Miner, Vice-President (Saint John) of the University of New Brunswick notes, "the university here is an active part of the strategy, whereas elsewhere [in larger cities], it is often taken for granted or ignored." (Miner, 1997)

The key component of the strategy is the realization that the campus cannot aspire to play the same role that larger universities play elsewhere. The campus strategy, based on the recognition that it is too small to do things by itself, seeks to identify niches in which the strengths of its faculty and curriculum match the needs of the city. Thus, the campus is aggressively and actively involved with partners in the community and abroad, thus ensuring stronger linkages in the community. This was seen in the initiatives and programs discussed above.

Other initiatives further the university's ties to the community. It recently began offering a major in computer science within its Bachelor of Science in Data Analysis. While similar to a computer science degree, the program enhances the analytical skills of graduates, making them more attractive to business, in that they possess computing skills and the ability to analyse pertinent information. Consistent with the city's efforts to raise the profile of the city abroad, the campus has launched a concerted plan for raising its international profile, aimed both at drawing international students to Saint John (in recognizing that this can forge future economic and social international linkages) and at exposing local students to the world outside of Saint John (as 80% of the campus' students are from the greater Saint John area).The pay-off has been immediate, as UNBSJ in 1997 received more applications at the undergraduate level than the much larger Fredericton campus of UNB. These applications were concentrated in business and engineering. (Miner, 1997) Thus, for example, a new international MBA program has been launched. The program, designed to differ from traditional two year MBA's, is an intensive 11 month program featuring international co-op placements and executive seminars. It also features a speaker's series featuring prominent leaders in the business community, such as the president of IBM Canada, and international academics. The program prepares graduates for immediate entry into the business world and develops links with the international business community. (For a description, see McLaughlin, 1997) Other international activities at the campus include the Eastern European Program, which organizes seminars and business missions for Canadian executives and arranges co-ops for students to work in countries such as Russia, Poland and the Ukraine. A research team has received a $1 million grant from CIDA in order to promote linkages with Cuba. The university also received $1.2 million from CIDA to train Vietnamese faculty and government officials in their efforts to shape a market economy, particularly in agriculture and food processing. Finally, the Electronic Commerce Centre received a $200,000 grant to undertake initiatives in electronic commerce in the Caribbean.

Given the relatively small size of the business and elite community in Saint John, the university has been able to enhance its role in the city's transformation efforts and raise its profile in the city. Consequently, this has solidified linkages between the academic research, business, and policy communities. Thus the campus, despite its small size, can be sold by city and business leaders as an asset. (Balcomb, 1997; Carson, 1997) Networking, both formally and informally, cements this role. For example, Dr. Miner is a member of the Board of Directors of the Saint John Board of Trade, ensuring regular contact with the business community. Simply being aware of concerns of the business community, or being knowledgeable of future strategies, allows the university to prepare for such changes and allows the university to proactively aid in these plans as opposed to simply reacting to them.

The local community college's curriculum is also geared towards preparing students for a knowledge-based economy. For example, its Computer Programming Technology programme is geared towards addressing a foreseen shortage of skilled IT workers. 95% of graduates obtain employment within six months of graduation, and 50-75% are placed before graduation, due to the work term component of the programme. Due to its success, the programme recently doubled in size. As well, the college offers programmes in Office Technology and Information Systems, training students in automated office systems, systems analysis and Internet development. These are offered in light of the province's emphasis on preparing a technology-literate workforce. Work term placements contribute to the high employment rate of graduates.

Networks

The Nordicity Group points to the importance of linkages through formal and informal networks in cluster development. Shared knowledge of information by city and business elites can assist in the growth and transformation process.

In the past decade, Saint John has experienced an energy surge in the goal of diversifying the economy and promoting the region as a desirable place for business. While at the provincial level Premier McKenna has served as the primary champion for New Brunswick and has set the tone for the local community, at the regional level a variety of organizations work to promote entrepreneurialism and the diversification of the economy. Housed in the aptly named Business Resource Centre and reflective of the teamwork approach adopted by the city's elite, ACOA, Uptown Saint John, the National Research Council, Enterprise Saint John, the Fundy Region Development Corporation, the Saint John Board of Trade and the New Brunswick Economic Development and Tourism ministry share resources and work together to assist in the growth process. The Business Resource Centre thus provides a physical location for much of Saint John's business networking and planning. As well, there is a significant overlap in the board of director membership of ESJ, the FRDC and the Saint John Board of Trade. (FRDC, 1996, p. 34)

The decision to house the various development agencies under the same roof was the result of the efforts of a few key individuals, such as Carson and Balcomb, to raise the profile of the development process in the city and region. The move greatly strengthened the networking process in that agency members are aware of the activities of other agencies. This minimizes the duplication of effort and allows for resource sharing, but more importantly, it allows for the sharing of knowledge. It also allows for visitors to the centre, be they potential new entrepreneurs or visiting investors or firms considering relocation, to rapidly learn about the entire business environment of the region.

While they all play an important role, two organizations will be discussed. Enterprise Saint John was created in 1993 as an independent economic development commission for the City of Saint John. Its role is to facilitate business development in the city and pro-actively seek new business opportunities. Thus, economic development was effectively removed from City Hall, allowing the private sector board of directors to become more involved in the development process.

ESJ thus has an important networking role as well, as seen for example in its annual retreat of the Board of Directors. Composed of representatives of the city's business and knowledge communities, members discuss the city's economic situation, economic trends locally and globally, and potential areas for action. In the past, these retreats have led to the creation of several task forces, including earlier in 1997 the creation of a medical task force, chaired by Larry Armstrong, a Vice-President for J.D. Irving Ltd. and a former provincial Deputy Minister. Members of this task force include Dave Carlin, CEO of the Atlantic Health Sciences Corporation, and Dr. Rick Miner, VP of UNBSJ. Premier McKenna met with the task force, which led to provincial support for the Institute of Health Sciences initiative. Not surprisingly, the proposed institute is coming closer to fruition.

The activities of ESJ are complemented by those of the Fundy Region Development Commission, one of twelve provincially-sponsored development boards. The FRDC provides business development services to firms and companies in the regions, helping them to create jobs.[43] In light of a rapidly changing economy and a significant number of people either unemployed or receiving social assistance, the FRDC developed in 1996 a new strategic plan, discussed above. The FRDC encourages new entrepreneurs to join either the local home-based business network or their local Chamber of Commerce, or to attend the Saint John Board of Trade monthly mixer. (Balcomb, 1997)

These formal networks are complemented by an active informal network. As the Nordicity Group notes, this type of networking is difficult to document but is instrumental in the clustering process. The relatively small size of the city increases opportunities for contact in non-formal settings. Word of mouth spreads quickly, minimizing the likelihood that key business or government people are unaware, for long at least, of new initiatives. This also allows for rapid action when new initiatives are undertaken, as it is much easier to get key players informed and onside. (Carson, 1997; Reich, 1991, pp. 234-240)

Government Support

The Nordicity Group suggests that pro-business government support assists the cluster development process. Again, this tone has been set by the provincial government, as discussed above. At the regional level, a pro-business tone has also been established. For example, the decision to create ESJ, thus in effect moving the city's economic development agency out of City Hall, was made in order to attempt to remove bureaucratic politics and in-fighting out of the planning efforts. (Balcomb, 1997) ESJ remains funded by the city, but it has a volunteer, private sector board of directors. Similarly, the FRDC's mandate is to develop entrepreneurialism in the region and assist in business creation. This is dictated by its sponsors, namely the federal, provincial and municipal governments. (See FRDC, 1996, pp. 4-6) Like the ESJ, its board of directors is from the private sector.

Time and Planning

The Nordicity Group suggests that cluster development can take many years, depending on the strategy adopted and the time required to build up a critical mass of firms. Nevertheless, planning is possible, although it should entail a situation analysis and establish goals and measurable objectives. (Nordicity Group, 1996, pp. 42- 44)

Major changes to the international political economy, including globalization, the rise of the knowledge economy, public sector cutbacks and changing demographics, have forced all levels of government to reassess their development strategies. In 1987, Saint John began the process of revising its strategy, following the recognition that the previous plan developed in 1973 no longer reflected current conditions.[44]

The process[45] stressed the importance of city leadership and business and community involvement in the development of the new plan, given the climate of continued cutbacks in federal and provincial expenditures. It stated that the resurgence and the transformation of the local economy, if it was to be successful, would need to be less dependent on those traditional sources of economic stability. The process recognized the fundamental changes occurring in the international political economy (information economy, new technologies) and in society (neo-conservatism; changing household structures), which set the terms for an evaluation of these changes on Saint John.

The background study recognized that growth was occurring in business management and the food and accommodation sectors, and that only tourism was showing great promise: other key sectors, such as the oil refinery and shipbuilding, were underperforming, in part due to the fact that the traditional manufacturing base had been slow to adopt new technologies. (See City of Saint John, 1987a, pp. 16-27, 37) However, the study also recognized the importance of integrating the activities of the community college and the university with the needs of employers and the development plan; taking advantage of the new spirit of entrepreneurialism and free trade (which dominated policy debates in the 1980s); and promoting the high quality of life and low cost of living as reasons for locating in Saint John. (City of Saint John, 1987a, pp. 37-38)

As a result, the city adopted four interrelated directions as the basis of its new development plan: promoting entrepreneurialism, high quality products and economic diversification; improving the environment and social development; and promoting community involvement. (City of Saint John, 1989) The plan set out short, medium and long term items of action, and the various roles in which the city and other actors should play in these activities. The plan demonstrated considerable foresight and awareness of changes occurring in the international political economy. Highlights of note include the focus on the tourism and hospitality sectors as growth opportunities; the need for continued revitalization of the city, its heritage and its infrastructure; the need for a concerted community-wide effort to support economic growth and diversification, including an "open for business" attitude at City Hall; and an emphasis on literacy and training of its citizens. Noticeably absent, though perhaps implied, is the need for a coordinated tracking mechanism of economic activity in the local economy.

The planning process at the city level has been mirrored at the regional level, as the FRDC released strategic plans in both 1991 and 1996. The FRDC's strategic plan described above has helped to ensure that important players in the economic transformation of the region are on the same page. The Business Resource Centre exemplifies the team approach that has been adopted by its tenants, as they are working together towards a common goal. (Balcomb, 1997) Similarly, Carson points to a "better synergy" with the community college and the university, in part due to explicit intentions of making them part of the city's transformation efforts. (Carson, 1997)

Efforts have focussed on promoting a climate of entrepreneurialism and attracting firms and investment from abroad. The FRDC helps entrepreneurs, mostly in the service sector, develop business plans, and in some cases, if an entrepreneur cannot afford private start up services, provides assistance.[46] It also assists existing businesses and entrepreneurs when required with technical, marketing and support information, and serves as an access point to various provincial and federal officials.

Despite the presence of several large New Brunswick-controlled firms, it is clear that attraction of firms and investment capital from abroad are required to instill a dynamism in the provincial economy. To this extent, the government and the Premier have set the tone, establishing a clear, pro-business environment. For example, the government highlights low labour costs, a stable work force (and one that is both relatively skilled and available, reflecting the high unemployment of the province) and a supportive government as key reasons for locating in the province. (See New Brunswick, 1995)

ESJ plays an active role in efforts to attract firms and investors from abroad. Firms are attracted to Saint John by the diversified manufacturing base, a significant pool of bilingual labour, and the R&D performed by NBTel. (Report on Business, 1994) The FRDC also reacts to leads from the Premier's Office and from the Ministry of Economic Development and Tourism (offering information on power, location, land, for example), and tries to find appropriate sites for the potential firm. (Balcomb, 1997) Finally, the city is also trying to take advantage of the federal duty deferral program and become a free trade zone. This plan, in the final development stages, would allow goods from abroad to be imported duty-free and then be re-exported.[47]

A key component of the planning process, however, is the presence of an adequate database or knowledge of local conditions which can allow local decision-makers to assess the possibility for local clustering and develop and implement a strategy for local economic growth and diversification. For example, in its study of West Sydney, Australia, the Nordicity Group (1996) noted, "desiring high technology development is not sufficient." As a result, the region undertook a detailed analysis of the region's strengths and weaknesses to assess the feasibility of such a strategy.

In Saint John, the lack of adequate data is problematic. For example, the FRDC does not possess detailed data on the performance of R&D in the region, even though it is clearly happening at Fundy Cable and NBTel. (Balcomb, 1997) Similarly, Carson points to a need to get a better handle on the skills set of the unemployed, particularly since there is a concern about those who get "left behind" as the economy becomes more knowledge-based. However, the local employment centre is examining this issue. (Carson, 1997) As well, ESJ intends to perform, in conjunction with UNBSJ, a semi-annual economic review by sector.

City leaders use the high quality of life in the city as a selling point of the region, an increasingly important factor in business location decisions, regardless of how quality of life is defined.[48] A key dimension of this exercise has been the effort to overcome Saint John's "feeling of continuously being slighted," a feeling that goes as far back as to when Fredericton was named the capital in 1784. (Balcomb, 1997) As well, the city is trying to overcome the perception that it is a polluted, blue collar city, despite clean-up efforts and the booming of the service sector.

This feeling has been changing, for as far back as the early 1980's, surveys revealed that Saint John was a good place to live,[49] confirmed recently by the Report on Business magazine. (Report on Business, 1996) In recent years, there has been a concerted effort amongst city elites to transform this image.

14.5 CONCLUSION

Some analysts have been pessimistic as to the economic future of Saint John. Milne, for example, has suggested that without adjustment towards a more service-oriented economy, Saint John will decline. In particular, retraining opportunities for the citizens are required. (Milne, 1996, 99-100, 103) As well, De Benedetti has pointed to Saint John's over-reliance on manufacturing as a potential hindrance in the transformation of the economy. (De Benedetti, 1994) Finally, concern has been expressed about the benefits of developing the call centre industry, for it has been suggested that once other countries with cheaper labour get their telephone systems established (ironically the same countries NBTel is exporting its technology to), Saint John might experience an exodus of call centres from the city and indeed the province. (Desserud, 1997)

At another level, doubts are surfacing about the continuation of the transformation of the province's economy and culture once Premier McKenna's successor is chosen in the spring of 1998. Concern has been expressed that while he was a politician with considerable vision and talent, McKenna did not groom potential successors. This can only create uncertainty and undermine his considerable efforts at transformation. (Desserud, 1997)

Still, the seeds for the successful transformation of the local economy have been planted. Optimism reigns that these efforts will pay off. The region is now entering the second decade of its transformation, yet it remains to be seen whether successful cluster development will occur.

Several lessons can be learned from this case study. First, while the region has established clear policy goals, it has relatively few policy instruments. In Canada, given the constitutional framework, municipalities are essentially creatures of the provincial governments. The latter are partial to wide-ranging economic development of the province, forcing municipalities to compete with each other in the development arena. Thus, Saint John is engaged in battle with Moncton and Fredericton for scarce investment, particularly in telecommunications.

Second, further research is required to determine what constitutes a cluster *in the Canadian context*. Given the population distribution in the country, focus might better be placed on "regional" as opposed to "local" systems of innovation. Thus, we might conceive of a southern New Brunswick innovation system, comprised of Saint John, Moncton and Fredericton. While each city might specialize in certain sectors of a cluster such as telecommunications, joint promotion of the three cities together would allow the region to be promoted to firms requiring minimal population or market sizes before considering relocation. This triad would be served by two competing fibre-optic networks connecting the cities, five university campuses (UNB-Fredericton,

UNB-Saint John, Mount Allison University, University of Moncton, St. Thomas University), and four campuses of the New Brunswick Community College. To this end, ESJ and the Greater Economic Development Commissions of Fredericton and Moncton are planning a joint promotion of the Southern New Brunswick triad. Thus a broader conception of the boundaries of the innovation system might allow the region to overcome problems associated with small size and a lack of access to capital.

Third, tools are required to help recognize when a potential cluster is emerging. This is important information for policymakers and economic development agencies in their efforts to promote growth. They need to enhance their knowledge of economic activity in the region, particularly in terms of local R&D performed in the key clusters and the extent of the linkages to other sectors of the economy. As well, given the emphasis on attracting firms to the region and the fact that development agencies speak of firms coming to the region because of the presence of NBTel, for example, it is a prime opportunity for local authorities to undertake an examination of business relocation attitudes. Why are firms attracted to an area? Is it because of a clustering dynamic, or is it simply due to "traditional" business considerations (low costs, etc.)? Such an endeavour would enhance Statistics Canada's redesign efforts.

Fourth, there is a role for Statistics Canada to develop "cluster surveys" to assist local planners measure economic activity. For example, ESJ's recent survey on the call centre industry, while providing valuable information, was a difficult undertaking, given ESJ's limited resources. Of particular value would be a longitudinal analysis which could help gauge clustering dynamics.

At this point in time, it is clear that a transformation is under way in Saint John, both in terms of the economic structure (a shift away from manufacturing) and in terms of attitudes (a team or community approach, a concerted planning effort, and optimism that change can be possible). Knowledge flow between the key actors is good, which is a benefit of a small region, and development is focussed on what is possible and attainable.

Of course, only time will tell if local efforts will be successful, which is one of the major problems of cluster analysis: it always looks backwards in time. Given the length involved in cluster development, statistical tools are required to assist planners in their efforts.

ENDNOTES

1.   On this issue, see Brodie (1990). This phenomenon has occurred consistently throughout Canadian economic history: for example, the decline of the once prosperous maritime economy following confederation; the economic boom in Alberta (and to a lesser degree Saskatchewan) driven by energy exploitation in the post - WW II era; the relative decline of the Ontario economy following the passage of the FTA and NAFTA agreements; and the dynamism of the British Columbia economy associated with the rise of the Pacific Rim.

2.   While this paper deals with the city of Saint John, where appropriate it will also address the greater Fundy Region, of which Saint John is a central component. The Fundy Region includes not only Saint John county but the neighbouring counties of Charlotte and King. This coincides with the provincial government's definition of the Fundy Region. The 1996 population of the Saint John CMA was nearly 130,000 (approximately 75,000 in Saint John proper), while the population for the Fundy Region is estimated to be at least 170,000 (official figures derived from the 1996 census are

not yet available). This allows the work of the Fundy Region Development Corporation to be included in this analysis. As well, it reflects the fact that regional economies and innovation systems, while perhaps geographically centred and concentrated in a major city, encompass a broader land space.

3.   The end of the federal frigate program in 1996 had a dramatic impact on the local economy, for it employed nearly 4,000 people at its height in 1991. Many of these jobs were high-skill and high wage. This is significant given that there are approximately 8,000 unemployed people in the area. Saint John Shipbuilding will in 1997 be undertaking a new, privately-sponsored ship construction program (worth $150 million), but it is expected that total employment associated with the program will only be 500 jobs unless new contracts can be obtained. To this end, it is significant that New Brunswick Premier McKenna, in his role as host of the 1997 annual Premier's conference, managed to obtain support for a resolution calling upon the federal government to undertake measures to make the Canadian shipbuilding industry more internationally competitive.

4.   For example, Enterprise Saint John, the city's economic development commission, now estimates that 75% of Saint John's labour force is employed in white collar jobs. (Carson, 1997)

5.   To this list can be added the proposition of Zoltan Acs (Statistics Canada workshop, March 12-13, 1997) that small cities with small universities (the Saint John campus of the University of New Brunswick has an enrollment of approximately 3,000 students and a full-time faculty of approximately 100) generally do not generate the dynamics necessary for "positive economic transformation." However, this is likely to be more true of the American experience than the Canadian experience, due to the substantial differences in the two university systems. American universities can essentially be classified as research or non-research universities, while almost all Canadian universities perform research to some degree. Moreover, their role in the local economy is different, given the particular population distributions in Canada. While this will not be explored in this paper, such a comparison is an important field of inquiry that requires more examination.

6.   Premier McKenna resigned from office in October 1997. His successor will be chosen in the spring of 1998.

7.   Premier Frank McKenna, "State of the Province Address," January 27, 1997.

8.   See *Globe and Mail*, "N.B. Realizing High-Tech Dream," December 24, 1996, p. B11.

9.   In 1996, the province reached an agreement with a private sector trainer, ITI Information Technology Institute, to establish post-graduate training facilities in Moncton, Fredericton and Saint John. Courses are to provide high end IT skills training for more than an estimated 200 students. On the recruitment drive, see *Globe and Mail*, "N.B. Hunts for Technology Specialists," January 24, 1997, p. A3. In part, this effort is to compensate for the out-migration of better-educated New Brunswickers which has contributed, Milne contends, to a higher provincial rate of unemployment. See Milne (1996), pp. 19-26.

10.   There were more than 10,000 applications. Consequently, the government again offered the program at the end of 1997.

11.   To visit the government's Itjobnet web site, the URL is: *http://itjobnet.gov.nb.ca/main.htm*

12.   For example, the provincial Ministry of Natural Resources is committed to creating 1,800 new jobs in value-added industries by the year 2000. See *Saint John Times Globe*, "Moncton Gets Value-Added Forest Firm," January 14, 1997, p. A10.

13.   *Saint John Times Globe*, "Plastic? Fantastic," June 13, 1997, p. B1..

14.   The bank was the Canadian Imperial Bank of Commerce. See *Saint John Times Globe*, "Second Survey Says N.B. Has Best Growth Rate," August 1, 1997, p. C8.

15.   See *New Brunswick Telegraph Journal*, "N.B. Tops Nation At Job Creation: StatsCan Survey," July 31, 1997, p. A3, and "Lowered Unemployment Rate No Sign of Good Times," August 9, 1997, p. B1.

16.   For example, Royal Bank economist Lise Bastrache recently said: "New Brunswick is seen as a province that is moving. People talk a lot about the province in a positive way but that kind of optimism is not validated in numbers (investment intentions)....New Brunswick's message is very much a pro-business message. It is considered the most dynamic province. The perception is good. But I am disappointed when I see the actual numbers." *New Brunswick Telegraph Journal*, "Lowered Unemployment", p. B1.

17.   For an illustration of the economic strength of the Irving family companies, see De Benedetti (1994, p. 225).

18.   For example, as the General Manager of Enterprise Saint John reported, Arthur Irving has told him that if he hears of a potential investor or company not coming to Saint John because of the Irving presence, then he is to arrange a meeting to discuss this matter in the hopes of attracting the firm or investor. (Carson, 1997)

19.   See *Saint John Times Globe*, "Shipyard's Coming Day of Reckoning Forces a Look at Diversification," July 12, 1995.

20.   See *Saint John Times Globe*, "You Gotta Eat," May 15, 1997, p. A1

21.   De Benedetti, (1994), pp. 223-224. Furthermore, the recent drop in after tax income is explained by the fact that the federal frigate program skewed income data upwards.

22.   See *New Brunswick Telegraph Journal*, "Workers Abandoning Saint John Market," February 3, 1996.

23.   For details of the competition between NBTel and Fundy Communications, see *Saint John Times Globe*, "On Your Mark...", May 1, 1997, p. A16; *New Brunswick Telegraph Journal*, "Second to None," May 3, 1997, p. D1; *Toronto Star*, "Fierce Competitors," July 21, 1997.

24.   See *Saint John Times Globe*, "Corporations Team up for Education," January 25, 1997, p. B1.

25.   At present, community college courses in call centre training are only offered at the Woodstock and Moncton campuses of NBCC.

26.   On systems development, see *New Brunswick Telegraph Journal*, "Technology Marriage Will Benefit Business," May 8, 1997, p. B1

27.   For example, call centres produce $40 million annually in long distance revenues for NBTel, or 25% of its total long distance revenue. See, *Toronto Star*, "New Brunswick Taps 1-800 Revolution," July 20, 1997. In terms of the potential employment impact that call centres now have, AT&T Canada in August 1997 announced that it would establish a call centre in Halifax, Nova Scotia, creating 1,000 jobs. If the firm had decided to locate in Saint John, the local unemployment rate would have dropped significantly.

28.   For an overview of this position, see *Toronto Star*, "New Brunswick Taps 1-800 Revolution," July 20, 1997.

29.   There are various dimensions to eco-tourism, including whale watching, fishing, hiking, bird watching, sailing, canoeing, kayaking and the historic and cultural dimensions of the region.

30.   See *Saint John Times Globe*, "Fundy and Friendly," August 14, 1997, p. A1.

31. The three new Bachelors in Health Sciences specialize in nuclear medicine, radiation therapy, and radiography. As well, the university has a bachelor's program in nursing.

32. For more details see *Saint John Times Globe*, "Hospital Wants Research Institute," July 18, 1997, p. A1.

33. See *Saint John Times Globe*, "More Doctors are Leaving Saint John," August 11, 1997, p. A1.

34. For a profile of McKenna, see *Maclean's*, "Fast Frank," April 11, 1994, p. 22.

35. See *Globe and Mail*, "N.B. Tries to Reverse Brain Drain," December 27, 1996, p. A3.

36. Indeed, shortly after assuming power, McKenna in 1988 hired a New York-based corporate image-maker to conduct focus groups on the province's image. The results showed that people outside of the province either did not know of the province or saw it as an economic wasteland. The latter belief was also held by people within the province. The recommended strategy was that the province identify areas of strength, build upon them, and publicize success stories. See Fowler (1996).

37. Statistics Canada, *Science Statistics*, vol. 20, no. 7, September 1996.

38. Statistics Canada, *Science Statistics*, vol. 20, no. 8, October 1996.

39. Statistics Canada, *Provincial Distribution of Federal Expenditures and Personnel on Science and Technology, 1994-95*, ST-97-02, February 1997.

40. Statistics Canada, *Science Statistics*, vol. 20, no. 9, October 1996.

41. Statistics Canada, *A Compendium of Science and Technology Statistics*, ST-97-01, p. 2, February 1997.

42. As de la Mothe and Paquet (1997, p. 2) note, emphasizing process rather than structure can help us understand innovation at the local level. For example, in the university sector, the Association of Universities and Colleges of Canada has devoted much energy promoting the role that universities play in local and regional economic development as key components of the national and local systems of innovation, rather than simply focussing on investments in HERD.

43. As such, it differs from ESJ in one respect in that it does not "seek and find" firms and investment.

44. For example, population and employment growth associated with federal and provincial programs did not occur. The 1973 Municipal Plan foresaw a population of 265,000 by the end of the 20th century. For an overview of the 1973 plan and how it was eclipsed by conditions, see City of Saint John (1987b).

45. The process began in 1987 with the establishment of a program for managing change. A background study was commissioned by the city and produced in December 1987. The city council created a Task Force for the Strategic Plan, which recommended adoption of a plan in which the city would follow four recommended "directions" for the city; finally, in April 1989, the city developed an action plan for implementing the strategic plan.

46. The FRDC assisted 157 new businesses which were created in 1995-96. (Balcomb, 1997)

47. See *Saint John Times Globe*, "Free Trade Zone for Area is Nearer Reality," June 3, 1997, p. A1.

48. Both Balcomb and Carson point to the *Report on Business*, which in 1994 and 1996 named Saint John as a leading place to do business because of its high quality of life, community spirit, and low cost of living. See Segedy (1997, pp. 62-63) for a discussion of factors that pertain to "quality of life" in business location decisions. Given the rise of the new information and communication

technologies, smaller and more remote communities tend to sell their strengths that contrast with major urban centres (e.g. less crime, "friendliness", cheaper cost of living, etc.) as reasons to locate. This is due to the rise of the service sector and the "near-universal availability of the classic variables of land, labour, and capital, as well as utility rates, taxes, and the elusive 'business climate and entrepreneurial environment.'" As a result, "the profit motive can be satisfied virtually anywhere." (Segedy, 1997, p. 59)

49.   See City of Saint John (1987a, p. 1) Factors in this transformation include urban renewal in the 1960's and as series of events in the 1980's: the restoration of the Imperial Theatre; the opening of Market Square in the heart of the uptown area; completion of the Regional Hospital, the largest health centre east of Montreal; the 1985 Canada Summer Games (which also led to the creation of the Aquatic Centre and the track and field stadium on the university campus; and the 1987 Youth World Cup of Soccer.

REFERENCES

Boyer, R. and Drache, D., (eds.) (1996). *States Against Markets: The Limits of Globalization*. London: Routledge.

City of Saint John (1989). *Goals for Saint John*.

City of Saint John (1987a). *Strategic Plan for Saint John: Background Study*.

Cohen, S.S., and Zysman, J. (1987). *Manufacturing Matters: The Myth of the Post-Industrial Economy*. New York: Basic Books.

De Benedetti, G.J. (1994). "Saint John: Are Its Strengths Its Weaknesses?" In George J. Benedetti and Rodolphe H. Lamarche, (eds.), *Shock Waves: The Maritime Urban System in the New Economy*. Moncton: Canadian Institute for Research on Regional Development, pp. 207-229.

de la Mothe, J. and Paquet, G. (1997). "Regional Systems of Innovation: Toward a Dynamic Evolutionary Perspective". Notes prepared for the *Workshop on Regional Systems of Innovation in Canada*, Ottawa, March 12-13, 1997.

Dick, B.R., and Hood, G.E. (1978). *Saint John Regional Plan: The Regional Economy*. New Brunswick Department of Municipal Affairs.

Foster, M. (1995). "The New Development Paradigm: Its Application to Small Regions." In Foster, M. and Davis, B. (eds), *Regions at the Crossroads*, University College of Cape Breton Press, Sydney, pp. 105-141.

FRDC (Fundy Region Development Commission) (1996). *The Fundy Region: Working Together Towards 2002*.

Grant, J. (1992). *A Handbook of Economic Indicators*. Toronto: University of Toronto Press.

Holbrook, J.A.D. (1997). "The Link Between the Natural Resources Sector and the Services Sector: The Case of Canada and the Pacific Rim" (CP 97-01). Vancouver: Centre for Policy Research on Science and Technology.

15 CANADIAN SCIENCE PARKS, UNIVERSITIES, AND REGIONAL DEVELOPMENT

Jérôme Doutriaux
University of Ottawa

15.1 INTRODUCTION

Research on technical innovation and the commercialisation of knowledge has clearly revealed the important role of regional high-technology clusters in the diffusion of technology, economic growth and job creation. Interest in high-technology clusters as knowledge-based growth poles came from the observation of the success of a few high-tech regions, in particular Silicon Valley in California, Route 128 around Boston, and the area around Cambridge in the U.K.. The observation by David Birch of the higher than average rate of job creation by small firms[1] (Birch 1987) in areas located around research oriented universities reinforced the interest of governments and regional development officers for promoting potential university-industry synergies in technological innovation.

Empirical evidence collected over the past 25 years has led to the identification of a number of characteristics which are generally associated with the development and growth of innovation-based local regions:

- a technology-oriented manufacturing base with a mix of large firms ("anchor organisations") and small entrepreneurial firms;
- a knowledge base made of a solid research infrastructure (public and private research laboratories, research universities), supported by an appropriate supply of qualified human resources (universities, post-secondary colleges, professional training organizations);
- a local quality of life attractive to highly skilled individuals;
- a local culture encouraging formal and informal networking, exchanges, communications and cooperation among local stakeholder (individuals, businesses, governments, educational institutions) and supportive of entrepreneurial activities; and

- an appropriate physical and business infrastructure (transportation, telecommunications, business services, venture capital, government rules, regulations and programs).

The relative importance of each of these characteristics depends on local conditions and on the manner in which local stakeholders interact. The impact of linkages between local organisations has been well analysed by Saxenian[2] (1994) who notes that technology-based local economic growth results from elements of cooperation as well as of competition: sharing expertise and knowledge in some domains, for example in the development of cooperative supplier networks, while specializing in specific niche areas. As stated in a recent OECD report, "...Assessments of the importance of collaborative enterprise activities in national innovation systems show that such co-operation can contribute considerably to firm innovative performance"[3] (OECD, 1997, p. 7).

For a number of years, research parks[4] have been considered an efficient way to create local conditions supportive of the innovation process. Designed for research and development organisations, high-tech firms and support services providers and often linked with a university, they are expected to create a "critical mass" of technology-based activities providing member organizations with an environment conducive to the cross-fertilization of technical and business ideas and supportive of the commercialisation process, thus contributing to regional economic development and job creation. Although the first formal science park was established in 1951 at Stanford University, science park creation really took-off in the 1980's in an effort to support economic development and job creation, and to encourage the growth of knowledge-based industries. In the USA alone, 91 science parks were established in the 1980s compared with 32 parks between 1951 and 1980[5] (Quintas, 1997). In Canada, 3 parks were created in the 1960's and 1970's, and 12 parks were created in the 1980's. According to the Association of University-Related Research Parks (AURRP)[6], there are now over 410 such parks at various stages of development in the world (136 in the USA).

It is on these research parks, their contribution to regional development, the effect of regional characteristics, and the importance of their linkages with local universities that this chapter is focussed.

The role of universities as complements to research parks cannot be neglected. Indeed, as noted by OECD, "Government-supported research institutes and universities are main performers of generic research and produce a body of basic knowledge for use and further development by industry. ... The general ability of industry to access that knowledge is important. This can be through patent data, published information,..., access to scientific networks and spin-off firms nurtured in technology incubators"[7] (OECD, 1997, p. 8) Not only do universities contribute directly to local research activities, human resource development and training, and local quality of life through their cultural activities; they also act as catalysts for activities and services offered by other groups and organizations. However, as noted by Acs, their importance must not be overstated because "...while world class universities are necessary for high technology economic development, they have not proven sufficient"[8] (Acs 1990, p. 315). However, they are still a key location factor for R&D intensive firms as shown recently by a Conference Board of Canada survey conducted for Technopolis 97: "In

North America, such companies are likely to search for areas with a significant pool of skilled workers and, to a lesser degree, a well-established, reputable university".[9]

The efficiency of research parks as a knowledge-based economic development mechanism was questioned in 1991 in a landmark study by Luger and Goldstein.[10] They found that about a quarter of the parks they observed did achieve their goals of attracting and encouraging research and development activities and contributing to job creation and economic growth. However, another quarter of the parks analysed had become pure real estate ventures with little effect on local socio-economic conditions, and the rest of the parks (50%) failed. They argued that in many regions, science parks may not be a wise investment[11] (Luger and Goldstein, 1991, p. 183-4). Their theory was that science parks were likely to be most successful in large metropolitan areas but that they would probably not function as growth poles there. Indeed, those areas tend to be naturally attractive to high-technology investments and to engineers and scientists because of their research intensive universities, their well developed research and development base, their business infrastructure, provided industrial development space (land, buildings) is available (thus the attraction of real estate-type business/research parks). They did judge that by attracting research and development activities, science parks could eventually contribute significantly to economic development in smaller metropolitan areas lacking a well recognized university if the right investments in long-term employment were made, but that it would take more time.

The heterogeneity of the research park population must be considered in the analysis. Donnelly recently noted that only about half of the approximately 150 research parks in the United States claim affiliation with a university[12] (Donnelly, 1996, p. 3). And among those, he found 41 parks "affiliated with a public university and utilizing land owned by that university or a university controlled entity..", 27 of these universities being "research universities".[13] Donnelly hypothesises that "the most privileged class of parks would appear to be those associated with large research universities located in metropolitan areas that are large enough to enjoy a critical mass of corporations, personnel, local entrepreneurs, and capital"[14] (p. 9). The success of parks associated with other types of universities would however be highly dependant on the size of their metropolitan areas, on the "energy coming from outside the institution", and on "special community assets".

It is on the relationship between Canadian research parks, their metropolitan areas, and the universities located there that this chapter is focussed. The analysis comes after an overview of the characteristics of Canadian research parks characteristics, of the main metropolitan areas in the country and of their universities.

15.2 RESEARCH PARKS IN CANADA, CURRENT STATUS[15]

The first Canadian business park with an industrial research orientation was created in the Toronto area in 1965. In 1990, in one of the first comprehensive studies of Canadian research parks, Bell and Sadlack identified 12 university-related research parks[16] (1992), all members of The Association of University-Related Research Parks, hosting from 1 to 65 firms and from 40 to 1200 employees. Their goal was to assess the effectiveness of research parks as a university-industry technology transfer mechanism. They therefore tried to look only at parks having formal ties with a university, a selection which, from their own admission, is not easy to make.[17] A more recent inventory of research parks was prepared by Robert Armit in 1993 and identified 14 university-related parks in Canada.[18] There are now 17 parks designed specifically for research intensive activities in the country (all but the University of Calgary Research Park are members of AURRP)[19], plus three in the planning stage (Table 15-1). Not all are university-related. They host from 2 to 153 firms (40 firms average, including firms in incubation), and from 40 to 7870 employees (1376 employees average). Comparable numbers for the USA are park averages of 35 firms and 1864 employees, one park having as many as 150 firms (Stanford Research Parks, 26,000 employees, the oldest US park created in 1951), and another having as many as 33,000 employees (Research Triangle, North Carolina, 57 firms, created in 1959).[20] Canada has 5 parks with more than 35 firms, compared with 14 in the USA. With an average area of 324 acres, Canadian parks are similar in size to the average North American park (average of 290 acres of "improved land" area, average of 572 acres for total land area).[21] Because of a higher rate of park creation in the USA in the late 1980's and again in 1994-95 (14 parks were created in 1994-95 in the USA compared with a total of 9 in 1991-93), the average age of parks in the US (12 years) seems lower than in Canada (15 years); however this may be misleading because of potential problems of interpretation of the words "planned", "created" and "developed".[22] Canadian park ownership/management is mixed, with 4 parks owned/operated by a university and most of the others being government-owned or not-for-profit operations. About half of the parks report having technology incubation facilities, a proportion significantly higher than in the US.[23]

Research parks in Canada form a relatively heterogenous population. For this analysis, they will be regrouped in two categories, the Industrial Research Parks and the University Related Research Parks, reflecting the formality of their links with a university. Selecting the category is easy for some parks (Sheridan Science and Technology park is clearly an industrial research park; Guelph and Western Ontario are clearly university-related), more difficult for others because of changes in ownership, management, or orientation since their creation. For university-related parks, our classification is congruent with that used by Robert Armit in his 1993 survey of Canadian parks.[24]

The Industrial Research Parks were generally created by land developers and/or regional governments to attract large corporations looking for an attractive area to locate their research facilities. Theses parks tend to be primarily real estate developments with plenty of green space, designed for scientific or industrial research laboratories. Services offered to park tenants tend to be limited, inter-firm linkages and relationships with universities and/or government laboratories being left to the discretion of each firm. In Canada, we identified five parks with a strong research orientation which seem to belong to that category. As noted below, a sixth park could be considered for that category if it is decided that it actually is a "research park".

Sheridan Science and Technology park, created in 1965 in the Toronto area to boost research activities in the region and owned by the Ontario Development Corporation, a provincial development agency. The parks has attracted a wide variety of R&D industries in a broad spectrum of sectors from energy electronics, chemicals,

Table 15.1 - Research Parks in Canada, Basic Information

R.P name (*: see note, bottom of table)	A CMA	B Year created	C # emp. (# organiz.)	D area (acr) / # build.	E Inc.?	F #spin-offs	G Owner/management
Discovery Park - SFU*	Vancouver	1980	150 (8)	74/1	MT	4	Private
- UBC *	Vancouver	1982	238 (23)	56/2	Y	23	Private
Discovery Place -BCIT*	Vancouver	1980	1500 (9)	80/9	N	1	Private
Calgary Research and Development Park	Calgary	Planned	---	1066/--	N	n.a.	not available
University of Calgary Research Park*	Calgary	1966/1983	1000 (72)	125/17	Y	n.a.	not available
Edmonton Research Park*	Edmonton	1983	1000 (35)	325/--	Y	n.a.	Not for profit / local government
Innovation Place*	Saskatoon	1977/1980	1500 (100)	120/21	N	12	Government (on leased university land)
Regina Research Park	Regina, Sask.	planned				n.a.	Local government
University of Manitoba Research Park*	Winnipeg	1980	50 (2)	108/1	N	n.a.	University of Manitoba
University of Western Ontario	London, Ontario	1989/1993	250 (21)	50/5	MT	3	University of W.O.
University of Guelph Research Park*	Guelph, Ontario	1982	400 (20)	37/5	MT	3	University of Guelph
Sheridan Science and Technology Park	Toronto	1965	2600 (18)	340/18	N	n.a.	Not for profit / provincial government
Ottawa Life Science Technology Park	Ottawa	1992	40 (8)	23/1	Y	n.a.	Not for profit / government, local association
Technoparc, St.Laurent campus (CITEC)	Montreal	1989/1993	230 (4)	690/1	N	n.a.	Not for profit
Technoparc, Montreal campus*	Montreal	1989	150 (2)	111/2	N	n.a.	Local government.
Laval Science and High-Tech Park*	Montreal / Laval	1988	2532(7)	227/3	Y	3	Municipal government
Technopole Saint-Hyacinthe	Saint-Hyacinthe	1984/1993	7870 (153)	810/--	Y	3	Local government
Quebec Metro High-Tech Park*	Quebec City	1988	2500 (102)	335/14	Y	n.a.	Several levels of governments
Parc scientifique, université de Moncton	Moncton, N.B.	Planned	n.a.		MT	n.a.	University of Moncton
Halifax Industrial Park, Ragged Lake Area*	Halifax, N.S.	1981	1316 (100)	2000 (60)	N	n.a.	Local Government

A: Census Metropolitan Area
B: year founded/ development
C: number of employees (number of organizations)
D: Surface (acre) / number of buildings
E: Presence of technology incubator in park (Yes / No / MT (multi-tenant but not a formal incubator)
F: Number of University/academic spinoffs in park
G: Ownership / Management
n.a.: Not available

Note: *: "University-related" parks included in Stephen Bell's 1991 study (re. S. Bell. J. Sadlack, "Technology transfer in Canada...", 1992, op-cit.
Sources: AURRP directory, telephone survey Fall/Winter 1996-97, Fax/telephones, November-December 1997

petroleum and metals, pharmaceuticals, motion pictures to forest products. Every park member normally buys his parcel of land. The park is managed by the owners association. This park is nearly full and can be considered quite successful. It currently houses 12 large industrial laboratories. The localisation in 1967 in the park of ORTECH, a 300-employees independent research organization founded in 1928 with contributions from industry with Ontario government matching grants has probably contributed to park success, even if linkages between park members are relatively limited. The park association can facilitate contacts between firms and with professors, students, researchers in local universities (University of Toronto, McMaster, York, Guelph, Waterloo), but the member firms are mostly left alone.

The *technopole Saint-Hyacinthe*, established in 1984 but actually developed in 1993, is the creation of the local economic development corporation in an agricultural zone not far from Montreal. The park houses 153 specialized firms in agribusiness and related areas including 17 public and private research centres doing over $20million of research annually. This park has a strong sectoral focus, employs 7870 persons and seems quite successful as a local growth pole. It also houses three technology incubators (CRDA, Cintech AA, Centre d'incubateur technologique maskoutain) and at least three spin-off corporations.[25] There are no special linkages with universities on the region although research laboratories in the park do receive a large number of academic researchers from the Université de Montreal (its faculty of Veterinary Medicine is located in Saint Hyacinthe) and Université Laval for doctoral or post-doctoral research.

Technoparc Saint Laurent, Technoparc Montreal, and the *Laval Science and High-technology Park* are three independent parks located in the Montreal metropolitan area. Technoparc Montreal, located only a short drive from downtown Montreal, was created in 1989 by the City of Montreal Economic Development unit. It is developing very slowly, has no special sectoral focus, and has no special links with local universities. Technoparc Saint Laurent was created in 1989 as a joint effort of the federal government, the provincial government, and the regional municipality. Its development actually started in 1993. It is specifically designed to foster the research and growth of high-tech companies. Although it has no special sectoral orientation, its strength are probably in aeronautics and biotechnology/pharmacy. Its Board of Directors includes representatives of the local universities. Park management is moderately proactive in developing university-park linkages by keeping in contact with the technology transfer offices of the Université de Montréal, McGill, UQAM, Concordia, organizing several university-industry forums each year (a full day meeting of academic/industry) and other networking/consultation activities. The Laval Science and High-Technology park is probably the most successful of the three, possibly because of the attraction created by its first tenants, the location in the park of the Institut Armand Frappier (Université du Québec), and the extensive networking done by park promoters with local university-industry liaison offices and venture capital firms. Its primary focus is on health-related industries, especially pharmaceuticals and biotechnological research. The quality of the physical environment is a strong point, as well as the recent creation of a pro-active technology incubator (Quebec Biotechnology Innovation Centre, 1995). As for the other parks in the Montreal area, formal university liaison activities are relatively limited and left to the initiative of the individual firms (which have links with Institut Armand Frappier (pharmaceuticals/medical research), Université de Montréal, McGill University...). Developed industrial land in the Montreal area being plentiful and relatively cheap; the comparative advantage of locating in these three research parks is not always very clear.

The *Halifax Business Park, Ragged Lake Area,* is the Halifax regions's first totally mixed use research/business park.. Its main but not exclusive focus is on medical and pharmaceutical technologies. The park was created and is managed by regional development authorities and has no link with Dalhousie University.[26] A business incubator and a technology centre are in the planning stage. The mixed-use orientation of this park qualifies it probably more as a business park than as a research park; this may be reflected in the fact that it is not any more a member of AURRP (1997).

A common characteristic of these industrial research parks is their relatively large size, their creation by governments/regional development entities, their lack of strong sectoral focus, and the absence (until now, at least) of technology incubators. They also tend to cater to the needs of established firms rather than start-ups, are physically removed from the universities, and are not very proactive in terms of university linkages, leaving each firm/research centre develop their own communication channels, exchanges, linkages with the universities in the region.

The *University-Related Research Parks* were created by universities or by local governments with strong university input. Their initial objectives were either to become regional development growth poles, and/or to be university income-generating ventures, and also to be a mechanism for the commercialisation of university technology through spin-offs. Eleven parks can be considered to be in that category.

Discovery Parks Incorporated (DPI) in Vancouver is the operating arm of the Discovery Foundation.[27] The Foundation was established by the government of British Columbia (B.C.) in 1979 to utilize an allocation of freehold land and leased land to promote and support the growth of the high technology research and development industry in B.C., particularly in association with the province's post secondary institutions. The Foundation was first managed by the province, then controlled by the universities.[28] It went bankrupt in 1990. Most of its growth has occurred since its restructuration in 1991 as an independent self-supporting organization with a new business plan with primary objective to support the development of research facilities at four academic institutions and at the Science Council of B.C. To date, it has been responsible for the development of 12 buildings housing 40 companies (including companies taking possession in 1998), employing 1900 scientists and technicians in a wide range of high technology sectors as diverse as marine navigational systems, digital telecommunications, computer software, value-added wood products and laboratory testing processes. This has been accomplished through the sale of land and lease of buildings at the 80 acre *Discovery Place* research park located next to the British Columbia Institute of Technology. These proceeds have been used to construct Multi Tenant Facilities (MTF) at UBC and Simon Fraser University with similar facilities planned for the future at other academic institutions. These facilities are open to university spin-off firms as well as to research-oriented non-university spin-offs. The two university sites pay their share of DPI's overhead costs as well as their direct operating costs, with the net revenues being returned to the universities for research purposes. DPI promotes corporate linkages with the universities by providing land and/or buildings required to house research and development companies who wish to locate adjacent to a post-secondary institution. With 28 spin-off firms housed in its facilities, DPI has proven to be a very efficient support mechanism for the university technology transfer activities through spin-offs.

The *University of Calgary Research Park*, owned by the University of Calgary, was created by the university in 1966 on government-owned land adjacent to its campus. Its marketing is done by the Calgary Research and Development Authority (CR&DA), " a [not-for-profit] tripartite initiative of the University of Calgary, the Chamber of Commerce and the City of Calgary created in 1981 to diversify the regional economy".[29] The park has no special sectoral focus. It houses several large private and government research laboratories as well as Discovery Place, a large university-owned multi-tenant facility. "The Authority's Technology Enterprise Centre operates two small business incubators",[30] one located in the park and the other in the same building as the Alberta Research Council in the centre of Calgary. These incubators benefit from the expertise of the university's New Venture Program and are considered among the most successful in Canada. Except for the incubator, there are no formal links between the university and firms in the park, but " the university works collaboratively with the tenants of the two parks to facilitate research and assist the transfer of technology..".[31] Contacts are facilitated by the close proximity of the University Research Park with the faculties of Science and Engineering, most companies having research projects with professors, students, academic researchers. Globally, the park has been quite successful in supporting university-industry interaction.

The *Edmonton Research Park* was created in 1980 on the outskirts of Edmonton as a regional development initiative by the city of Edmonton in close cooperation with the University of Alberta. Its first facility was open in 1983, a telecommunications research company originally employing 30 people. The park is now full and managed by Economic Development Edmonton, a not-for-profit organisation partially subsidized by the city of Edmonton which also manages the Edmonton Convention Centre. Its Board of Directors represents the University of Alberta, the governments of Edmonton and of Alberta, and the private high-technology sector. It now houses a number of public and private laboratories in a variety of sectors, but its strengths are in line with the regional industrial base in the areas of medical biotechnologies, telecommunications, electronics, petroleum research, frontier engineering, and computer software. Its success can be attributed in part to its very successful Advanced Technology Centre, a state of the art incubator facility (also full), and to the strong cooperation between the city of Edmonton, the university and local research labs.

Innovation Place in Saskatoon, created in 1977 with actual development in 1980 is another success story. It was created on land owned by the University of Saskatchewan and leased to the government, with responsibility for its development given to the Saskatchewan Economic Development Corporation, a provincial crown corporation.[32] Most (federal and provincial) research institutes in the region are located in the park or adjacent to it, as well as many agricultural biotechnology firms. The park currently hosts 100 organizations, 65 from the private sector. Part of its success as a growth pole for this agricultural region comes from its well defined focus on agricultural biotechnology and related areas (close to 40% of its activity is in agricultural research, 25% in information technologies, the rest being divided among resource research, environment and other sectors), from the strong cooperation between the various levels of government, National Research Council agricultural laboratories (federal), the University of Saskatchewan, and the local post-secondary colleges. It resulted from the selection of a narrow niche, world class in that field, and aggressive marketing. Some significant spin-offs firms in the park had their origin in the university. The fact that they did not seem to have made much use of the services of UST Inc. (there is no technology incubator in the park or at the university) led Meg Barker to conclude that "our results do not point to the University of Saskatchewan as being the leader in regional technology-based economic development, rather, it appears to be an important catalyst in that development".[33] Innovation Place does not organize formal university-industry activities.

The *University of Western Ontario,* in London, Ontario, is a university owned and managed park developed on university land next to the main campus. The park was created in 1989 to "promote and support private- and public-sector research and development appropriate to the research activities and interests of the University and to the economic and social goals of the region..".[34] Its first building was built in 1993. It currently hosts 21 organizations (17 private enterprises) employing about 250 persons. It does not currently have a clearly defined sectoral orientation but hopes to increase its biotechnology and advanced manufacturing focus in the future.[35] This plan will benefit from the recent opening in the park of NRC's Integrated Manufacturing Technology Institute in a building which is a university - city of London joint venture. To maximize park-university synergy, the university has recently named the same person as director of its industry liaison office and of its research park. It is too early to assess the success of this park which seems to suffer from the cost its university-financed infrastructure (which includes a multi tenant building and a hotel). Increased cooperation with the city of London in promoting the park as a regional research growth pole and as a partner in further infrastructure development is seen as a future success factor.

The *University of Manitoba Research Park,* created in 1980, is also university-owned. It houses two firms with a total of 50 employees (AURRP data) in biotechnology and blood products. The park does not have a technology incubator. Tenant relationships with the university tend to be informal and include research contracts with university professors as well as the funding of a research chair.

The *University of Guelph Research Park* was created in 1982 as an academic initiative by the University of Guelph with Ontario provincial government funding. It sits on university land next to the campus. Construction of the first building began in 1986 (Agriculture Canada "Health of Animals" laboratory).Two other buildings came in 1987 and 1991 (Semex Canada (artificial insemination for cattle) and Agriculture Canada regional headquarters), followed by a successful multi-tenant building, the Research Park Centre built by the university (which is 90% occupied). The park is now home to more than 20 organisations (400 employees) including 3 university spin-off companies.[36] Its main focus in on agricultural research. Tenants have good relationships with the university research office located on campus. The recently created Guelph University Alumni Research and Development corporation (GUARD) working out of the university may further contribute to spin-off creation. Park development is linked to local industrial receptor capacity which takes time to develop.

The *Ottawa Life Sciences Technology Park* was created in 1992 as a University of Ottawa initiative backed by the provincial government (which provided the land) and the City of Ottawa. It failed because of its dependence on the financial support of the university and the local hospitals. It is now owned by the Ontario Development Corporation (ODC), a government development organization, and managed jointly by the ODC, the Life Sciences Council (a public/private partnership representing the health sector's interests and including the ;local universities, local hospitals, National Research Council (federal) and a number of private firms), and the city of Ottawa. The park is located next to the university of Ottawa medical school and Faculty of Health Sciences and to Ottawa's major hospitals. It currently has one multi-tenant facility hosting 8 firms with 40 employees. It is too early to evaluate its success as a growth pole in life science industries, a $300 million local industry which is small compared with the region's telecommunication and software sectors and its public sector activities (seat of Canada's federal government), but growing at a fast pace. Given the private support it receives through the Life Science Council its likelihood of success are good.

The *Parc de Haute Technologie de Québec Métropolitain*, located next to the campus of Université Laval in the Quebec city metropolitan area, was created in 1988 as a joint venture between the Federal Government (30% of current operating costs), the provincial government (25%), and the local municipalities. It is managed by the Corporation du park technologique which organizes a large number of support activities (information sessions, conferences, workshops, networking activities, marketing), linkages with the university are quite strong. More than half of the members of the park's Board of directors come from the university. The university technology transfer office also has a permanent office in the park. At least 10 of the park's tenants are Université Laval spin-offs, and 7 are spin-offs of the local Institut National d'Optique. The park does not have a strong sectoral focus, firms being active in domains related to the region's traditional industries (forestry) or to its research base (optoelectronics, information technologies and telecommunications, new materials, nordicity). The success of this park can be attributed to the high level of cooperation between the federal and provincial governments, the municipalities and the universities (Université Laval, Institut National d'Optique).   the park has also a successful technology incubator benefiting from the university business faculty exceptional expertise in entrepreneurship and small business management. In fact, the 35 of the 38 firms that went through the incubator since 1988 are still in operation.

If success is measured in terms of *park employment* and *number of organizations*, four of these university related parks can be considered very successful (Calgary, Edmonton, Innovation Place, Quebec). The others were recently reorganized (The Discovery Parks, reorganized in 1993 after difficult times and which now seem to be doing quite well), have been experiencing slower growth (Guelph, Manitoba, Western Ontario) or are too young to tell (Ottawa).

A common characteristic of the four "successful" parks which sets them apart from the slower-growth parks is the quality of their linkages with the universities, the involvement of regional and/or municipal authorities from park inception, and their goals to be tools of regional economic growth. They are also managed by for profit or not-for-profit organisations which are separate from the universities although with good university representation on their Boards of Directors. By comparison, the "slower growth" parks tend to be university-driven with less proactive regional inputs and support.

15.3 RESEARCH PARKS, REGIONAL CHARACTERISTICS AND UNIVERSITIES

In the previous section, Canadian research parks were re-grouped in two categories:
- 5 industrial research parks (IRP)
- 11 university-related research parks (URRP)

As shown in Table 15-2, the IRPs and the URRPs are located in distinctly different Census Metropolitan Areas (CMAs):
- Canada's two largest CMA, Toronto (1st in terms of population) and Montreal (2nd) have only IRPs. And, expect for Guelph, there are no formal land-based research parks in the other large CMAs located within 120 kilometres of these two very large CMAs, Hamilton (9th), Waterloo/Kitchener (11th), St.Catharines/Niagara (12th), Oshawa (16th), Sherbrooke, in spite of the presence –in some cases-- of excellent research-oriented universities (McMaster, Waterloo).
- Each of the following 7 CMAs ranked by size of population has at least one URRP (except Hamilton which is close to Toronto area). Vancouver (3rd CMA by size) and Ottawa (4th) are two special cases: Vancouver's research parks were at one time formally linked with the universities, but are now privately held and managed, although in close cooperation with local universities; and Ottawa's research park is a recent creation in a sector in emergence (health sciences), the industrial high-tech development of the region (telecommunications, microelectronics, software) having occurred in the absence of any land-based research park.
- The other CMAs in Canada (ranked 11th and up in terms of population) have no land-based research parks. Some are located in the greater Toronto regions, other are important regional centers with excellent research universities (Halifax with Dalhousie University) or smaller universities (Windsor, Regina, Victoria). There are two exceptions, Guelph and Saskatoon with highly focussed universities/parks.

Canadian CMAs can therefore be regrouped in three categories:

15.3.1 The Largest CMAs (Toronto, Montreal) with Non University-Related Industrial Research Parks.

These CMAs are characterized by Tables 15-2 and 15-3:
- the size and diversity of their population;

Table 15.2 - Research Parks by Census Metropolitan Areas (CMA)

	CMA	CMA 1996 population (thousands)	CMA, rank in Canada (by pop.)	Number of firms among Canada's top 100 / next 100 in terms of sales	Number of firms among Canada's top 50 / next 50 R&D spenders	% labour force in manufacturing / in Nat. Sc., Eng., Math.
I. Industrial Research parks						
Sheridan Science and technology	Toronto	4 216.3	1	34 / 41	22 / 21	9.5 / 5.1
- Technopark St. Laurent - Technopark Montreal - Laval Science and Technology	Montreal	3 258.9	2	21 / 16	10 / 11	11.8 / 4.6
Saint Hyacinthe	(Montreal)*	77.0	>25			
II. University Related Parks						
Discovery Parks (SFU, UBC, Disc. Place)	Vancouver	1 811.9	3	11 / 14	2 / 2	7.8 / 4/0
Calgary	Calgary	827.6	6	18 / 10	1 / 4	7.1 / 7.1
Edmonton	Edmonton	869.0	5	2 / 0	1 / 1	8.7 / 4.5
Saskatoon	Saskatoon	221.0	17	2 / 0	0 / 0	
Manitoba	Winnipeg	669.6	8	3 / 2	0 / 0	11.0 / 3.5
Western Ontario	London**	406.4	10	0 / 2	0 / 0	
Guelph	Guelph**	104.2	> 25	0 / 0	0 / 0	
Ottawa Life Sciences Technology	Ottawa	1 005.2	4	1 / 1	5 / 3	4.7 / 7.45
Quebec Metro High Tech	Quebec City	677.1	7	0 / 0	0 / 0	7.2 / 5.9
III. Large Metropolitan areas without formal land-based research parks (some have very successful technology incubators)						
	Hamilton**	627.7	9	2 / 0	1 / 1	
	St.Catharines / Niagara**	379.3	12	0 / 0	0 / 0	
	Oshawa**	262.7	16	0 / 0	0 / 0	
	Victoria (B.C.)	306.8	14	0 / 0	0 / 0	
	Windsor	278.9	15	0 / 0	0 / 0	
	Waterloo / Kitchener**	387.6	11	0 / 4	1 / 0	16.5 / 4.25
Halifax Industrial Park, Ragged Lake Area	Halifax	336.3	13	0 / 1	0 / 1	
	Regina	196.7	18	0 / 0	0 / 0	

Notes   * Saint Hyacinthe is an agricultural community located about 50 kilometres from Montreal.
** Distances from Metro Toronto (kilometres) : St. Catharines (110), Guelph (75), Hamilton (70), London (185), Oshawa (60), Waterloo (90).

Source: - CMA populations: Canadian Markets 1996, published by The Financial Post Data Group.
- Largest 200 Canadian firms in terms of 1995 sales: Blue Book of Canadian Businesses, 1997, Canadian Newspaper Services International Ltd.
- Largest R&D performers: " Canada top 100 corporate spenders, 1995", Evert Communications, http://www.evert.com/
- Percentage of workforce in manufacturing (processing, machining and related activities, product fabrication, assembling and repair) , percentage of workforce in natural science, engineering, mathematics: Market facts (Financial Post).

Table 15.3 - University research activities in major Census Metropolitan Areas, 1995

CMA (In the same order as in Table 15-2)	Major universities	Total number of full time students (graduate students only)	Sponsored research, Total ($ millions)	Industrially sponsored research ($millions)	Number of doctorates granted (mean 1994,1995)
Toronto	University of Toronto York university Ryerson Polytechnic	36088 (7811) 24344 (2184) 10583 (0)	272.7 19.6 3.7	97.6 3.6 1.5	515 89 0
Montreal	Université de Montreal McGill University Concordia University université du Quebec a Montreal Ecole Polytechnique	19487 (3754) 22472 (6058) 13768 (1602) 18374 (1305) 3801 (928)	191.6 120.4 13.4 25.9 32.1	84.0 23.7 1.2 6.0 13.3	242 277 56 55 55
Vancouver	University of British Columbia Simon Fraser University	24449 (6680) 10279 (2018)	146.4 20.5	44.8 4.3	306 74
Calgary	University of Calgary	19557 (2673)	64.0	20.7	106
Edmonton	University of Alberta	26489 (3422)	102.8	24.5	280
Saskatoon	University of Saskatchewan	14881 (1627)	45.6	11.9	74
Winnipeg	University of Manitoba	16755 (2637)	50.5	16.2	110
London**	University of Western Ontario				133
Guelph**	University of Guelph				103
Ottawa	University of Ottawa Carleton University	16291 (2604) 14941 (1662)	55.9 20.2	19.6 5.0	92 74
Quebec City	Université Laval	23805 (4955)	84.9	20.7	228
Hamilton**	McMaster University				127
St.Catharines**	–no local university--				
Oshawa**	–no local university--				
Victoria (B.C.)	University of Victoria				51
Windsor	University of Windsor				34
Waterloo / Kitchener**	Waterloo University Wilfrid Laurier University	16981 (1638) 6438 (421)	49.5 2.3	11.2 1.1	160 0
Halifax	Dalhousie University				68

Notes   * Saint Hyacinthe is located about 50 kilometres from Montreal
** Distances from Metro Toronto (kilometres) : Ste. Catharines (110), Guelph (75), Hamilton (70), London (185), Oshawa (60), Waterloo (90).

Source: estimated from data received from the Association of Universities and Colleges of Canada; data for 1994-95.

- their diversified industrial base, well developed industrial infrastructure, and attractiveness to business, illustrated by the presence of the headquarters of most of the largest private corporations in Canada;
- a very high level of total R&D activity estimated at $800 per person in 1996 (total $3.7 billion) in Toronto and $1000 per person (total $3.3 billion) in Montreal
- a strong *industrial research* intensity, 26.4% of all R&D performed by industry in Canada being done in the Toronto CMA ($440.00 of industrial R&D per capita) and 23% being performed in Montreal ($496.00 per capita), compared with per capita averages for industrial R&D expenditures of $390.00 for the province of Ontario, $280.00 for the province of Quebec, and $234.00 nationally.[37] Most of the largest industrial performers of R&D in the country are also located in those CMAs (Table 15-2);
- the presence of a solid higher education sector with a number of post-secondary colleges and universities, including the four most research-intensive universities in Canada (highest ratio of sponsored research to total university operating budget): University of Toronto with a sponsored research budget of $272 million representing 28.5% of the university's operating budget, Université de Montréal ($159.9 million (26%), McGill ($120 million (23%), Ecole Polytechnique ($32.1 million (28.3%) in particular (Table 15-4).[38] The high percentage of sponsored research financed by industry (53% at Université de Montréal in 1994, 38% at the University of Toronto, 41% at Ecole Polytechnique) shows also the strong industrial research orientation of those institutions (Table 15-4).
- a large number of federal and provincial government organizations and laboratories performing intramural R&D activities in the CMA.

Because of their size, well developed industrial infrastructure, established business and financial networks, strong research base (industrial, government, academic), renowned academic institutions, these CMAs are natural growth poles. Past studies have shown that the large research-oriented corporations located in those CMAs know how to access university research and do not really need the special services and infrastructure normally offered in university-related parks to tap academic knowledge[39] (Doutriaux and Barker, 1995). Access to university research is facilitated by pro-active industry liaison offices with well established industrial networks. In these large urban regions, access to land and appropriate physical facilities is a greater challenge than access to knowledge and to business services, thus the appeal and success of the industrial research/business parks. The multiplicity of sources of knowledge (universities, public laboratories, research institutes) and their geographic dispersion does not either favours one specific location for high-tech activities. The lack of direct involvement of local universities with "university-related" research parks is congruent with the findings of a recent survey of seventeen universities (sixteen respondents) organized by the Ottawa Carleton Research Institute as part of a study for the National Research Council of Canada, research parks being judged not very important as a university-industry technology transfer mechanism.[40]

Table 15.4 - Major Canadian Universities per CMA (1994-95 numbers of students or millions of dollars)

		A Total number of full time students.	B Ful time graduate as % of total	C % grad in E,M,P	D % grad in Health	E Value of sponsored research, $million (and as % of total university budget)	F Indust. R., as % of total research	G Number of patents (patent income, $1000)	H Type of university
Vancouver	Simon Fraser U.B.C.	10279 24449	19.4 28.1	19.2 25.9	0.0 15.3	20.5 (8.1) 145.4 (19.3)	22.0 32.4	2 ($15.7) 56 ($758)	C MD
Calgary	U. of Calgary	19557	13.0	22.7	23.6	64.0 (18.1)	33.0	6 ($741)	MD
Edmonton	U. of Alberta	26489	12.9	24.2	27.0	102.7 (18.1)	24.3	6 ($400)	MD
Saskatoon	U.Saskatchewan	14881	10.9	25.9	17.7	45.6 (17.1)	26.1	-n.a.-	MD
Winnipeg*	U. of Manitoba	16755	15.5	24.6	20.9	50.4 (15.1)	34.0	6 ($126)	MD
Waterloo/ Kitchener	U. of Waterloo Wilfrid Laurier	16981 6438	10.9 6.5	54.0 0.0	1.1 0.0	49.5 (18.1) 2.2 (2.6)	25.5 44.0	14 ($1413) -n.a.-	C PU
Toronto	U. of Toronto York U. Ryerson.	36088 24344 10583	21.6 9.3 0.0	18.0 7.5 --	38.6 0.0 0.0	272.6 (28.5) 19.6 (5.6) 3.6 (2.1)	38.0 20.5 41.0	11 ($953) -n.a.- -n.a.-	MD C -n.a.-
Ottawa	U. of Ottawa Carleton U.	16291 14941	16.0 10.9	15.8 23.8	23.0 0.0	55.7 (18.0) 20.2 (11.0)	30.7 35.2	-n.a.- 5 ($11.6)	MD C
Montreal	McGill U. U. de Montreal Concordia UQAM Ecole Polytechnique	22472 19487 13768 18374 3801	26.9 19.2 11.6 7.1 24.4	21.0 5.9 24.4 8.9 100	23.7 35.1 0.0 0.0 0.0	120.4 (23.0) 159.9 (26.0) 13.4 (6.0) 25.9 (9.0) 32.1 (28.3)	22.8 53.2 11.2 23.3 41.4	-n.a.- -n.a.- -n.a.- -n.a.- -n.a.-	MD MD C -n.a.- -n.a.-
Quebec	Université Laval	23805	20.8	17.9	17.0	84.9 (17.0)	26.7	-n.a.-	MD

A: # Full Time students, 1995
B: Full Time Graduate Students, % of Total F.T. students
C: % F.T. Graduate Students in Engineering, Math, Physics
D: % Full Time Graduate Students in Health related fields
E: Total sponsored research expenditures, 1994-95 , $ million (sponsored research as % of total university budget)
F: *Industrial* research as % of total sponsored research
G: Average # of US licences issued, 1992-94 (average income from licenses, 1992-94, $1000s)
H: Type of University, McLeans classification, ("The fifth annual ranking, Maclean news magazine, November 24, 1997);

    MD (medical doctoral): universities with a broad range of PhD programs as well as medical schools (other universities in that category include Western Ontario, Queens, and McMaster in Ontario, Sherbrooke in Quebec, and Dalhousie in Nova Scotia);
    C (comprehensive): universities with a significant amount of research activities and a wide range of graduate and undergraduate programs;
    PU (primarily undergraduate): mostly focussed on undergraduate education

15.3.2   The Large CMAs with University-Related Research Parks.

Geographically, most of these CMAs (Table 15-2) are relatively isolated (200 kilometres or more from Montreal or Toronto), except for Guelph which is only 75 kilometres from Toronto. They are all the seat of one or more research-oriented university (sponsored research budget equal to 17% to 19% of total university operating budget,[41] industrial funding of research accounting from 25% to 33% of total research.

Two types of parks can be observed:
- The parks with a general sectoral orientation of Vancouver (the three Discovery Parks), Calgary, Edmonton, Winnipeg, London, and Quebec City;
- The parks with a strong sectoral focus of Saskatoon, Guelph and Ottawa (new).

In the first category, two elements of success: the date of creation of the park, older parks tending to be more successful than the more recently created parks, and the level of cooperation between regional development authorities and the university, success being strongly correlated with the level of cooperation between local stakeholders.

In the second category, a strong sectoral focus for the parks is an added element of success: This strong sectoral focus is needed to overcome the negative impact of the relative geographic isolation of the CMA (Saskatoon), the pull-effect of Toronto (Guelph), or the dominat sectoral image (communication/microelectronics vs. health sciences) of the Ottawa CMA. This strategy has worked for Innovation Place in Saskatoon and seems to be well on the way to success in Guelph. The Ottawa experience is different. Thanks to its very intensive research orientation (due to the presence of a large number of federal government laboratories[42] and of 5 of the top 50 industrial R&D spenders in the country, including NORTEL (Table 15-2)), and to the actions of a local university-industry consortium, the Ottawa Carleton Research Institute (OCRI) and of the regional development corporation (OCEDCO), the region has become a very active telecommunication-microelectronics-software growth pole (with an estimated 40,000 high-tech jobs in 1997). That development took place in the absence of any formal research park, and the region as a whole could be considered as a "distributed Research Park".[43] Technology incubators were recently created by the National Research Council and the Communication Research Centre to support new start-ups, and another private company, Newbridge, is planning to develop a private research park to support its activities in telecommunication/software. Only recently was a sectorially focussed research park created to support the development of the region's health science industry.

15.3.4   The Large CMAs with No Land-Based Research Parks

As noted earlier, a number of the large CMAs with no land-based research parks are in the greater Toronto region. Only two have research universities, Hamilton (McMaster University) and Kitchener/Waterloo (Waterloo, Wilfrid Laurier), and could be expected to have research parks. Hamilton has a very successful technology incubator. And the Waterloo/Kitchener region which did at one time consider developing a research park, has evolved very successfully as a high growth high-technology region through the joint actions of the municipalities of Kitchener, Waterloo and Cambridge and of the University of Waterloo, and through the university technology commercialisation programs and its spin-off firms. The complete "Technology Triangle Region" (TTC) of Kitchener / Waterloo / Cambridge is in fact a "distributed research park". There have been talks about the creation of a formal land-based research park on land adjacent to the University of Waterloo, but by January 1998, no formal decision had been made.

In another part of the country, Halifax, where Dalhousie University is located, has a very proactive and successful technology incubator;[44] as noted previously, however, the Halifax Business Parks are mixed-use industrial parks with a limited research/university orientation which may be partially due to the physical distance between the parks and the university. In Victoria (British Columbia), university spin-offs get the support of the university industry liaison office, but the desire of the university to develop/be associated with a successful research park has never materialized.

Although there are no research parks in CMAs without at least one "research university", the reverse is not true. Of the 15 Canadian universities labelled "Medical-Doctoral" by McLeans magazine,[45] three are in the Toronto/Montreal areas (Table 15-4), eight are in a CMA with a university-related research park. The other four are in CMAs with no research parks: Queens University in Kingston (Ontario), McMaster in Hamilton (Ontario), Dalhousie University (Halifax) and Sherbrooke (90 kilometres south east of Montreal). Queens University, located in a region with limited industrial receptor capacity, has a productive technology transfer office (PARTEQ) working with firms in a larger region which includes Ottawa, Toronto and Montreal.[46] Hamilton and Halifax have pro-active incubators, and Sherbrooke benefits from Montreal's infrastrucuture.

15.4 CONCLUSION

As outlined in this chapter, and in tune with literature reports for other countries, research park characteristics and park-university relationships are very much related to their regional environments. What may be different in Canada compared with other countries is that regional research intensity (both in term of total and of per capita R&D expenditures) and metropolitain area size (in terms of population) are directly related, this country's largest metropolitan areas being also its most research intensive. The "size"and the "research intensity" effects cannot therefore be analyzed separately.

Our analysis had led to the following observations which may be useful to regional development officers and university technology managers:

- In the largest Canadian metropolitan areas which are also the country's most research-intensive regions (Toronto, Montreal, and, because of its research intensity, Ottawa):
- The most important function of research parks seems to be as providers of serviced land in a quality environment for research intensive organizations.

Linkages with the local research intensive universities are mostly left to each firm's initiative and the universities are not involved in park activities or management.

- With their strong research orientation and research and manufacturing activities spread geographicaly throughout the region, Canada's largest CMAs tend to operate as "distributed research parks" with no direct need for specific university-related land-based parks.

- Local networks and established linkages reduce the need for the type of services traditionally offered in university-related research parks to facilitate university-industry research cooperation. Unavailability of developable land in the direct vicinity of research universities may also be a factor in the lack of direct university involvment in research parks.

- The specialized needs of start-ups and other emerging high-tech firms are met by independant incubators or specialized services sometimes described as "virtual research parks".

- Because of the existing critical mass of research activities, high-tech firms, academic institutions, business services, risk capital, these metropolitain areas are attractive to research intensive organizations and these regions operate naturally as "growth poles".

- Formal research parks have however a role to play in these regions as growth poles for the development of research/manufacturing activities in new or emerging sectors not in the regional main stream (health sciences in Ottawa, biotechnology in Guelph in the greater Toronto region). The role of the research park is then to give visibility to these "new activities", create a localized critical mass of professionals and services, and provide localised access to the specialized technical and business services needed in the emerging sector.

- In the other large metropolitan areas with research/"medical-doctoral" universities,

- The most successful parks tend to have been *created/managed by regional development authorities with the strong cooperation/involvement of local universities.*

- Parks created primarily as an instrument for university-industry cooperation and for university-industry technology transfer did not seem to be as successful as the parks created with high-tech *regional development as first objective* and university-industry cooperation in second place.

- Sectoral focus depends on the region and its historical industrial base, providing a minimum of industrial receptor capacity for local research activities. In some cases, there may be no predetermined direct sectoral focus (Calgary, Edmonton, Quebec); in other cases, the need to create a world-class niche led to a successful specialisation (Innovation Place in Saskatoon, University of Guelph).

- When no land-based park has been developed, active and successful technology incubators acting as "virtual research parks" have generally been created to nurture high-tech start-up corporations.

- Some of these regions are in fact "distributed research parks", research and high-tech activities being integrated into the local urban fabric.
- In the other large metropolitan areas with no research universities, there are no research parks.
- This observation is congruent with past research findings on knowlwedge-based regions which require a minimum of local research activities, a research university being one of the usual component of a regional research infrastructure.

Not directly included in our analysis is the role of technology incubators: are research parks and technology incubators two independant regional infrastructures or are they complementary? In very large metropolitan areas operating as "distributed research parks" are they a substitute for formal university-related research parks? In other large metropolitan areas with a research university, are they a logical first step in regional development prior to the development of a research park or are they a logical complement for start-ups and SMEs who could then become park tenants?[47] While it is easy to make the case for the creation of incubators in research parks to support and nurture future park tenants, is not as clear at the regional level which one should come first and how that relates to local characteristics and the need to create receptor capacity.

ENDNOTES

1. David Birch, Job Creation in America: How Our Smallest Companies Put the Most People to Work, New York, The Free Press, 1987.

2. Saxenian, Annalee, *Regional Advantage: Culture and Competition in Silicon Valley and Route 128*, Harvard University Press, 1994.

3. "National Innovation systems: Background Report", OECD, Directorate for science, technology and Industry, DSTI/STP/TIP(97)2, Feb. 27, 1997, page 7.

4. In this chapter, the words science park, technology park or research park will be used as synonym to describe a campus-like development designed specifically for high-tech companies, private and public research organizations, and related support services.

5. Paul Quintas, " Fresh Growth in Science Parks", *Physics World*, Feb. 1997, p. 47.

6. http://www.siue.edu/AURRP/

7. "National Innovation Systems: Background Report", op.cit., page 8.

8. Acs Z.J., 1990,"High technology networks in Maryland: a case study", *Science and Public Policy*, 17,5, October, p.315.

9. "Janus Zieminski and Jacek Warda, "What makes Technopoles Tick?, a Corporate Perspective", The Conference Board of Canada, September 1997; presented by Gilles Rhéaume at the Technopolis 97 conference, Ottawa, Canada, September 10, 1997.

10. Michael I. Luger and Harvey A. Goldstein, "Technology in the Garden, Research Parks and Economic Development", The University of North Carolina Press, 1991, pp. 74,75.

11.   Ibid., pages 183-84.

12.   Brian Donnelly, "US Research Parks and their Role in Technological Entrepreneurship", paper presented at the INFORMS conference, Atlanta, Georgia, November 6, 1996, page 3.

13.   According to the classification established by the Carnegie Foundation for the Advancement of Teaching, a research university offers a full range of bachelor, master and doctoral programs, gives high priority to research, awards at least 50 doctoral degrees each year, and receives at least [US] $15.5 million in federal support annually (Donnelly, op.cit., page 5).

14.   Ibid., page 9.

15.   The contribution of Marc Racette to the preliminary date collection (Fall 1996) for this section is acknowledged.

16.   Bell Stephen and Jan Sadlack, " Technology Transfer in Canada: Research Parks and Centres of Excellence", Higher Education Management, 4, 2, July 1992, pp. 227-244.

17.   They excluded Sheridan Research Park (Toronto area) with no formal university linkages but lots of R&D activities but included the Halifax Business Park (Ragged Lake Area) in spite of the complete absence of university involvement (Bell and Sadlack, op.cit, "conclusions").

18.   Armit, Robert, *Partners in Progress, University Related Research Parks in Canada,* AURRP, 1993; two of the 14 "parks" are actually incubators (The Greater Hamilton Technology Enterprise Center in Hamilton (Ontario), and the Carleton Technology and Training Center (now closed, at Carleton University, Ottawa (Ontario))

19.   Association of University Related Research Parks (AURRP) 1997 membership directory.

20.   Derived from AURRP data reported by Sarfraz Mian, "Technology Business Incubation and the Development of Technology Incubators", paper presented at the Thematic Workshop on Technology Incubators, Committee for Scientific and Technology Policy, OECD, Paris, June 1997.

21.   Estimated from data reported by Sarfraz Mian, op.cit. In Canada, if the 2000 acres mixed use Halifax Business Park (Ragged Lake Area) is removed, average park land area comes to 219 acres.

22.   According to the Association of University-Related Research Parks, there were 136 parks in the USA, 4 in Mexico and 18 in Canada. In 1996.

23.   About 25% of US parks have such facilities, but a recent study of University parks done by Coopers and Lybrand for AURRP shows a growing trend towards the establishment of technology incubators in existing and new parks (as reported by Sarfraz Mian, 1997, op.cit.).

24.   Robert Armit, *Partners in Progress...,* op.cit.

25.   Communication from Mario De Tilly, Commissaire Industriel, Corporation de développement économique et industriel de la région de Saint-Hyacinthe Inc., December 02, 1997.

26.   Halifax Business Park, telephone conversation, November 25, 26th, 1997: there is a network of 14 business parks in the Halifax region, none dedicated exclusively to research activities. There have been discussion for the creation of a Nova Scotia high-technology park, but no definite plans yet. The lack of university insolvent in the Halifax Business Parks was noted in 1991 by Stephen Bell and Jan Sadlack *(op.cit).*

27.   Most of this text on Discovery Park was adapted/copied from a writeup supplied by Craig Aspinall of Craig Aspinall and Associates, suite 302-116 Homer St,. Vancouver, B.C., December 11, 1997.

28.   Interview with Peter Kimoff, December 03, 1996.

29. *Technology Enterprise Centre*, Calgary Research & Development Authority, May 18, 1993, page 44.

30. Ibid.

31. Ibid.

32. Meg Barker, "Modalities of U-I Cooperation in the APEC Region, Country Report for Canada", report prepared for the Association of Universities and Colleges in Canada, August 1996, section 3.1.3.

33. ibid, section 3.1.5.

34. http:/www.uwo.ca/research/ResPark/park.htm

35. Informal telephone conversation with Tim Walzak, Director, Office of Industry Liaison and UWO Research Park, December 05, 1997.

36. Telephone conversation with Barbara Reid, University of Guelph Research Park, November 26, 1997

37. 1994 data; estimated from Statistics Canada cat. 88001, vol.20, n.9, table 6, "Provincial Distribution of R&D by Performing Sector, 1994", and cat. 88001, vol.20, n.5, table 5, "Total Intramural [Industrial] R&D Expenditures by Province".

38. The university with the next highest sponsored research to operating budget ratio is the University of British Columbia (ratio of 19.3% in 1994; Table IV), most of the other large universities having ratios in the 18% (Calgary, Alberta, Ottawa, Waterloo)) or 17% range (Laval, Saskatchewan).

39. Doutriaux, Jerome, Margaret Barker, "The University-Industry Relationship in Science and Technology", occasional paper n.11, Industry Canada, August 1995, page 94.

40. TTC University survey, *http://www.ocri.ca/initiatives/bronsur1.html*, question 31: 9 respondents felt that university-related science parks were a factor "over-rated in importance" in technology transfer, 4 rated it as "important", 3 as "among the most important".

41. Except the University of Manitoba in Winnipeg at 15% (Table IV).

42. In 1994, the federal government alone performed $736 million of R&D activities in the National Capital Region (Ottawa CMA) which comes to about $750.00 per capita. When local industrial research activities are added, the Ottawa CMA is by far the most research-intensive CMA in Canada. Source, federal R&D : Statistics Canada cat. 88-001, v.20, n.9, Table 6.

43. Research Park Task Force, final report, August 1996, page 3; report prepared for the Ottawa Carleton Research Institute (OCR) and the Ottawa Carleton Economic Development Corporation (OCEDCO).

44. Robert Armit, informal communication, November 27 , 1997.

45. "Medical Doctoral (MD) ", definition at the bottom of Table IV.

46. Kingston is located 175 kilometres South of Ottawa, 260 kilometres East of Toronto.

47. Rutam Lalkuka, Jack Bishop, " Technology Parks and Business Incubators: The potential for Synergy", in *The Economics of Science Parks*, edited by M. Guedes and P. Formica, AURRP, 1996.

REFERENCES

OECD (1997). *National Systems of Innovation.* Paris, OECD.

Saxenian, A. (1994). *Regional Advantage: Culture and Competition in Silicon Valley and Route 128,* Cambridge: Harvard University Press.

PART E

QUO VADIS?

16 SOME LESSONS AND CHALLENGES FOR MODEL BUILDERS, DATA GATHERERS AND OTHER TRIBES

John de la Mothe and Gilles Paquet
University of Ottawa

16.1 INTRODUCTION

The preceeding wide ranging tours that make up this volume make of no claim to comprehensiveness. But they have the merit of having surveyed a vast territory. From this voyage, one must now derive some lessons.

The eclectic approaches used by our authors are bound to frustrate model builders who have been in search of the definitive template for modelling the innovation process and local/regional dynamics in a knowledge-based economy. The statistical data collectors can only be challenged by the lack of agreement on a canonical data set and by the variety of data sources used. The academic and professional analysts are undoubtedly puzzled by the seemingly heterogeneous and widely varied *outillage mental* and perspectives used by the different experts in tackling their specific regional or local systems. And, the policy-makers may reasonably wonder how this sample of experiences can help those charged with the responsibility of designing innovation and regional/local development policies.

There is no easy way to serve well all these different families of readers with one set of concluding comments. But this is what we would like to attempt in the next few pages.

We proceed in three stages. First we make some general comments of general applicability. Second, we identify some important threads woven through the fabric of the different papers that might usefully be pulled in order to reveal some important patterns. Finally, we suggest that a few major challenges remain that need to be addressed directly by all those parties interested in innovation and local/regional development. These are translated into a set of general recommendations to each of the different families of readers.

16.2 TWO BASIC POINTS

One of the major points that emerges from the studies in this book is the ill-specified nature of both the innovation process and of the local/regional development process. If there is a common denominator to the papers, it is the insistence of the different authors on the   idiosyncratic nature of most of the clusters that have been studied. Whether this specificity is ascribed to culture, infrastructures of one sort or another, or some other form of particular synergies, it is omnipresent.

This poses a major challenge, for it makes each cluster more or less exceptional, and nothing is more difficult to craft that a theory of exceptions.

Moroever, this scattered approach has proved more useful in the characterization of the clustering process than in the definition of the underpinnings of the innovation process. Indeed, what might be one of the principal findings that have emerged from this ensemble of papers is the much greater difficulty in grappling with the innovation process than with the clustering process. While there have been disagreements among the authors about the nature of an optimal/viable/vibrant cluster, there is no such convergence on the innovation process front. Indeed, it may even be suggested that in many of the papers the focus on clustering has often led to the focus on innovation being sacrificed almost entirely.

A provisional result of these investigations has been the realization that what constitutes a "territorialized system of innovation" remains an essentially contested concept. This could only lead to a blurred notion of local/regional systems. While there would appear to be some convergence of opinions on the fact that the capturing of the benefits of innovation best takes place at the "meso-level" and in a "territorialized milieu"– not within national structures – the process of innovation *per se* has not been mapped out yet and its dynamic at the meso-level has not been explicated.

A second major point, related to the first, is the characterization of innovation as an emergent phenomenon. While the word is not used explicitly, this is an omnipresent reality. The authors have proposed an array of innovator success factors that are present in the specific clusters that they have studied, but they have almost always noted that the degree of innovativeness or creativity that has evolved is the result of some obscure alchemy of those factors.

Whether this sort of quasi-organic emergence of innovation can really ever be understood fully is not a matter on which the authors agree. Some would appear to be satisfied with the existence of this sort of mysterious penumbra around the process of innovation. Others feel that this mystery can be probed and anticipatory tools can be developed. But it is fair to say that no paper in this book has disclosed the dynamic structure of the emergent innovation process.

Even on the matter of what might be the engine of the learning process that underpins innovation, there has been little willigness to raise the issue front and center. This is particularly surprising in view of the fact that learning has become the central core of the new canonical thinking about the source of the wealth of nations. While there are some insights to be derived from the "blurred firm" of Annalee Saxenian for it is a system in the making, and from the "region" as the learning system in Richard Florida's work, or about Cooke and Morgan's (1993) densely networked region (as used by Gertler *et al*), it is still unclear at this point in time how these new "units of analysis" are learning.

In the absence of some serious reflections both on the epistemology and the economics of the basic relationships on which the innovation and "meso-territorial" dynamics are built, progress is bound to be slow and uneven. This work has progressed somewhat in the recent past, but it remains very much beyond the frontiers of the conventional work in the economics of innovation and local/regional development. Indeed, given the very stylized nature of the analytics that are in good currency, it may appear to be no one's professional duty to delve into these murky waters. Yet, it is *une tâche incontournable*. For without a conceptual refurbishment that brings about the "recovery of practical philosophy" and the development of a much more refined conceptual apparatus to handle cognitive networks, the inquiry on both fronts (i.e. in the processes of clustering and emergence in local and regional systems of innovation) may well by fundamentally handicapped (Toulmin 1990; Laurent and Paquet 1998).

16.3   KEY DIMENSIONS

Despite these broad reservations, and despite the fact that there is still no canonical syncretic approach that effectively integrates all the results of all the papers at the end of this voyage, it should be clear that the chapters presented above have made a substantial contribution to the clarification of the issues in the worlds of innovation and local/regional development. They have contributed to define a *problématique en devenir* and have explored, in all sorts of useful ways, many dimensions of this broad approach.

Without meaning to impose on the contributors an overall framework they would feel most uncomfortable with, we have underlined some of the factors that have been emphasized by the contributors and are likely to remain key dimensions of the broad portrait of local and regional systems of innovation in the process of emerging.

a) A general search has been launched by all the contributors for a *meaningful unit of analysis*. While many have developed their own for the purposes of their chapters, there has been an explicit longing for a more satisfactory unit for comparative purposes. It is clear that it is unlikely to be the conventional firm. Whether it might be the "firm and its circumstances" that may qualify as the new relevant unit of analysis remains also unclear. Saxenian's blurred firm and talent pool, Florida's learning region, Landry and Amara's and Wymbs' business networks, Acs *et al*'s specific spill-over areas, Gertler *et al*'s densely networked region, and Voyer's clusters are all different versions of a new meso-unit which remains to be analyzed in much greater detail before we can hope to suggest that it might become the new unit of analysis. What is required at this time is a second-round of analysis to refine these different concepts and find some core features that might underpin all these provisional probing tools.

b) More than in most other books of this sort, the contributors have tended to focus on *processes* rather than structures. Structures remain omnipresent because they are much easier to describe and to use as presumed explanatory variables, but most contributors have tried their hand at identifying processes at work both in innovation *per se* and in the development of local and regional clusters. Yet, this gambit has not necessarily yielded the benefits that were anticipated. Most often, the interaction or transmission processes have simply been identified and very loosely described, or have been used in questionnaires without much precision. As a result, it is not unfair to say that we are extremely far from being able to decompose the innovation and clustering processes into sub-processes to attempt some useful re-engineering. 'Process' remains mainly a label. This points to a need for modeling processes in an experimental way in the next phase of work on local and regional innovation systems.

c) The notion of *learning* has also permeated most of the papers, but it has probably remained too much of a closed concept. It is as if reference to the cognitive space *à la* Ziman (1991) or to cognitive economics (McCain 1992; Moati and Mouhoud 1994; Paquet 1998 a, b) was still perceived as being too risky. And yet the learning issue is resonating centrally in most of the papers. Even though there is already some work in progress in this area, and even some effort to deal with learning in territorialized innovation systems (Guilhon et al 1997), this remains very much exploratory work and additional investment is probably also one of the great priorities for those interested in studying territorialized innovation system as learning systems.

d) *Governance and culture* are also omnipresent dimensions in the chapters in this volume. They are defined in fundamentally different ways, but they are often used in the same breath, and are invoked as determining factors in explaining the success and failure of territorialized innovation systems. Governance has to do with the ways in which a system or a complex organization steers itself. Culture "is the shape a place takes when it's inside the heads of its peoples: all the habits, attitudes and values they take for granted" (O'Toole 1998: 61). Both dimensions are clearly integrated. Culture shapes the habitus (or the system of dispositions and inclinations), and this in turn impacts on the pattern of governance; but governance is also contributing to the shaping of culture and ethos through the learning that it generates.

In the chapters presented above, this dynamic interaction between governance and culture does not play itself as fully as it might. Rather, culture is taken more or less as a set of static features that bind and program the style of governance. Lawton-Smith *et al.* and Roy are more explicit than other contributors - even though this is done descriptively in separating culture and governance and in showing how they interact, but even they do not explore as fully as they might have this nexus of forces.

e) The ethnographic description of the various linkages binding together and defining the territorialized innovative clusters, provided by the contributors, has generated a broad portrait of the rich array of interactions making up a working *innovative milieu*. The interfirms links are explored in great details by Gertler *et al*, and in more specific ways by Shuetze, Landry and Amara, Saxenian, and Lawton-Smith *et al*. But too little is said about the centrality of interpersonal resources (Foa 1971) and about the development of new forms of social glue binding the clusters into quasi-communities.

The analysis of the linkages has taken a mechanical bent which may turn out to be a disservice to the analysis of clusters. Schuetze comes closest to probing idiosyncratic relationships when he examines how small and medium sized firms appropriate knowledge, White Acs *et al*, Lawton-Smith *et al* and Nimijean make use of proximate spill-over and cultural proximity in their analysis. But this is done in an oblique way. Little is done to probe the sources of increasing returns beyond the traditional aglomeration economies; and yet, it is clear from the analyses above that most of the contributors feel that there is more to clustering than agglomeration economies, and much more to innovation than sheer proximity.

f) The notion of *entrepreneurship* (private, public, civic) is central to the innovative cluster and yet not much effort is dedicated to probing this corner of *terra incognita*. Roy and Voyer make reference to the phenomenon front and centre, but most of the other authors only refer to it obliquely without making any attempt to factor this collective phenomenon in the innovative process. Yet, the central importance of entrepreneurs in disclosing new worlds or in reframing the representations of the world of production has been explored with sufficient case that it should by now be part and parcel of any exploration of local and regional systems of innovation (Spinosa *et al* 1997).

g) It has also become clear from the chapters presented above that *statistical data cannot be collected in a one-form-fits-all way* if one is to be able to analyze the particular circumstances of the different innovative clusters in a useful way. The most illuminating insights have emerged from tailor-made surveys and from minutely crafted case studies which have tried to collect information matching as closely as possible the particular circumstances of the cluster or of the transversal dimensions of greatest interest for the particular local and regional system of innovation. This poses a major challenge to statistical agencies that may have to modularize somewhat their data collection in order to provide the users with the possibility of constructing tailor-made indicators from broad easy-to-re-assemble data sets.

16.4 MODEST PROPOSALS

Readers will generate their own senses of the directions in which the research program on local and regional systems of innovation should proceed. Not all those interested in this general area of study will derive the same priority list. However, a few general points would appear to be percolating with such force that few will probably feel that they do not belong to the top portion of any research agenda.

We present very briefly some of those priorities not with the ambition of imposing a research agenda on our colleagues, but with the hope that it might at least serve as a basis for discussion in the preparation for our next symposium.

a)   For Model Builders

Models are important as cognitive maps. But there is a real danger that, in the study of local and regional systems of innovation, there is not sufficient attention being given to middle-range phenomena. A number of decades ago, in the field of urban sociology, some practitioners complained that one could only find either studies of street-corner societies or of urbanization in the Western world. The study of territorialized innovation systems may face the same challenge. What is needed is more work at the meso-level, more work also focusing of the basic procesess at work, and more work on the learning process in particular.

b)   For Data Gatherers

Data sets are essential as corroborative evidence to test interesting hypotheses. Up 'till now, much attention has been devoted to the production of standardized national data sets, mainly of use for international comparison. The emerging interest in local and regional systems of innovation will force a need onto statistical agencies to reframe their approach. Data sets will have to be designed with meso-level users in mind. And since these users have quite diverse approaches and quite different data needs, the only way will be for national statistical agencies to allocate more of their scarce resources to the development and construction of both elaborate and diverse surveys.

This is quite a challenge, both methodologically and substantially, for statistical agencies, but it would appear to be the only way in which they can ensure that both data sets with a certain degree of integrity are produced centrally and particular differentially-focused analyses are conducted locally.

c)   For Academic and Professional Analysts

Academic and professional analysts have an equally daunting challenge. They have to develop an all out strategy of exploration of the different territorialized innovation systems with the view of generating a cumulatively more heuristically powerful conceptual framework. Models may vary but a conceptual framework of interpretation likely to serve as an infrastructure for all the modelling activity is badly needed.

Nothing less than to sort of effort that has led to the development of evolutionary economics, and what is in the process of developing in cognitive economics, can provide the intellectual capital necessary for a fruitful next step. The real danger at this time is a balkanization of research and the over-investment in case studies. A division of labour is necessary and substantial intellectual resources must be invested in the

R&D of the territorialized innovation system industry. This would mean the development of analytical frameworks that explore the various processes of learning, coordination, proximity, entrepreneurship, and so on, and an effort to bring these partial views together in an effort to produce a broad interpretative framework that is capable of guiding the research agenda in the next phase.

d) For Policy Makers

Policy-makers are by definition impatient. They are "solutionists". They are less interested in processes than in cures. At this time, they may have to be persuaded to have patience. There is no merit in advocating solutions before one is clear about the ways in which it is likely to work out in practice. At this time, we know much about meso-innovation systems, but we know little about the way to grow them. What one may hope to develop out of the action proposed under a., b. and c. above is the emergence of modest general propositions that are likely to be useful to policy-makers. But, it is unlikely that such propositions can emerge in the very short run.

Does it mean that policy-makers have to be inactive in the meantime? Certainly not. What is required is that they undertake only planned non-permanent experiments at the meso-level in order to both accumulate additional information on the dynamics of meso-innovation systems and to provide a laboratory for those analysts in search of testing grounds.

16.5 CODA

It does not make much sense to close a book like this one without indicating the road ahead for the group that has just completed the first leg of a long voyage toward a better understanding of the dynamics of local and regional systems of innovation.

The road ahead is branching in two directions that we wish to follow daringly at the same time.

The first one is the development of a set of templates to look at local and regional systems of innovation on the basis of the existing literature and of the new insights generated in this volume in order to help design the sort of data sets necessary for empirical work to progress. This is the basis of a future symposium to be held and published in 1999 (de la mothe and Paquet 1999a).

The second one is the development of a template for a comparative framework to analyze the similarities and differences of territorialized innovation systems in the different continents. This is a project already under way and will appear in book form in 1999 (de la Mothe and Paquet 1999b).

REFERENCES

Cooke, P. and Morgan, K. (1993). "The Network Paradigm: New Departures in Corporate and Regional Development", *Environment and Planning D: Society and Space*, 11, pp. 543-564.

de la Mothe, J. and Paquet, G. (eds) (1999a). *Innovation Systems in Middle Power Countries* London: Pinter.

de la Mothe, J. and Paquet, G. (eds) (1999b). *Territorial Innovation Systems: Firms, Technologies and Networks,* (forthcoming).

Foa, U. (1971). "Interpersonal and Economic Resources" *Science*, 171, 3969, 29 January, pp. 345-351.

Guilhon, B. *et al.* (eds) (1997). *Economiede la connaisance et organisations*. Paris: L'Harmattan.

Laurent, P. et Paquet, G. (1998). *Economie et épistémologie de la relation – coordination et gouvernance distribuée*. Lyon: Vrin.

McCain, R.A. (1992). *A Framework for Cognitive Economics*. Westport: Praeger.

Moati, P. et Mouhoud, E.L. (1994). "Information et organisation de la production: vers une division cognitive du travail" *Economie appliquée*, XLVI, 1, pp. 47-73.

O'Toole, F. (1998). "The Meanings of Union" *The New Yorker*, April 27/May 4, pp. 54-62.

Paquet, G. (1998a). "Evolutionary Cognitive Economics" *Information Economics and Policy,* Elsevier.

Paquet, G. (1998b). "Lamberton's Road to Cognitive Economics" in S.Macdonald and J. Nightingale (eds) *Information and Organization: A Tribute to the Work of Don Lamberton*, Amsterdam: Elsevier.

Spinosa, C. *et al* (1997). *Disclosing New Worlds*. Cambridge: The MIT Press.

Toulmin, S. (1990). *Cosmopolis,* Chicago: University of Chicago Press.

Ziman, J. (1991). "A Neural Net Model of Innovation" *Science and Public Policy, 18*, 1, pp. 65-75.

INDEX

Economics of Science, Technology and Innovation

1. A. Phillips, A.P. Phillips and T.R. Phillips:
   *Biz Jets. Technology and Market Structure in
   the Corporate Jet Aircraft Industry.* 1994   ISBN 0-7923-2660-1
2. M.P. Feldman:
   The Geography of Innovation. 1994   ISBN 0-7923-2698-9
3. C. Antonelli:
   *The Economics of Localized Technological
   Change and Industrial Dynamics.* 1995   ISBN 0-7923-2910-4
4. G. Becher and S. Kuhlmann (eds.):
   *Evaluation of Technology Policy Programmes
   in Germany.* 1995   ISBN 0-7923-3115-X
5. B. Carlsson (ed.): *Technological Systems and Economic
   Performance: The Case of Factory Automation.* 1995   ISBN 0-7923-3512-0
6. G.E. Flueckiger: *Control, Information, and
   Technological Change.* 1995   ISBN 0-7923-3667-4
7. M. Teubal, D. Foray, M. Justman and E. Zuscovitch (eds.):
   *Technological Infrastructure Policy. An International
   Perspective.* 1996   ISBN 0-7923-3835-9
8. G. Eliasson:
   *Firm Objectives, Controls and Organization. The Use
   of Information and the Transfer of Knowledge within
   the Firm.* 1996   ISBN 0-7923-3870-7
9. X. Vence-Deza and J.S. Metcalfe (eds.):
   *Wealth from Diversity. Innovation, Structural Change and
   Finance for Regional Development in Europe.* 1996   ISBN 0-7923-4115-5
10. B. Carlsson (ed.):
    Technological Systems and Industrial Dynamics. 1997   ISBN 0-7923-9940-4
11. N.S. Vonortas:
    Cooperation in Research and Development. 1997   ISBN 0-7923-8042-8
12. P. Braunerhjelm and K. Ekholm (eds.):
    The Geography of Multinational Firms. 1998   ISBN 0-7923-8133-5
13. A. Varga:
    *University Research and Regional Innovation: A Spatial
    Econometric Analysis of Academic Technology Transfers.*
    1998   ISBN 0-7923-8248-X
14. J. de la Mothe and G. Paquet (eds.):
    Local and Regional Systems of Innovation   ISBN 0-7923-8287-0

KLUWER ACADEMIC PUBLISHERS — BOSTON / DORDRECHT / LONDON